In Defense of Japan

In Defense of Japan

FROM THE MARKET TO THE MILITARY IN SPACE POLICY

Saadia M. Pekkanen and Paul Kallender-Umezu

Stanford University Press
Stanford, California

Stanford University Press
Stanford, California

Publication assistance for this book was provided by the Job and Gertrud Tamaki Professorship at the University of Washington, Seattle.

Printed in the United States of America on acid-free, archival-quality paper

Library of Congress Cataloging-in-Publication Data

Pekkanen, Saadia M.
In defense of Japan : from the market to the military in space policy / Saadia M. Pekkanen and Paul Kallender-Umezu.
p. cm.
Includes bibliographical references and index.
ISBN 978-0-8047-0063-4 (cloth : alk. paper) 1. Astronautics and state—Japan. 2. National security—Japan. 3. Astronautics, Military—Japan. I. Kallender—Umezu, Paul. II. Title.
TL789.8.J3P45 2010
623'.690952—dc22 2010013900

Typeset by Westchester Book Group in 10/14 Minion

For Robert, ever aloft
and
for Ina and Yuko, with love, forever

CONTENTS

TABLES AND FIGURES

PREFACE

THIS BOOK REPRESENTS a journey that began at the end of 2002, after a series of rather animated breakfasts at the Hotel Okura in Tokyo. After well over five years of research and writing, this journey has taught us as much about the state of Japan's space politics, policy, and technology as about what their combination may mean for Japan's security directions. This book surprised us in many ways, the most important of which is that it was not the one we started out to write. As industry analysts, we began very modestly with a focus on the major pillars of Japan's space technology—space launch vehicles, satellites and spacecraft, and emerging technological niches—as well as the set of public and especially private actors who make them.

In connecting the dots for this book, we learned that, as in other countries, space technology in Japan did not proceed in a vacuum. The more we studied the sector's twists and turns, the more we learned to place Japan's space technology in its proper evolutionary context—that is, in the flow of technical and paradigmatic changes worldwide, as well as the currents of political and social changes in Japan. Early on, we came to the joint conclusion that the entire tenor of Japan's space program was shifting from what we could best describe as *the market-to-the-military*. By this we mean that it is no longer commercial but national security paradigms that are ever more critical in driving Japan's space policy forward.

Of course, only the excellence of Japan's civil space program could have brought Japan to consider making such a shift. We are not unmindful of the many criticisms that have been levied against Japan's civilian space program—some of which are well deserved—but there is also much there with which to be impressed. No matter what the criticisms, we are certainly impressed: with the sheer ingenuity of its engineers, with the stunning technological progress on a dime and with few huge disasters, with the subtle ways in which the legal and institutional structure was stretched to fit the technological realities, and also with the dogged persistence with which its leaders stressed autonomous and independent access to space at every turn.

Because not everybody shares our enthusiasm for the minutiae of Japan's space technology and policy, we also worked to make this a book of wider interest to space analysts, academics, policymakers, Japan and Asia specialists, technology experts, corporations, and businesses. The book is aimed specifically at those seeking to understand the broad contours of Japan's space-related history and trends, those looking to understand the role of Japanese public actors and, especially, private corporations in advancing it, and also those seeking to understand what the discrete components of Japanese space developments may potentially mean for the country's security policy in this new century.

Stripped to its essentials, our point is simple: Japan has the technical wherewithal to be marked as a military space power and now has placed national security as the centerpiece of its space development strategy. Having developed a range of military space technologies that came about in its civilian space program, Japan is actively advancing space as a keystone of its security and diplomatic strategy. This reality will affect the substance and direction of the country's national security in this new century. The militarization of space assets in Japan is not so much a game of numbers as it is an issue of technological realities, and now also the legal and institutional orientations that are beginning to reflect them. The story of how and why these technological realities came about and what they mean concretely for Japan's defense glues our narrative together. There has certainly been news about Japan's space programs in the Japanese media. But much of it has blandly focused on the big successes and failures of Japan's rockets and satellites, as well as the activities of various Japanese astronauts. There has not been much analysis of the military angle, excepting the case of Japan's spy satellites. Most foreign observers also report on Japan's space program much in the same way, saying very little, if anything, about its relation to military and defense realities.

Of course, for most of the postwar period, the Japanese government has itself never even remotely hinted at anything other than a civilian space program—until the spy satellites, until Ballistic Missile Defense, and, more important but much less well known, until the Kawamura initiative, and, most transformative of all, the new Basic Space Law all came along. To make our case clear, we show how distinct components of Japan's space technology fit historically in the militarized aspects of space technologies in other countries at every turn possible.

We are deeply grateful to the following people, who may well agree or disagree with the contents and interpretations presented in this book, but without whose time and keen insights over the years our analysis of Japanese space policy would not have been possible: Yasunori Matogawa, former director of the Japan Aerospace Exploration Agency (JAXA); Masakazu Iguchi, former chairman of the Space Activities Commission (SAC); and Masakazu Toyoda, secretary-general at the Secretariat of the Strategic Headquarters for Space Policy (SHSP); as well as Kazuto Suzuki, Setusko Aoki, Akira Kubozono, Hirohisa Mori, and the countless other government and corporate officials interviewed over the years by Paul Kallender-Umezu. We also thank the following people who provided comments on part or all of the manuscript: Chris Hughes, Andrew Oros, T. J. Pempel, Ken Pyle, Robert Pekkanen, John Pekkanen, and Richard Samuels, as well as the very helpful reviewers for Stanford University Press. We thank our research assistants for their timely and superb work: Jacob Brown, Vitaliy Pradun, Hiro Sasada, Jessica Leitham, and Tasuku Watanabe. We also thank Stacy Wagner, our editor at Stanford University Press, for her incisive and thoughtful commentary on this manuscript and for taking it forward.

Above all, we thank each other as good friends and colleagues over the years: SMP thanks PKU for tolerating her tendency to speak academically with such great good humor back in the real world and for so generously sharing his expertise; PKU thanks SMP for following along in his tracking of the actual twists and turns in the everyday news when we were seeking the big picture, and most of all for listening carefully back in the ivory tower. Finally, and most important, we thank our spouses and families for giving us the love, time, and space that made our long journey possible.

SMP
Seattle, Washington, United States

PKU
Tokyo, Japan

ABBREVIATIONS

AAW	Anti-Aircraft Warfare
ABL	Airborne Laser
ABM	Anti-Ballistic Missile Treaty
ADEOS	*Advanced Earth Observing Satellite (Midori)*
ADRTS	*Advanced DRTS*
AFOSR	U.S. Air Force Office of Scientific Research
AFRL	U.S. Air Force Research Laboratory
AFSPC	Air Force Space Command (U.S.)
AIAA	American Institute of Aeronautics and Astronautics
AIRBOSS	Advanced Infrared Ballistic-Missile Observation Sensor System
AISS	Alternative Infrared Satellite System
ALFLEX	*Automatic Landing Flight Experiment*
ALOS	*Advanced Land Observing Satellite* (*Daichi*)
AMRAAM	Advanced Medium-Range Air-to-Air Missile
AMRC	Advanced Mission Research Center
ANGELS	Autonomous Nanosatellite Guardian for Evaluating Local Space (U.S.)
AOARD	U.S. Asian Office of Aerospace Research and Development
APG	Aviation Program Group

AP-MCSTA	Asia Pacific—Multilateral Cooperation in Space Technology and Applications
APRSAF	Asia-Pacific Regional Space Agency Forum
APSCO	Asia-Pacific Space Cooperation Organization
APSCC	Asia-Pacific Satellite Communication Council
ARTEMIS	Advanced Relay and Technology Mission Satellite (Europe)
ASAT	anti-satellite (weapon or system)
ASBC	Advanced Space Business Corporation
ASCM	Anti-ship Cruise Missiles
ASDF	Air Self-Defense Force
ASER	Advanced Satellite Engineering Research Project
ASM	air-to-surface missile
ASNARO	*Advanced Satellite with New System Architecture for Observation/Sasuke*
ASR	*Advanced Solid Rocket/Epsilon*
ASTER	Advanced Spaceborne Thermal Emission & Reflection Radiometer
ASTRO	Autonomous Space Transport Robotics Operations (U.S.)
ASTRO	*Astronomical Observation Satellite* (*Astro-EII/ Suzaku, Astro-F/Akari, Astro-G, Astro-H*)
ATSML	Anti-Terrorism Special Measures Law
AUGI	(Basic Act on the) Advancement of Utilizing Geospatial Information
AVNIR-2	Advanced Visible and Near Infrared Radiometer Type-2
AVSA	Avionics and Supersonic Aerodynamics
AWACS	Airborne Warning and Control System
AWST	*Aviation Week & Space Technology*
BADGE	Base Air Defense Ground Environment
BMC4I	Battle Management Command, Control, Computers, and Intelligence
BMD	Ballistic Missile Defense

BMDO Ballistic Missile Defense Organization (U.S.)

BMDR Ballistic Missile Defense Research

BS *Broadcasting Satellite* (*Yuri* series)

Bus Satellite chassis, or frame

C4ISR Command, Control, Communications, Computers, Intelligence, Surveillance, Reconnaissance

CAD Computer Aided Design

CELLSAT *Cellular Satellite*

CEOS Committee on Earth Observation Satellites

CFRP Carbon-fiber Reinforced Plastic

CIA Central Intelligence Agency (U.S.)

CIRO Cabinet Intelligence and Research Office

CLB Cabinet Legislation Bureau

CMOS Complementary Metal-Oxide Semiconductor

CO Cabinet Office (Prime Minister)

CoCOM Coordinating Committee for Multilateral Export Controls

COTS commercial-off-the-shelf

COMETS *Communications and Broadcasting Engineering Test Satellite* (*Kakehashi*)

CPSDU Committee on Promotion of Outer Space Development and Utilization

CRCSS Cooperative Research Center for Satellite Systems (Australia)

CRL Communications Research Laboratory

CS *Communication Satellite* (*Sakura* series)

CSDC Council on Security and Defense Capabilities

CSICE Cabinet Satellite Intelligence Center

CSTP Council on Science and Technology Policy

CTBT Comprehensive Test Ban Treaty

CX-OLVE ConeXpress Orbital Life Extension Vehicle (UK)

DARPA Defense Advanced Research Projects Agency (U.S.)

DART	Demonstration of Autonomous Rendezvous Technology (U.S.)
DASH	*Demonstrator of Atmospheric Reentry System with Hyper Velocity*
DCS	Defensive Counterspace
DERA	Defense Evaluation and Research Agency
DGPS	Differential Global Positioning System
DIH	Defense Intelligence Headquarters
DOD	Department of Defense (U.S.)
DPRK	Democratic People's Republic of Korea
DPJ	Democratic Party of Japan
DRC	Defense Research Center
DRTS	*Data Relay Test Satellite* (*Kodama*)
DRTS-X	*Data Relay Test Satellite-X* (code name for a follow-on to *Kodama*)
DS2000	Melco standard satellite bus
DSCA	Defense Security Cooperation Agency
DSP	Defense Support Program (U.S.)
DTSI	Defense Trade Security Initiative
ECS	*Experimental Communication Satellite* (*Ayame* series)
EDT	electrodynamic tether
EGS	*Experimental Geodetic Satellite* (*Ajisai*)
EKV	Exoatmospheric Kill Vehicle
EO	Earth Observation
EOC	Earth Observation Center
EORC	Earth Observation Research Center
ERSDAC	Earth Remote Sensing Data Analysis Center
ESA	European Space Agency
ESRO	European Space Research Organization
ETS	*Engineering Test Satellite* (*Kiku* series)
EU	European Union

EUTELSAT	European Telecommunication Satellite Organization
EXPRESS	*Experiment Reentry Space System*
FAS	Federation of American Scientists
FMCT	Fissile Material Cut-Off Treaty
FREND	Front-end Robotics Enabling Near-term Demonstration
FSX	Fighter Support Experimental
GALEX	Galaxy Express Corporation
GALILEO	Global Navigation Satellite System (Europe)
GCOM	*Global Change Observation Mission* (*GCOM-W* [*Water*], *GCOM-C* [*Climate*])
GEO	Geosynchronous Orbit
GEOS	Global Earth Observation System
GIS	Geographic Information Systems
GLONASS	Global Navigation Satellite System (Russia)
GOJ	Government of Japan
GMD	Ground-based Midcourse Defense
GMMSS	Global Multimedia Mobile Satellite Communications System
GMS	*Geostationary Meteorological Satellite* (*Himawari* series)
GOSAT	*Greenhouse Gases Observing Satellite* (*Ibuki*)
GPS	Global Positioning System (U.S.)
GSDF	Ground Self-Defense Force
GSI	Geographical Survey Institute
GSO	Geostationary Orbit
GTO	Geosynchronous Transfer Orbit
GX	*Galaxy Express* (*rocket*)
HOPE	*H-II Orbiting Plane*
HOPE-X	*H-II Orbiting Plane–Experimental*
HSFD	*High Speed Flight Demonstrator*
HTV	*H-II Transfer Vehicle*

HYFLEX	*Hypersonic Flight Experiment*
IA	IHI (Ishikawajima-Harima) Aerospace
IAEA	International Atomic Energy Agency
IAT	Institute of Aerospace Technology
ICBM	intercontinental ballistic missile
ICO	Intermediate Circular Orbit
IFSEC	U.S.–Japan Industry Forum for Security Cooperation
IGS	*Information Gathering Satellites*
IGY	International Geophysical Year
IHI	Ishikawajima-Harima Industries
IKAROS	*Interplanetary Kite-craft Accelerated by Radiation of the Sun*
IMU	Inertial Measurement Unit
INDEX	*Innovative-technology Demonstration Experiment Satellite* (*Reimei*)
INMARSAT	International Mobile Satellite Organization (formerly International Maritime Satellite Organization)
INS	Inertial Navigation System
INTELSAT	International Telecommunications Satellite Consortium
IRBM	Intermediate-Range Ballistic Missile
ISAS	Institute of Space and Astronautical Science
ISL	Inter-Satellite Link
ISR	Intelligence, Surveillance, and Reconnaissance
ISS	International Space Station
ISS	*Ionosphere Sounding Satellite* (*Ume* series)
ISTS	International Symposium on Space Technology and Science
IT-RMA	Information-Technology Revolution in Military Affairs
JADGE	Japan Aerospace Defense Ground Environment
JAEC	Japan Atomic Energy Commission
JAMIC	Japan Microgravity Center
JAROS	Japan Resources Observation System and Space Utilization Organization

JAS	*Japan Amateur Satellite* (*Fuji* series)
JAXA	Japan Aerospace Exploration Agency
JDA	Japan Defense Agency (now MOD)
JDAM	Joint Direct Attack Munitions
JEM	*Japanese Experiment Module* (*Kibō*)
JEMRMS	Japanse Experiment Module Remote Manipulator System
JERS	*Japanese Earth Resources Satellite* (*Fuyō*)
JMA	Japan Meteorological Agency
JPL	Jet Propulsion Laboratory (U.S.)
JRANS	*Japanese Regional Advanced Navigation Satellite*
JSDF	Japan Self Defense Force
JSI	Japan Space Imaging Corporation
JTAGS	Joint Tactical Ground Station
KARI	Korea Aerospace Research Institute
KDD	Kokusai Denshin Denwa
KEI	Kinetic Energy Interceptor
Keidanren	Japan Business Federation
KHI	Kawasaki Heavy Industries
KOMSAT-3	Korea Multipurpose Satellite-3
KSPC	Kakuda Space Center
KSRC	Kashima Space Research Center
LAN	Local Area Network
LANDSAT	Land Remote-Sensing Satellite
LCT	Laser Communication Terminal
LDAR	Large Deployable Antenna Reflector
LDP	Liberal Democratic Party
LDR	Large Deployable Reflector Antennas
LDREX(-2)	Large-scale Deployable Reflector Experiment Model (-2)
LE	liquid engine

LEO	low Earth orbit
LITVC	Liquid Injection Thrust Vector Control
LNG	Liquid Natural Gas
LOD	Launch on Demand
LOX/LH2	Liquid Oxygen/Liquid Hydrogen
LOX/LNG	Liquid Oxygen/Liquid Natural Gas
LRB	Liquid Strap-On Boosters
LRST	Long-Range Surveillance and Track
LSR	Laser Ranging Equipment
LUCE	Laser Utilizing Communication Equipment
MAFF	Ministry of Agriculture, Forestry, and Fisheries
MCA	Management and Coordination Agency
MD	Missile Defense
MDA	Missile Defense Agency (U.S.)
MDS	*Mission Demonstration Satellite*
MEADS	Medium Extended Air Defense System
Melco	Mitsubishi Electric Corporation
METI	Ministry of Economy, Trade, and Industry (formerly MITI)
MEXT	Ministry of Education, Culture, Sports, Science and Technology (previously MOE)
MHI	Mitsubishi Heavy Industries
MIC	Ministry of Internal Affairs and Communications (formerly MPHPT)
Micro-OLIVe	*Micro OMS Light Inspection Vehicle*
MIRV	Multiple Independently Targetable Reentry Vehicles
MITI	Ministry of International Trade and Industry (now METI)
MKV	Multiple Kill Vehicle
MLIT	Ministry of Land, Infrastructure, and Transport
MNC	Multinational Corporation

MNTVC Movable Nozzle Thrust Vector Control
MOD Ministry of Defense
MOE Ministry of Education (now MEXT)
MOF Ministry of Finance
MOFA Ministry of Foreign Affairs
MOS *Marine Observation Satellite* (*Momo* series)
MOX Mixed Oxide
MPHPT Ministry of Public Management, Home Affairs, Posts and Telecommunications (now MIC)
MPT Ministry of Posts and Telecommunications (later MPHPT)
MRBM Medium-Range Ballistic Missile
MRJ Mitsubishi Regional Jet
MSAS MTSAT-based Augmentation System
MSAT Mobile Satellite
MSDF Maritime Self-Defense Force
MSLS Micro Satellite Launch System
MTCR Missile Technology Control Regime
MTSAT *Multi-functional Transport Satellite* (*continuation of Himawari* series)
MUSES *Mu Space Engineering Spacecraft* (*MUSES-A/Hiten, MUSES-B/Halca, MUSES-C/Hayabusa*)
NAL National Aerospace Laboratory of Japan
NASA National Aeronautics and Space Administration (U.S.)
NASA ADS NASA Astrophysics Data System
NASDA National Space Development Agency of Japan
NDIA National Defense Industry Association (U.S.)
NDPO National Defense Program Outline
NEC Nippon Electric Corporation
NEDO New Energy and Industrial Technology Development Organization

NeLS	*Next-generation Low-Earth Orbit System*
NFIRE	Near Field Infrared Experiment (U.S.)
NHK	Nippon Hōsō Kyōkai (Japan Broadcasting Corporation)
NIC	National Intelligence Council
NICT	National Institute of Information and Communications Technology
NIES	National Institute of Environmental Studies
NISTEP	National Institute of Science and Technology Policy
NOAA	National Oceanographic and Atmospheric Administration (U.S.)
NPO	non-profit organization
NPT	Non-Proliferation Treaty
NRO	National Reconnaissance Office (U.S.)
NSC	National Security Council (U.S.)
NSDI	National Spatial Data Infrastructure
NSIO	National Space Intelligence Office (U.S.)
NSSI	National Security Space Institute (U.S.)
NSSPG	National Space Strategy Planning Group
NTT	Nippon Telegraph and Telephone
NTWD	Navy Theater-Wide Defense System
OBC	Onboard Computer
OCS	Offensive Counterspace
OCO	Orbiting Carbon Observatory
OICETS	*Optical Inter-orbit Communications Engineering Test Satellite* (*Kirari*)
OMS	Orbital Maintenance System
OPS	Optical Sensor
OREX	*Orbital Reentry Experiment* (*Ryūsei*)
ORS	Operationally Responsive Space

OSR	Orbital Servicing Robot
OTA	Office of Technology Assessment (U.S.)
PAC-3	Patriot Advanced Capability-3
PALSAR	Phased Array type L-Band Synthetic Aperture Radar
PARC	Policy Affairs Research Council (later renamed Policy Research Council)
PETSAT	*Panel Extension Satellite*
PLANET	*Planetary Satellite* (*Planet-A/Suisei, Planet-B/Nozomi, Planet-C/Akatsuki*)
PNT	Positioning, Navigation, and Timing
PPP	public-private partnership
PPR	Peaceful Purposes Resolution
PRISM	Panchromatic Remote-sensing Instrument for Stereo Mapping
PRISM	*Picosatellite for Remote-sensing and Innovative Space Missions*
PSI	Proliferation Security Initiative
PTE	Performance Test Engine
QZSS	*Quasi-Zenith Satellite System* (*Michibiki*)
RBR	Reconfigurable Brachiating Robot
RCAST	Research Center for Advanced Science and Technology
R&D	Research and Development
REM	Reentry Module (*USERS*)
RESTEC	Remote Sensing Technology Center of Japan
RFT	Radio Frequency Terminals
RIPS	Research Institute for Peace and Security
RITE	*Remote Inspection Technology Experiment*
RLV	Reusable Launch Vehicle
RMA	Revolution in Military Affairs
ROSETTA	Robot-oriented Space Evolution Technology Task Force
RSC	Rocket System Corporation

RSGIS	Remote-Sensing Geospatial Infrastructure System
RSS	Reconfigurable Space System
RTF	Radio Frequency Terminal
SAC	Space Activities Commission
SAM	Surface-to-Air Missile
SAO	Smithsonian Astrophysical Observatory
SAR	Synthetic Aperture Radar
SBIRS	Space-Based Infrared System (U.S.)
SBSS	Space-Based Space Surveillance System
SCC	Space Communications Corporation
SCC	Security Consultative Committee
SCD	SM-3 Cooperative Development Project
SCOPE	*cross Scale COupling in the Plasma universe*
Scramjet	Supersonic Combustion Ramjet
SDF	Self-Defense Force
SDI	Strategic Defense Initiative
SDIO	Strategic Defense Initiative Organization
SDS	*Small Demonstration Satellite*
SEEDS	*Space Engineering Education Satellite*
SELENE	*Selenological and Engineering Explorer* (*Kaguya*)
SEM	service module (*USERS*)
SERVIS	*Space Environment Reliability Verification Integrated System* (*Servis-1, Servis-2*)
SFU	*Space Flyer Unit*
SHSP	Strategic Headquarters for Space Policy
SJAC	Society of Japanese Aerospace Companies
SLR	Satellite Laser Ranging
SLBM	submarine-launched ballistic missile
SLV	Space Launch Vehicle
SM-3	Standard Missile-3

SMART-OLEV SMART Orbital Life Extension Vehicle (Sweden)
SOD Space on Demand
SOHLA Space-Oriented Higashioka Leading Association
SOL Space Open Laboratory
SOLAR *Solar Physics Satellite* (*Solar-B/Hinode*)
SPAC Satellite Positioning Research and Application Center
SPICA Space Infrared Telescope for Cosmology & Astrophysics
SPIE International Society for Optical Engineering
SPOT Satellite pour l'Observation de la Terre (France)
SRB Solid Rocket Booster
SRBM Short-Range Ballistic Missile
SSA space situational awareness
SSB Solid Strap-on Boosters
SS/L Space Systems/Loral
SSO Sun Synchronous Orbit
SSM surface-to-surface missile
SSSAT Solar Sail Demonstration Satellite
SST Scaled Experimental Supersonic Transport
SSTG Smart Satellite Technology Group
SSTO Single-Stage to Orbit
STA Science and Technology Agency
STDRC Space Technology Demonstration Center
STS Space Transportation System (U.S. space shuttle)
STSRDC Space Transportation System Research and Development Center
STSS Space Tracking and Surveillance System
S&T Science and Technology
TAO Telecommunications Advancement Organization
TDRS Tracking and Data Relay System Satellite (U.S.)
THAAD Terminal High Altitude Air Defense

TMD	Theater Missile Defense
TNSC	Tanegashima Space Center
TPAE	Three Principles on Arms Export
TRDI	Technical Research and Development Institute
TRMM	Tropical Rainfall Measuring Mission
T-SAT	Transformational Satellite Communication System (U.S.)
TT&C	Tracking, Telemetry & Command
TVC	Thrust Vector Control
UAV	Unmanned Aerial Vehicle
ULA	United Launch Alliance
UN	United Nations
UNISEC	University Space Engineering Consortium
UNOOSA	United Nations Office for Outer Space Affairs
USAF	United States Air Force
USC	Uchinoura Space Center
USDOD	United States Department of Defense
USEF	Institute for Unmanned Space Experiment Free Flyer
USERS	*Unmanned Space Experiment Recovery System*
USTR	United States Trade Representative
VEP	Vehicle Evaluation Payload
VLBI	Very Long Baseline Interferometry
WAAS	Wide Area Augmentation System (U.S.)
WEOS	*Whale Ecology Observation Satellite System*
WESTPAC	Western Pacific
WINDS	*Wideband InterNetworking Engineering Test and Demonstration Satellite* (*Kizuna*)
WMD	Weapons of Mass Destruction
WTEC	World Technology Evaluation Center
XSS	Experimental Satellite System (U.S.)

In Defense of Japan

1 THE MARKET-TO-MILITARY TREND

THIS BOOK FOCUSES ON JAPAN'S CAPABILITIES IN SPACE, with a view to understanding their progression over time.[1] In the face of repeated commercial disappointments and continuing scientific uncertainties, Japan has managed to develop and maintain an indigenous space industry—one that today marks it as a military space power. Certainly, these developments in the industry have taken place in plain sight of the public within a steadily advancing civilian space program.[2] Were all this part of a long-term coherent national strategy, of course, it would be understandable and perhaps play a part in the current debates refashioning our understanding of Japanese grand strategy. Inconveniently for the theorists, however, Japan's commitment to the space industry long predates the very recent formulation of anything like a coherent national strategy, which came with the passage of Japan's first ever Basic Space Law in 2008, and the subsequent Basic Space Plan in 2009.

How, then, can we understand the long-term and significant commitment the country has made to space? How do we begin to understand the shift from what we call the market-to-the military in space developments? This latter question is all the more important because the new Basic Space Law matters in a fundamental and transformative way. It means that Japan's space policy has officially transitioned from one that exists only for peaceful purposes (a distinct definition that originally limited Japan from the development of any space technology that could be used for military purposes) to one with a strong—and, at long last, visible—emphasis on national security and the use of military space as a critical component of Japan's strategic defense.[3]

Important answers lie in the market. To be clear, the market was the driver not because of corporate success, but because of corporate setbacks. Through a conjunction of historical accidents rather than overarching purposeful design, corporations found their choices narrowing over time. With investments already in the commercial space industry that were not turning a profit, corporations looked to salvage or bolster their bottom lines by pushing their allies in the government to develop military space projects. Because of the unusual prevalence of dual-use technology in the space industry, this could be economically profitable and, as it turned out, politically attractive and legitimately possible over time as Japan faced rising external security challenges.[4] These elements make up the essence of the market-to-military trend. In the meantime, through small twists and turns, the *strategy* of militarizing Japanese space assets—as observable through formal laws, institutions, reports, plans, policies, and so on—continues to reflect the economic interests and especially *capabilities* of the private makers of space technology.

Why might a focus on space development be important at all? This is a dual-use sector, with assets that yield both civilian and military value and that are difficult to distinguish neatly across these very dimensions.[5] Whether right or wrong, desirable or not, governments and militaries around the world increasingly mark space as a strategic asset and see it as a primary provider and enabler of of war-fighting capabilities. For most of the postwar era, because of tightly held constraints on the kinds of technology it could develop for space use, Japan never remotely hinted at the militarization of its space program. By this, we mean the recognition, value, and use of space assets for military, or national security, purposes. In the context of space-based military capabilities, we follow the key mission areas identified by the United States Joint Chiefs of Staff as well as the United States Air Force Space Command (AFSPC).[6] These include the abilities of a government to provide *space support* (i.e., to be able to get to and maneuver in space with functioning launch vehicles and spacecraft) and *space force enhancement* (i.e., to be able to increase the combat and success potential of a combat force). Both of these have long been seen without controversy, at least outside Japan, as force multipliers in that the technology increases the potential of traditional forces across a range of operations. But then there is the possibility of weaponization of space assets, which has also been carefully excised from mention in Japan's civilian space program. This takes us into the nebulous—and, we believe, largely indistinguishable—realm of *space control* and counterspace (i.e., to be able to reap advantages of space assets while others cannot through surveillance, protection, prevention, and negation); and also *space force application* (i.e., to be able to overtly engage in weaponization), which is highly controversial in terms of cost and effectiveness in the long run.

The dimensions above are certainly relevant to an analysis of Japan's space developments, and this is essentially what we undertake in this book. The market-to-military trend in Japan's space sector is very much in keeping with worldwide trends, which showcase the importance of land, naval, air, and increasingly space, as dominant theaters of operations. As the most prominent example, since 2001, there has been a corresponding concern at the highest echelons of the U.S. government about protecting the nation's ever-burgeoning reliance on space assets for both commercial and, especially, military purposes—a trend we will show is also reflected in Japan's case in successive stages and across technologies.[7] Other observers have of course provided commentaries on the shifting priorities of Japanese space policy from the late 1990s onward, foreseeing either a continuing and deepening national security role or an opportunity to be managed by major space players like the United States.[8] To date, however, detailed analyses of the concomitant space systems—rockets and missiles, satellites and spacecraft, guidance, reentry, command, control systems, and so on—needed for Japan to develop an independent strategic military capability have not been available.

We take steps in this general direction, with the goal of showing the range of systems Japan has and is developing with its space technologies that can be used for its national defense goals. From a global perspective, this is hardly controversial. The fact is space is not a true sanctuary from military activity by governments around the world—Japan now explicitly included.[9] There are, of course, caveats in our Japan-centered narrative—failures, underdeveloped technologies, wrong turns and twists. Nevertheless, we maintain that Japan's cutting-edge space technologies, and now its institutional and legislative changes, mark Japan as a military space power. To put it in the more well-known dimensions above, we show how Japan long ago traversed space support, is now deeply engaged in space force enhancement, and may well have discrete elements of both space control and space force application well under way.

UNDERSTANDING JAPAN'S SPACE POLICY

In this book we focus on space policy in Japan in a thematic and chronological fashion—the legal and institutional context, the players, the industry, and especially the technology—to show how and in what ways Japan has been able to develop military capabilities in space. This is pertinent at a time when there are changes in Japan's security policies that fuel debates about the country's security directions and, popularly put, re-militarization ambitions.[10] As recently as the start of 2008, it seemed that the debate over the scope and contents of Japan's defense had emerged forcefully from its postwar shadows, with even nuclear options no longer as taboo in Japanese public discourse as they once were.[11] As some suggested outright, it may well be true that Japan's military posture has not been this robust since before the Pacific War.[12] A number of small but concerted institutional and legal moves have interacted with geopolitical developments to move Japan steadily down the path to ever more assertive security postures.[13] In early 2007 these moves resulted in the transformation of the Japan Defense Agency (JDA) into the Ministry of Defense (MOD)—the symbolic upgrading of Japan's defense concerns most visible to an audience both at home and abroad.

But the complicated fact is that such discrete elements in domestic politics, and the din associated with them, are now coalescing across people, parties, institutions, and laws in new struggles over how best to secure Japan in this new century.[14] Nobody is quite sure what all this tumult means, or where it may be headed. In the meantime, these struggles continue to highlight ongoing controversies on a number of significant dimensions—the constrained but increasing nature of Japan as a military actor, the push and pull of Japan's continued dependence on the alliance with the United States, and possibly the residual but eroding anti-militaristic sentiments across Japan.[15] They also certainly continue to energize debates about retrenchment and radicalism in Japan's security policy.[16] Even with the gathering storm over Japan's re-militarization debates, however, the case for Japan as a military space power is difficult to make all around. There are at least three reasons for this, the last of which segues into the analytics.

First, a focus on Japan's actual military capabilities is a small part of the debates over Japanese security policies. Seen in this light the very idea of Japan as a military space power seems misplaced given disputes over Japan's status even as a conventional military power. However, earlier works have broken considerable ground in correcting impressions of Japan as a military pygmy.[17] Even a very cursory examination we undertake below in line with these works shows that Japan continues to be on par with other countries in terms of military expenditures and even, to some extent, force capabilities. Table 1.1 illus-

Table 1.1. Japan's Defense Expenditures in Comparative Perspective

	2002		2007	
Country	*Value (US$, in millions)*	*As Percentage of GDP*	*Value (US$, in millions)*	*As Percentage of GDP*
United States	348,555	3.3	552,568	3.99
United Kingdom	38,141	2.4	63,258	2.28
Japan	39,200	1.0	41,039	0.93
France	36,492	2.5	60,662	2.37
China	68,963	1.6	46,174	1.42
Russia	50,800	4.8	32,215	1.54

SOURCES: 2002 figures: "Table 44—Comparative Defense Expenditures and Military Manpower, 2002–2004," *The Military Balance* 106(1), 2007, pp. 398–403; 2007 figures: "Table 38—International Comparisons of Defense Expenditures and Military Manpower," *The Military Balance* 109(1), 2009, pp. 447–452.

trates Japan's defense expenditures over the course of the 2000s. Putting aside the United States, whose expenditures dwarf those of all others, it shows that Japan's known value of military expenditures is comparable to that of other dominant players.[18] Despite the 1 percent ceiling, estimates over roughly a five-year period place the country's military expenditures among the highest in the world.[19]

Table 1.2 shows Japan's military capabilities in comparison to other dominant military players, especially in the Asian region. In keeping with other estimates, it presents a mixed picture of Japan's defensive and offensive capabilities across the services.[20] Japan's Ground Self-Defense Force (GSDF) continues to lack any serious offensive capabilities for land warfare abroad, but can certainly defend its own turf.[21] The Air Self-Defense Force (ASDF) may also be somewhat stronger on defense, being able to protect its airspace using its fighters for defense and its Airborne Warning and Control System (AWACS) aircraft for coordination to improve situational awareness. Although Japan does not field offensive ground-attack missiles or air-to-ground munitions, ASDF pilots have begun to practice dropping live bombs, done first in Guam in July 2007, using F-2s that are jointly produced by the United States and Japan.[22] The ASDF has also moved to improve its aerial refueling capabilities, and to modernize its own fleet.[23] Finally, the Maritime Self-Defense Force (MSDF) has considerable defensive and offensive powers and is the strongest part of Japan's military capabilities. It has a significant number of modern cruisers/destroyers, with the Kongō- and Atago-class destroyers also to be

Table 1.2. Japan's Conventional Military Forces in Comparative Perspective

Service	*Measurements*	*United States*	*Japan*	*South Korea*	*North Korea*	*China*	*Russia*	*India*
Personnel	Army	488,944	148,300	560,000	950,000	1,600,000	395,000	1,100,000
	Army Reserves	675,450	40,000	4,500,000 (all services)	600,000	800,000	20,000,000	960,000
	Navy	376,750	44,500	63,000	46,000	255,000	142,000	55,000
	Air Force	346,150	45,900	64,000	110,000	400,000	160,000	161,000
Ground Forces	Main battle tanks	8,023+ (all modern)	980 (all modern)	2,390 (1,080 modern)	3,500+ (0 modern)	7,580 (1,280 modern)	22,991 (4,931 modern)	3,978 +1,133 reserve (344 modern)
	Light tanks	0	0	0	560 PT-76 (all modern)	1,150 (all modern)	150 PT-76 (all modern)	190 (all modern)
	Artillery pieces	8,041	1,980	10,774+	17,900+	17,700+	30,412+	10,360+
	Attack helicopters	1,359 total (all modern)	90 (0 modern)	60 (0 modern)	24 Mi-24 (all modern)	39 (8 modern)	139 (all modern)	52–62 (all modern)
	Conventional short-range ballistic missiles	0	0	0	650? (0 modern)	725 (all modern)	0	20? (0 modern)
	Conventional medium-range ballistic missiles	0	0	0	~10 (0 modern)	21 DF-21/21A (all modern; carry conventional and nuclear warheads)	0	0
Maritime Forces	Aircraft carriers	11 (10 modern)	0	0	0	0	1 (modern)	1 (non-modern)
	Cruisers/destroyers	72 (all modern)	44 (40 modern)	6 (all modern)	0	28 (25 modern)	25 (14 modern)	8 (3 modern)
	Frigates/corvettes	31 (all modern)	9 (all modern)	37 (9 modern)	8 (0 modern)	48 (all modern)	45 (3 modern)	49 (27 modern)
	Attack submarines, [illegible]	59 [illegible]	0	0	0	4 [illegible]	23 [illegible]	0

	Attack submarines, conventionally powered	0	16 (all modern)	20 (all modern)	63 (0 modern)	59 (23 modern)	19 (all modern)	16 (14 modern)
	Anti-submarine warfare/maritime patrol aircraft	279 (221 modern)	80 P-3C (all modern)	16 (8 modern)	0	4 (0 modern)	70 (35 modern)	37 (6 modern)
Air Forces	Flying hours, per year	189+	150	n/k	20	80–180 (180 on modern craft)	20–25	180
	Fighters	541 (533 modern)	260 (190 modern)	210 (all modern)	299 (20 modern)	1,179 (178 modern)	765 (735 modern)	137 (74 modern)
	Strike aircraft	1,978 (1,853 modern)	0	295 (165 modern)	211 (all modern)	1,242 (73 modern)	793 (2 modern)	674 (170 modern)
	Naval strike aircraft	656 (567 modern)	0	0	0	642 (48 modern)	117 (49 modern)	15 (all modern)
	Medium-range bombers	0	0	0	80 H-5	222 total (0 modern)	0	0
	Long-range bombers	180 (86 modern)	0	0	0	0	80 (16 modern)	0
	Naval bombers	0	0	0	130 (0 modern)	58 (all modern)	0	0
	Tankers	594 (59 modern)	1 (modern; 4 more on order)	0	0	13	20 (all modern)	6 (all modern)
	Surface-to-air missiles (SAMs)	1,281+	832	340+	10,340+	1,862	4,365+	3,500+

SOURCE: Information on the ground, maritime, and air forces for each country is based on that found in *The Military Balance* 107(1), 2007, as follows: for the United States, pp. 28–42; for Japan, pp. 354–357; for South Korea, pp. 359–361; for North Korea, pp. 357–359; for China, pp. 346–350; for Russia, pp. 195–205, for India, pp. 314–319. North Korea's conventional missiles information is taken from "North Korean Missiles," available online at www.globalsecurity.org (accessed 18 July 2007). The chart represents the numbers present in the original sources, although not necessarily organized or added in the same fashion. For example, in any given country, SAMs operated by all services are added for a single numerical value. Because different militaries organize the same assets (like SAMs) under different services, to facilitate side-by-side comparisons, assets in the chart are organized according to their role, rather than what service they are under in their original countries. Hence, instead of using the more conventional army, navy, and air force rubrics, the chart uses the ground, maritime, and air forces rubrics.

NOTE: Modern equipment is defined as a design that is not outclassed by systems widely deployed throughout the world but is not limited to cutting-edge. Reserve equipment is included in the counts.

equipped with the Aegis Combat System that allows them to integrate and coordinate air defense networks over naval task forces.[24] The MSDF also has 80 P-3C Orion aircraft for anti-submarine operations and maritime patrol in the region.[25]

Looking more specifically at its aggregate defense expenditures as well as its standing across land, air, and naval mission capabilities, there is strong evidence that Japan's military power is often greatly underestimated. Our purpose in highlighting them is to suggest that, like conventional ones, Japan's space-based military capabilities also deserve attention as a cohesive whole. This is because, in line with the Revolution in Military Affairs (RMA)—which only underscores the importance of space assets and technologies to national defense—the SDF no doubt recognizes that its advantage over regional rivals will come not so much in the quantitative upgrading of conventional capabilities across its services.[26] Rather, it is bound to come in the technological and qualitative enhancements in space-based resources.

Such an analysis is especially valuable today as Japan is thought to have a much more effective military than it had in the postwar period—that is, in terms of internal management and organization, foreign integration, and work with the United States, and, of particular interest to us, use of intelligence and technology.[27] As we attempt to show in the rest of the book, Japan's considerable military capabilities are made more considerable by the country's already advanced space technology prowess. Features like its reconnaissance and surveillance capabilities, as well as the country's burgeoning ballistic missile defense (BMD) system, will continue to be upgraded and integrated into its military space infrastructure by domestic players. Therefore, like its conventional capabilities, Japan's space capabilities should not be underestimated, and deserve to be better understood by allies or rivals.

A second reason why it has been difficult to appreciate Japan's military capabilities in space is because any such cohesive emphasis has been absent at the national level, both in terms of strategy and organizational coherence. As we document more thoroughly in Chapters 2 and 3, from the time that Japan launched its historic *Pencil* rocket in 1955 to the passage of the Basic Space Law in 2008, there was much contestation over and many changes in the governance even of civilian space activities, let alone military oriented ones. Indeed the one element that was supposed to color all of Japan's space actitivites in relation to defense was the Diet's Peaceful Purposes Resolution (PPR) in 1969, which pointed in the opposite direction to space militarization in terms of official policy. Great care was taken by the government to express all its subsequent interpretations and policy changes under this rubric, and the emphasis was generally on the commercialization of space assets and the scientific explorations of the heavens. Independent access to space was certainly part of the official rhetoric and policy, but any hint of its relation to boosting Japan's defenses was absent from public documents and statements. To be sure, the saga of Japan's spy satellites and BMD drew much attention, but these were treated almost as isolated developments in the larger debates about the peaceful purposes resolution and Japan's re-militarization rather than as discrete elements of a military space infrastructure.

By and large, what attention there was focused on the negative facets of Japan's civilian and commercial space activities: the lack of progression and failure to commercialize Japanese space technologies in global markets, the lack of space planning at the highest levels, mixed signals from top committees supposedly controlling space activities, endless reviews following depressingly familiar failures, a devastating downgrading of Japan's space activities to the second rank of national scientific priorities, and attacks on Japan's international space activities by a budget-conscious Ministry of Finance (MOF). Compounding these negatives was the fact that the research, development, and testing of specific space technologies, was spread across a morass of public institutions and private corporations that themselves went through successive bouts of changes. All this made it difficult to pinpoint, much less assess, the potential military applications of legitimate civilian space technologies. While this morass of players, especially on the public side, was often pinpointed as a source of national incoherence in space policy, it also had the effect of shielding the development of technologies from negative budgetary, political, or public attention. Component by component, even highly controversial militarized space technologies, such as potential reentry warheads or anti-satellite (ASAT) systems, continued to be researched, manufacturted, and tested under legitimate civilian uses and goals.

Third, as we now turn to showing more specifically in terms of the analytical framework of the project, the construction of academic and policy debates on Japan's security policies has not exactly allowed a comprehensive focus on Japan's space capabilities. From a political economy perspective, the debates have generally favored analyses based on ideas and institutions, but not so much interests. Specifically, existing debates on Japan's militarization have focused attention on the importance of theoretical paradigms and constitutional changes. As discussed below, controversies about such approaches are broadly important in framing our market-to-military thesis concerning Japan's space technologies. However, we argue that, by and large, a focus on the concrete private interests behind Japan's space technologies lends greater specificity to explaining Japan's space policy over time. We therefore concentrate as much as possible on the activities of defense-related corporations that are central to advancing the actual state of Japan's space technologies—the kinds of products they are able to produce and what those products suggest for the market-to-military thesis over the postwar period.

This, then, is a book about Japan's space-related technologies and policy and not a book about Japan, international relations theory, or legal change. Nevertheless, below we situate and relate our focus to broader questions of Japan's security along these very dimensions to show how and in what ways they might be relevant to our question. This clarifies our own positions on some of the existing controversies and puts our specialist focus in broader contexts. Our emphasis allows us to assess whether and how realist approaches (emphasizing power and techno-security concerns) and constructivist frameworks (emphasizing pacifist and anti-militarist norms) were relevant to the course of the space technology that came into play in the postwar period. It also allows us to get a clear sense of whether constitutional norms constrained the militarization of space assets over the same period.

Analytical Approaches

This section draws out three possible ways of analyzing the market-to-military trend: realist and constructivist approaches, as well as constitutional concerns, both of which have figured prominently in larger debates about Japan's militarization, and interest-based approaches that have not. We do not provide any definitive or competing tests of these theoretical paragdigms or analytical approaches; rather, we assess their interplay in the evolution of Japan's space saga. Our focus is largely on the interest-based approach that we think is most relevant to the advancement of space technologies in national defense.

Contested Theories In some of the more prominent debates over Japan's militarization, scholars have focused on a long-running debate that pits theoretical perspectives in the study of international relations against each other. At the center of the controversy is the school of realism, with its historical emphasis on self-interest, relative power, competition, and security as the prime determinants of outcomes among states.[28] These concerns are reinforced by the persistent presence of anarchy in the international system, which has itself led to a fundamental divergence in contemporary realism, whether of the neorealist (focused on international outcomes) or neoclassical realist (focused on foreign policy strategies) variety. Simply put, offensive realism holds that anarchy provides strong incentives for offensive action as a means to security; however, defensive realism gives greater weight to moderation and prudence in the search for security.[29] Such intra-realist divisions are not necessarily clear-cut conceptually or practically as states' motivations stem from a range of interrelated security, power, and expansion concerns.

But for realists of all stripes, the imperative for Japan (as other states) in contemporary international politics is clear: militarize or else! This is not an issue of autonomy or international collaboration, which is indeed a false dichotomy, whether in conventional or space assets.[30] As the Council on Security and Defense Capabilities (CSDC), a highly influential advisory group to the prime minister, noted pragmatically at the end of 2004, whether a state like Japan militarizes by direct acquisition of military goods from abroad or by indirect indigenization of related technology at home is less important than the fact that it must do so in its own interest.[31] Following realist premises, this is because the consequence of anarchy in world politics for Japan, like all other positional states, is that it cannot ultimately count on anyone; that it must above all seek to survive by evaluating its power relative to other states; that the only certainty is uncertainty; and that there are never iron-clad guarantees of security for any state even from allies or alliances in a fluid geostrategic environment. This fluidity is evidenced today in the relationship between the United States and the former Soviet Union: their strict bipolarity is gone, and other strategic competitors are rising.

Not surprisingly, as discussed earlier, scholars have shown a great deal of interest in understanding the potential impact of these structural changes on Japan's security policy. Although it is fair to say that there is ever more support for a realist framework in the latest emerging literature on Japan's security policy, it is still difficult to pinpoint definitive backing for either offensive or defensive realism across the board because it is difficult to categorize security postures neatly across these dimensions. Perhaps more on point is whether Japan's acquisition of military-related—specifically space-related—technology has implications for the offense-defense balance.[32] Although alluring in its simplicity (balance shifts based on technology affect the likelihood of conflict) and predictive power (technology favoring the offense increases the likelihood of conflict), the offensive-defensive balance of military technology also remains controversial. Chief among the controversies is that scholars have found it very difficult to come up with objective and robust criteria for distinguishing between offensive and defensive technologies (and postures)—which in turn means an emphasis on both tangible and perceptual elements. Such problems are especially acute in dual-use space technologies, and for this reason we stick to the more nuanced earlier emphases on space support, space force enhancement, space control, and space force application in our assessment of Japan's space technologies.

Contrary to realist impulses, scholars have used constructivism to argue that Japan is and will remain disinclined toward militarism. At its core, the constructivist paradigm accords centrality to ideological, ideational, and social processes in world politics and from that persepective asserts that they can have a causal impact on actors' behavior even in what has long been analyzed as a materially oriented realpolitik realm with self-interested actors.[33] Put simply, the "international" is not some given and absolute reality but one that is socially structured and unstructured by interacting actors with shared identities and interests that are themselves socialized over time across institutions and issues. Collective beliefs and values have a profound impact on the behavior of actors, the way they see the world, and the way they wish to see it.

From this vantage point, then, even realpolitik is constructed. It is constructed not in the heads of self-interested actors or in material capabilities but, more importantly, in practices, processes, and normative expectations in the international system. Nor is all of this some reactive slam on rationalist approaches. After all, at any given time, actors may well strategize rationally to reconfigure existing sets of preferences and identities and even the broader social context. From studies asserting the importance of state identity and cultural-institutional context as having an independent impact on national security policies to those stressing the ways in which these elements determine defense politics at a given time, the focus is on carefully tracing processes and empirics to provide ammunition for central claims about the importance of norms and changes in social construction.

Central to constructivist-oriented paradigms about Japan is its alleged "culture of anti-militarism," with its roots in antipathy toward the collective memories of uncontrollable militaristic actions during the Pacific War.[34] Given the realities of Japan's postwar defeat and surrender, analysts focus on a set of institutionalized norms that have shaped and, some believe, continue to constrain postwar Japanese security policy—strict civilian control, penetration by economic ministries, constraints of Japan's peace constitution, and anti-militaristic public skepticism. In short, pacifist and anti-militarist norms pervade Japanese society.[35] Whether sympathetic to the school of constructivism or not, others have also often lent indirect support over the years to such constructivist claims with their broad-based assertions that Japan is not militarizing commensurate with its economic might in the global system as realism might predict or that it continues to remain a political or security pygmy.[36]

Like the realist paradigm, the constructivist one is compelling but also questionable in practice. As they cannot use it to explain both continuity and change, those who helped pioneer studies on the culture of antimilitarism in Japan are, for example, more cautious about its continued importance in the future of Japanese military security.[37] This is not to say that the normalization of Japan's defense posture commensurate with the use of force exercised by powers such as France or Great Britain will take place anytime soon, but that both Japan's military and its military operations are now accepted across domestic politics. Others have also acknowledged that, although long-standing anti-militarist constraints continue to be at play, decisions such as the development and deployment of surveillance satellites do ultimately reflect a progression toward realist-oriented security policies.[38] Even those who downplay potential shifts in Japan's security postures or identity on the grounds of tepid or incremental responses in legislation, policies, technologies, and direction are mindful of the realities of change in Japan.[39] At the very least, their works suggest that the social construction of defense politics for Japan (independent of actors and material capabilities as constructivists stress) has changed domestically and internationally. To put it in constructivist terms, then, the point is that Japanese actors may now be engaged in the social reconstruction of a new defense paradigm for Japan. The slower recognition and acknowledgment in this direction thus far may be because constructivism overemphasizes the role of social structures and norms and does so at the expense of understanding the role of agents who can create and change them in the first place—which segues into our own argument about the importance of following the interests of private actors.[40]

For the purposes of simply framing our central concern with the market-to-military trend, the bottom line from these nuanced academic works is this: if realist paradigms (of any stripe) are valid, we should expect to see a greater emphasis in the direction of military capabilities in space and to see such an emphasis come to the fore more visibly with the increasing geopolitical uncertainties for Japan; but if constructivist paradigms are valid, we should expect to see that the specific set of norms considered important significantly swayed or checked the market-to-military trend over the postwar period.

Contested Constitutionalism Parallel to the academic debates, there are also controversies over constitutional limits and interpretation that are more pragmatically relevant in the real world. Although such controversies serve as the fount for the constructivist paradigm related to Japan, they stand on their own for most lawyers and legal scholars. The most well-known controversy concerns Article 9 of the 1947 Japanese constitution, by the existing provisions of which the Japanese people not only forever renounce war as a sovereign right but also the threat or use of force to settle international disputes. To give effect to this aspiration, the provisions also make clear that Japan will never maintain land, sea, and air forces, as well as potential war materiel.[41]

From its inception, the revision of the American-imposed constitution has been a cornerstone of conservative politics in Japan.[42] Formal activities to revise the constitution date back as early as 1955, the same year that Japan launched its historic *Pencil* rocket. Even the Americans backtracked under the exigencies of the Cold War early on, beginning with calls in 1950 for Japan to support U.S. efforts during the Korean War. Successive LDP governments were also committed, to varying degrees, to what can best be described in the pragmatic interest of Japan's defense as constitutional bypassing. By the mid-2000s, a number of concerted steps by conservative leaders signalled strong interest in constitutional change—researching the right to collective self-defense, pushing through a new educational law focused on instilling patriotism in schools, upgrading the JDA to ministry status, revising the Self-Defence Force (SDF) law, and most directly on point, suggesting a timetable for constitutional revision.[43] The question perhaps is no longer whether or not the Japanese constitution will be changed. Rather, the critical question may well turn out to be in which direction, to what extent, and with what ultimate purpose constitutional changes will come about, as the various legislative changes as well as public proposals show.[44]

The pragmatic politics of Japan's contested constitution came to the fore when the Japanese government dispatched the SDF personnel for the multinational force in Iraq in December 2003.[45] In turn, this triggered several high-profile responses from the United States and Japan, with the United States suggesting that Article 9 hindered U.S. as well as Japanese defense interests at several levels. Diplomatically, the United States' support for Japan's efforts to become a permanent member of the United Nations Security Council was contingent on the willingness and ability of Japan to deploy military forces in the interests of the international community. Domestically, the presence of constitutional interpretations prohibiting the exercise of Japan's right to collective self-defense also needed to be revised in order to strengthen the military alliance between the two countries, pressures for which built as Japan geared up for missile defense deployment in the 1990s, as we detail in Chapter 6.[46] In fact, the old domestic taboos and regional reservations about dispatching Japanese forces had already been melting away at that point, as incrementally the SDF began to play a greater visible role in foreign operations since 2001.[47] At that point, in the wake of 9/11, Japan dispatched ships of the Maritime Self Defense Force (MSDF) to the Arabian Sea to provide rear-area logistical support for the U.S. military operations, both against Al-Qaeda and against the Taliban in Afghanistan. To make that possible, in October 2001 the Anti-Terrorism Special Measures Law (ATSML) was enacted, and it expanded the scope of permissible noncombat operations for the SDF.

The LDP and the DPJ have moved to consolidate their party positions on issues of national security. In contrast to most of the postwar period, they can do so quite visibly in a domestic political environment in which there is growing public support for revision or amendment of a U.S. imposed constitution.[48] Politically, structural changes in Japan's electoral system toward a hybrid mixed-member electoral system, institutional reforms that have increased the prime minister's crisis and security policymaking capacity, as well as a decrease in inter-party differences over security issues bode well for Japan's militarization.[49] Efforts at distinguishing policies across parties are certainly important in that they continue to gain media visibility around the world when they do crop up. The July 2007 Upper House election, in which the DPJ came out ahead and the LDP was trounced, was a forceful reminder that the passage of further constitutional or defense-related legislation is not exactly assured.[50] Of particular interest was that although many insiders believed that the proposed space bill would be signed into law without much trouble, the loss of LDP control delayed its passage.[51] Such legislative battles might be thought to be even more relevant as of August 2009, because the DPJ is in power, and Japan appears to have gone beyond a one-party system. One view is that as genuine party differences on legislation and policies come increasingly to the fore, electoral politics across parties and individuals may well become a more viable force to contend with in shaping defense policies over the years.[52] However, on the front line of current space development, ideological or party differences on Japan's military space policy appear to be less relevant. The Basic Space Law, which enacts Japan's military space development over the next ten years, was passed with strong support by the DPJ. Another clear measure of this can be found directly in the space budget passed in January 2010, which has kept all the dual-use and military space programs mandated by the Basic Space Law.[53]

At this stage it can safely be said that differences on security issues have become less ideological on the left-right divide, are far more pragmatic in character, and now run primarily within rather than across political party lines.[54] Since 1994, when the voters effectively dismantled the Left in Japanese politics as the Socialists accepted the premiership for their principles (strict interpretation of Article 9 and opposition to the U.S.–Japan alliance and the SDF forces), it has been clear that there is currently little of the Left remaining in Japanese politics. There are, in other words, few credible adherents of the strict postwar pacifism now left across the full spectrum of Japan's political parties; and, judging from the patterns thus far in the new security environment, there are probably few pacifists left who are not willing to compromise on boosting Japan's defense, maybe even through some sort of revision of the existing constitution. In the end, rather than high-profile electoral maneuvering, what needs to be watched with regards to Japan's defense are incremental moves within and across party lines.[55]

Formal steps toward revision of the Japanese constitution have been taking place since 2000, and that trajectory to date is worth noting.[56] In January 2000, Constitutional Research Commissions were set up in both the Upper and Lower Houses of the Japanese Diet. By December, basic principles and drafts plans for a new constitution began to be circulated by political parties like the Liberal Party (which went on to merge with the DPJ in 2003) and the LDP. In October 2001, discussions by the Constitutional Research Commission in the Lower House revealed that there was broad consensus in favor of constitutional revision, regarding Article 9 and the exercise of collective defense, among the specialists invited by both the LDP and DPJ. A year later, in November 2002, the same Constitutional Research Commission submitted an interim report calling for constitutional revision in light of the shifting domestic and international environment. In March 2005 the Lower House Constitutional Research Commission passed the draft for its final report, which enjoyed the support of the LDP, DPJ, and New Kōmeitō, although both the LDP and DPJ continued to circulate their own textual revisions and draft constitutions. By April 2005 the Upper House also issued its final report.

Given the economic stakes evident in the militarization of space-related technologies that we stressed earlier, business interests in Japan have also weighed in on the constitutional revision debate. In April 2003, Japan's highly influential and very powerful business lobby group, Nippon Keidanren, called for constitutional revision and legal restructuring in areas like foreign policy and national security; in January 2005, it went on to produce a report that advocated the revision of the constitution and the enactment of legislation to allow a national referendum on revisions to the constitution. All this activity led to an even more concrete result, as far as enabling procedures for formal revision. In April 2007, the LDP and New Kōmeitō used a majority vote to push forward the national referendum bill to amend the constitution in the Lower House; the bill then also cleared the Upper House a month later in May over the express objections of the then DPJ opposition.[57] At the very least, the procedural way is now paved for Japan to mould the U.S.-imposed constitution from a bygone era into one that speaks to Japanese interests in the unfolding new geopolitical game. As of 2010, constitutional revision in Japan remains an open-ended story, fueling controversy over changes such as the newly enacted anti-piracy law that allows the government to dispatch the MSDF (along with the Japanese Coast Guard) abroad to combat piracy, and the coming into effect of the National Referendum Law.[58]

The constitutional dance has always generated much controversy and interest because of what it implies about Japan's reassertion of its sovereign military rights in the international and regional system. We cannot say for sure whether and how Japan's constitution and Article 9 in particular, will indeed change even with the procedural steps locked in place.[59] From our vantage point as industry analysts, we also have a slightly different perspective on the matter. Seen through the prism of development of space assets, the controversies over constitutional revision are, in our judgment, less a barometer of Japan's remilitarizing ambitions; instead, we believe they are far more instructive today in showing that the Japanese government and industry have come a very long way from that episode in Japan's postwar history in which they consciously chose to avoid association with defense production.[60] The bottom line here is that, given the wording of Article 9, we should expect to see the constitutional text, norms, and interpretations check the progress of militarization of space assets.

Focusing on Economic Interests Framing our approach with existing theoretical paradigms and constitutional controversies is important because they not only provide a rich context but also ensure against excessive dogma. While mindful of their importance, however, we believe that far more relevant for the advancement of Japan's space technologies are the industrial and electronics companies, who are also Japan's primary defense contractors, and who have also operated under the same social and political exigencies as other actors in postwar Japan.

The starting point for understanding their activties is the contemporary geopolitical context, which has been uppermost in explanations of Japan's ever more visible realist orientations. The type of threats Japan faces at the tumultuous start of the twenty-first century, with two nuclear-armed neighbors in northeast Asia, means that, more than ever, Japan's military defense requires a sophisticated technological response—one that has long been entrenched and is now openly being acknowledged, across sectors and over time.[61] In an open interview, Gen Nakatani, former JDA director-general and later member of the House of Representatives, pointed correctly to the changed nature of threats for Japan's security in the region.[62] These have moved from concerns about outright invasion to ones that highlight the more sophisticated link between technology and security, such as ballistic missile attacks, surveillance and intelligence, and the presence of spy ships in Japanese waters. In laying out its visions for Japan's security needs in the new century, the CSDC began a 2004 report by noting that the complex nature of threats Japan now faces range from terrorist attacks by nonstate entities to traditional warfare.[63]

The unstable external environment was also the launching theme for the National Defense Program Guideline released in 2005 by the then JDA.[64] The Special Committee on Space Development, under the Policy Affairs Research Council of the LDP, took its cue from the rapid development of missile launches by North Korea and China's burgeoning space program to heavily criticize the existing institutional and strategic space policy structure for its failure to consider integrated space development and utilization in a way that spoke to Japan's national security.[65] The timing of nuclear testing by North

Korea in 2006 played into the hawkish orientations of the incoming Abe government and further underscored the necessity of an integrated riposte such as the institutionalization of a Japanese version of the U.S. National Security Council (NSC), which has not come to fruition, and the upgrading of the JDA to a full-fledged MOD, which has.[66] As of 2009, a range of events and opinions suggest that regional military realities are coloring Japan's security discourse—chief among them North Korea's continued nuclear and missile tests, as well as China's rising power over the long run.[67]

But the external still has to be filtered through the domestic lens. We believe that ultimately, of course, the specific turns and twists in Japan's defense will be determined by domestic factors, including the fate of institutionalized norms that already, contrary to expectations even a decade ago, appear to be more belligerent.[68] Thus our line of reasoning is simple but solidly situated in the flow of geopolitical realities in which Japan finds itself: Largely because they can profit from the shifting parameters of Japan's security concerns by pushing the militarization of space technologies, the interests and capabilities of private actors have become increasingly critical to the future of any general debates about Japan's militarization.

This focus in the broader security picture of Japan is not new. In fact, private industry and defense contractors have long been embedded in and were central to Japan's militarization saga across historical periods.[69] Our emphasis that Japanese space-related corporations have moved to advantage themselves economically in the contemporary shifting geopolitical landscape has resonance in the interwar period as well. Nor is our emphasis analytically controversial from the perspective of political economy in which, relative especially to diffuse interests, the greater influence of concentrated interests on economic policymaking is considered to be a well-established proposition.[70] Generally, political economists expect policy to be biased in favor of special or concentrated interests because publics or electorates may well be confounded by the effect or technical complexity of such policies. However, this is not to discount the importance of diffuse interests, which can have a powerful impact given factors such as preference intensity on a particular issue or policy, patterns of group organization, and domestic political institutions. Thus although we emphasize the importance of space-related enterprises in the context of Japan's space policy, we too remain attentive that the anti-militaristic sentiments of the Japanese public has thus far constrained the outright militarization of Japan's space technologies, whether by public or private enterprises.

Thus, to sum up, our simple underlying contention is that space-related technologies can, have been, and are increasingly being shifted *from the market to the military* in the interest of Japan's national defense. As Japan's external security concerns rose to the fore of public discourse, we contend that in seeking to survive over time, a specific set of Japanese corporations has shifted more visibly than ever before from commercialization to militarization of space-related technologies—a process that has had and will continue to have important implications for Japan's space military capabilities for national defense. We also contend that as the security discourse in Japan has changed, this market-to-military trend, vocalized by the corporate sector, began to be reflected in the national space strategy. Developments as recently as 2005, such as the Kawamura initiative discussed later in this book, made clear that the policymaking structure for Japan's space development would shortly reflect the market-to-military trend.

If our underlying reasoning is correct, we can expect to see several interrelated things: that the underlying space technologies are transposable from civil to military uses; that a key set of corporations with concrete interests remain pivotal in advancing space technologies on both fronts; that the more commercial prospects seem shaky, the more likely it is that the military angle looms large over civil space-related ventures; and that, finally, the fabric of political and legal structures reflect military space concerns. Although not doctrinaire and certainly mindful of other elements that come into play, we will trace these essential themes through the narrative. Over the course of the first decade of the 2000s, we have seen the *market-to-military trend* come to center stage in Japan's space policy, with important consequences for Japan's defense and security debates at a pragmatic level.[71] At the end of 2002, as our research for this book got under way, the trend was barely detectable, even in an integrated analysis of the technological and legal developments in Japan's space sector. In 2009, as we went to press, the covers of this trend were being lifted more quickly than we had anticipated, revealing a much more overt militarized stance in official Japanese space policy.

Mindful of the historical complexities, as well as theoretical and constitutional controversies outlined above, we want to be clear about what we are *not* saying: We are not claiming that corporations were the only players in shaping Japan's space policy or the broader nature of Japan's defense politics. In addition, we are not claiming that the development of space technologies, the functionality of which can be more critical in military systems, is the solution to all of Japan's defense concerns in the future. But we do maintain that corporate interest and competence are critical elements in the market-to-military trend. By concentrating on the space industry, we seek to show that the technological state of Japan's militarization is far more advanced than appreciated, given the range of space-related technologies that are already at play within the portfolios of relevant corporations and government agencies. To that end, we are focused, first and foremost, on carefully tracking the incremental but concrete changes in specific space technologies across the postwar period.

If the historical experience of other space powers is a guide, we should fully expect Japan to also exploit the dual-use strategic nature of space technology. Perhaps more significant, Japan's own historical experience attests to this leaning as military production was dual-use right from the start, and private contractors have long been central to technological diffusion across private and public spheres.[72] Such changes and advances in Japan's space technologies have not come about overnight; rather, as we show, the technologies were painstakingly acquired by Japanese corporations, often in conjunction with government agencies, over the course of several decades. As discrete elements and to the untrained eye, these technologies do not themselves appear to be anything other than what they are—solid rockets to launch scientific satellites, liquid rockets to launch heavier satellites, communication and Earth observation satellites, highly precise positioning systems, pod de-orbiting ability, satellite inspection technologies, and so on.

But the critical point, as stressed earlier, is that many space technologies are inherently dual-use, making it difficult even for professionals to distinguish their civilian and military uses from each other. The most salient recent example came in early April 2009 with North Korea's rocket launch. When North Korea asserted that its rocket launch was only a satellite mission, Japan and the West charged that any such rocket could also be linked to a ballistic missile program that had the capability to deliver a nuclear weapon.[73] The same logic can also be brought to bear on assessing Japan's civilian space program. To those familiar with space technology and applications, those very same precise positioning systems, pod de-orbiting systems, and, especially satellite inspection technologies just mentioned, for example, are also basically military applications that Japan can legitimately say it developed for expressly peaceful purposes.

At the most basic level, it is this duality embedded in Japan's technoeconomic security that allows us to posit the basis for a comprehensive Japanese security policy in the space sector in the long run.[74] Japan's well-established doctrine of comprehensive security—one that pragmatically recognizes the twin importance of economic and military security—as a means of advancing its national interests is critical.[75] As a nation beholden to the U.S.–Japan alliance—where a weaker state like Japan faces the attendant risks of entrapment or abandonment from its more powerful partner—Japan's quest for technological autonomy through defense production has undoubtedly allowed much of its strategic military capability to be couched within the postwar goal of economic growth.[76] By focusing on what private actors have been able to achieve, we show how this line of reasoning resonates, albeit unevenly, across a range of Japan's space activities.

A ROADMAP FOR THIS BOOK

The story of the militarization of Japan's space technology is inextricably bound with unfolding geopolitical realities. Within them Japan's defense-related corporations have moved successfully to exploit the changing external and internal security atmosphere in the interest of their own commercial survival and profitability. Thanks to their efforts, there is little question that space-related assets have made progress in boosting Japan's military capabilities and, as we enter the next decade, military space will become an essential, fundamental, deployed, and strategic component of Japan's defense. When the changes in the space sector in Japan are examined cumulatively and carefully over time—in particular, the products from private actors—they suggest that Japanese militarization has moved well beyond the theoretical and constitutional dimensions. Concerns with space-sector funding notwithstanding, we aim to show that taken together, the incremental market developments within this sector over the postwar period are coalescing into a strategic military capability.[77] Japanese space development, in short, has moved from the first phase of catch-up, through the second phase of attempted commercialization, and well into the third phase in which the market-to-military trend is increasingly evident.

With this as background, empirics take center stage in the remainder of the book: What are the notable activities and products in Japan's space industry? What do space-related technologies imply about Japan's military capability over time? What, in turn, does all of this imply about Japan's tangible militarization prospects in the near future? The remainder of the book illustrates the market-to-military trend in a thematic and chronological fashion. Chapter 2 sets out the actual evolution of Japan's space policy that has brought efforts to militarize space technologies over the course of the 2000s into sharp relief. It provides an account of space development in Japan from the postwar period to the close of the 2000s, focusing on the historical, institutional, and corporate elements that have coalesced to push Japan further down the path from the market to the military. Chapter 3 turns to the Japanese players in the space militarization saga, laying out the key players, their motivations, and tactics. The next three chapters then turn directly to the historical development of specific space assets, drawing out links to the ways in which they have been or are being militarized in the interest of defending Japan. Chapter 4 focuses on one of the stated official pillars of Japan's space program, namely rockets. This includes both liquid and especially solid rockets that have implications for Japan's acquisition of ICBM technology. Chapter 5 focuses on the second official pillar of Japan's space program: satellites and spacecraft, which are critical for reconnaissance and, less appreciated, for supporting military communication networks and counterspace, particularly anti-satellite (ASAT), capabilities. Chapter 6 focuses on specific cutting-edge space-related technologies that are currently under development, particularly with Operationally Responsive Space (ORS), Command, Control, Communications, Computers, Intelligence, Surveillance, Reconnaissance (C4SIR), and counterspace technologies. Chapter 7 concludes by assessing Japan's capabilities, drawing out the analytical and policy implications of the country's space capabilities. It also reexamines our findings under the newly minted Basic Space Law and the new Basic Space Plan, which for the first time in Japan's postwar history give the country a comprehensive national space strategy.

2 EVOLUTION OF JAPAN'S SPACE POLICY

IN THIS CHAPTER, we turn to the historical onset of space development in Japan, paying close attention to the country's institutional structure governing space policy, as well as the ups and downs the Japanese space industry faced in global competition in the postwar decades. During this period, there have been a range of legal and political constraints put in place to govern the development of space assets as shown in Table 2.1, and some of these were designed specifically to guard against the military uses of space. Over time, however, the militarization of space-related technologies became an ever steadier component of the overall space policy rubric in terms of rhetoric, planning, and industry demands. Using the principal constraints on Japan's space policy as set out in Table 2.1, this chapter focuses on tracing such developments in two parts, one focusing on government-centered changes and the other on industry-related demands.

GUIDING PRINCIPLES FOR SPACE ACTIVITIES

It is helpful to begin with a clear understanding of the fundamental legal and institutional context in which the Japanese government has operated. This section provides a brief discussion of the main instruments set out in Table 2.1, with a particular focus on those restraining the military uses of space that have long been of pragmatic interest to defense-related actors in the public and private spheres.

In 1969, Japan's industrialization of space began with two events, both of which indicated a clear and unequivocal official commitment to peaceful-only, nonmilitary development.[1] The first of these was the well-known Peace-

Table 2.1. Relevant Instruments for Japan's Space Policy, 1947–2009

	Year	*Instrument*[1]
Domestic	1947	Constitution of Japan
	1954	Article 9 Interpretation (exclusively defensive defense)
	1967	Three Non-Nuclear Principles (no possession, production, passage/introduction) Three Principles on Arms Export (partial ban)
	1968	Space Activities Commission (SAC) Establishment Law Statement of Minister of State Concerning the Term "Peace"
	1969	Resolution on Space Development for Exclusively Peaceful Purposes National Space Development Agency (NASDA) Establishment Law
	1975	*Promotion of Space Science* (Science Council, PM's Office; Ministry of Education)
	1976	Unified Government View of Arms Export (total ban) Nuclear Non-proliferation Treaty Cabinet order on Defense Spending Freeze at 1 Percent of GDP
	1978	*Fundamental Policy of Japan's Space Activities* (SAC)
	1981	Resolution as to the Problem of Arms Exports and Others
	1983	Statement of Chief Cabinet Secretary on Transfer of Military Technologies to the United States
	1984	*Fundamental Policy of Japan's Space Activities* (revised, SAC)
	1985	Resolution on Utilization of Satellites for "Peaceful Purposes" by the Self-Defense Force (SDF) Common Use Principle
	1986	Five Principles for Japanese Participation in the SDI Project
	1989	*Fundamental Policy of Japan's Space Activities* (revised, SAC)
	1995	Science and Technology Basic Law
	1996	*Fundamental Policy of Japan's Space Activities* (revised, SAC) First Science and Technology Basic Plan 1996–2000

(*continued*)

Table 2.1. *(continued)*

	Year	*Instrument*[1]
	1998	Statement of the Chief Cabinet Secretary Regarding Cooperative Technical Research with the United States on Ballistic Missile Defense (BMD)
	1999	Japan–U.S. Memorandum of Understanding on Cooperative Research on BMD
	2001	Second Science and Technology Basic Plan 2001–2005 *Promotion Strategy by Area* (Council on Science and Technology Policy, [CSTP])
	2002	*Basics of Future Space Development and Utilization* (CSTP) Law Concerning Japan Aerospace Exploration Agency (JAXA)
	2004	*Basic Strategy for Space Development and Utilization* (CSTP) *Japan's Visions for Future Security and Defense Capabilities* (Report by the Council on Security and Defense Capabilities [CSDC; Araki Report])
	2005	*Toward Establishment of New Space Development and Utilization System* (Report by the National Space Strategy Planning Group [NSSPG; Kawamura Initiative]) *National Defense Program Guideline, FY 2005*
	2006	*Towards the Establishment of a New Space Development and Utilization System: Japanese Space Policy as a Peaceful Nation* (Report by Special Committee on Space Development, Policy Affairs Research Council of the Liberal Democratic Party [LDP]) Third Science and Technology Basic Plan 2006–2010
	2007	Basic Act on the Advancement of Utilizing Geospatial Information (AUGI)
	2008	Basic Space Law Establishment of a State Minister for Space
	2009	*Basic Space Plan* (Strategic Headquarters for Space Policy [SHSP]) Establishment of a cabinet office for space development Ministry of Defense (MOD) Mid-Term Defense Plan, including military space policy
Bilateral	1962	Japan–U.S. Exchange of Notes Constituting an Agreement Relating to the Intercontinental Testing of Experimental Communications Satellites

1968	Japan–U.S. Exchange of Notes Constituting an Agreement Regarding the Establishment and Operation of a Satellite Tracking Station in Okinawa
1969	Japan–U.S. Exchange of Notes Constituting an Agreement Concerning Cooperation in Space Activities for Peaceful Purposes
1972	Japan–European Space Research Organization (ESRO) Exchange of Notes Concerning the Cooperation in Space Exploration
1974	Japan–U.S. Exchange of Notes Constituting an Agreement Relating to a Tracking Station on Kwajalein
1975	Japan–U.S. Exchange of Notes Constituting an Agreement Relating to the Furnishing of Satellite Launching and Associated Services
1978	Japan–Canada Exchange of Notes Constituting an Agreement Concerning the Establishment of a Temporary Satellite Support Facility at Churchill Research Range
1980	Japan–U.S. Exchange of Notes Constituting an Agreement Relating to Space Co-operation in Shuttle Contingency Landing Sites
	United States of America and Japan: Exchange of Notes Constituting an Agreement Relating to Space Launch Assistance
1983	Japan–U.S. Memorandum of Understanding Relating to the Operation of the LANDSAT System
1985	Japan–U.S. Memorandum of Understanding for A Cooperative Program Concerning Design (Phase B) of A Permanently Manned Space Station
1987	National Space Development Agency of Japan (NASDA)–European Space Agency (ESA) Memorandum of Understanding for the Direct Reception and Distribution of MOS-1 Data
1988	Japan–U.S. Agreement on Cooperation in Research and Development in Science and Technology (extended repeatedly)
1989	Japan–U.S. National Aeronautics and Space Administration (NASA) Memorandum of Understanding on Cooperation in the Detailed Design, Development, Operation and Utilization of the Permanently Manned Civil Space Station
	U.S.–Netherlands Exchange of Notes Concerning the Joining of Japan to the Arrangement Concerning Application of the Space Station Intergovernmental Agreement of 29 September 1988
1990	Japan–U.S. Exchange of Notes Concerning the Policy and Procedure of the R&D and Procurement of Artificial Satellite
	Japan–Australia Exchange of Letters constituting an Agreement concerning Co-operation on the Project for the Geostationary Meteorological Satellite-4 System

(*continued*)

Table 2.1. *(continued)*

	Year	*Instrument*[1]
	1991	NASDA–ESA Memorandum of Understanding concerning mutual access to data from the Japanese ERS-1 and the European ERS-1 missions
	1994	NASDA–Indonesia National Institute of Aeronautics And Space (LAPAN) Cooperative Project of Utilization for Japanese Earth Resources Satellite-1 Data
		Institute of Space and Astronautical Science (ISAS)–ESA Memorandum of Understanding concerning cooperation on the Infrared Space Observatory Satellite Program
		NASDA–NASA Memorandum of Understanding for Cooperation in the Advanced Earth Observing Satellite Program
		NASDA–LAPAN Collaborative Research Agreement for Studies for Production of Fundamental Datasets for Earth Science and Technology Researches
		Japan–U.S. Agreement Concerning Collaboration in the Space Flyer Unit (SFU) Program
		NASDA–ESA Memorandum of Understanding Concerning an Optical Link and S-band Experiment between the ARTEMIS and the Optical Inter-orbit Communications Engineering Test Satellite
	1995	Japan–U.S. Agreement Concerning Cross-Waiver of Liability for Cooperation in the Exploration and Use of Space for Peaceful Purposes
	1996	NASDA–France Centre National d'Etudes Spatiales (CNES) Inter-Agency Agreement concerning the Preparation of a Long Term Cooperation in the Field of Space Programs
	1997	NASDA–ESA Agreement Concerning the Direct Reception, Archiving, Processing and Distribution of ERS-2 SAR Data
		NASDA–ESA Memorandum of Understanding Concerning the Launch of ARTEMIS
		Japan–Australia Exchange of Notes Constituting an Agreement Concerning Cooperation on the Geostationary Meteorological Satellite-5 System
	1998	Japan–U.S. Joint Statement on Cooperation in the Use of the Global Positioning System
International[2]	1963	Treaty Banning Nuclear Weapon Tests in the Atmosphere, in Outer Space and Under Water
		Declaration of Legal Principles Governing the Activities of States in the Exploration and Uses of Outer Space, United Nations (UN) General Assembly Resolution 1962 (XVIII) of 13 December 1963
	1967	Treaty on Principles Governing the Activities of States in the Exploration and Use of Outer Space, Including the Moon and Other Celestial Bodies (Outer Space Treaty)

1968	Agreement on the Rescue of Astronauts, the Return of Astronauts, and the Return of Objects Launched into Outer Space (Rescue Agreement)
1971	Agreement Relating to the International Telecommunications Satellite Organization
1972	Convention on International Liability for Damage Caused by Space Objects (Liability Convention) Agreement on the Establishment of the Intersputnik International System and Organization of Space Communications
1975	Convention on Registration of Objects Launched into Outer Space (Registration Convention)
1976	Convention on the International Mobile Satellite Organization
1977	Convention on the Prohibition of Military or Any Other Hostile Use of Environmental Modification Techniques, UN General Assembly Resolution 31/72 of 10 December 1976
1979	Convention on the Transfer and Use of Data of Remote Sensing of the Earth from Outer Space
1982	Principles Governing the Use by States of Artificial Earth Satellites for International Direct Television Broadcasting, UN General Assembly Resolution 37/92 of 10 December 1982
1986	Principles Relating to Remote Sensing of the Earth from Outer Space, UN General Assembly Resolution
1988	Japan–U.S.–ESA Member States–Canada–Russia Agreement on Cooperation in the Detailed Design, Development, Operation, and Utilization of the Permanently Manned Civil Space Station)[3]
1992	International Telecommunication Constitution and Convention Principles Relevant to the Use of Nuclear Power in Outer Space, UN General Assembly Resolution 47/68 of 14 December 1992
1996	Declaration on International Cooperation in the Exploration and Use of Outer Space for the Benefit and in the Interest of All States, Taking into Particular Account the Needs of Developing Countries, UN General Assembly Resolution 51/122 of 13 December 1996
1998	Japan–U.S.–ESA Member States–Canada–Russia Agreement Concerning Cooperation on the Civil International Space Station

SOURCES: For the historical information use is made of "'Heiwa' to iu Go no Imi ni kan suru Nihonkoku no Kokumu Daijin no Genmei (1968 Nen Dai 61 Kokkai Kagaku Gijutsū Shinkjō Taisaku Tokubetsu Iinkai Giroku)" [Declaration of Japan's Minister of State Concerning the Term 'Peace' (Minutes of the Special Committee on Science and Technology Promotion Policy, 61st Diet, 1968]; "Waga Kuni ni okeru Uchū no Kaihatsu Oyobi Riyō no Kihon ni kan suru Ketsugi [Resolution Concerning Japan's Basic Development and Utilization of Space," (Plenary Session of the House of Representatives, 5 May 1969); and Uchū Kaihatsu Jigyōdan Hō ni Tai Suru Kokkai no Futai Ketsugi [Supplementary Resolution by the Diet Concerning the National Space Development Agency Law]," (Special Committee on Science and Technology Promotion Policy, House of Representatives, 13 June 1969); with all three instruments available via the official index on Uchūhō (Space Law) online through the library at JAXA at www.jaxa.jp (accessed 5 November 2009). In

(*continued*)

Table 2.1. *(continued)*

addition to information from JAXA's Uchūhō (Space Law) (accessed also earlier 30 June 2008) information is from the Ministry of Defense (MOD), *Defense of Japan 2008*, available online at www.mod.go.jp (accessed 3 March 2009), pp. 411–412; Ministry of Foreign Affairs (MOFA), "Japan's Policies on the Control of Arms Exports," "Japan–U.S. Cooperation in Equipment and Technology," "On the Three Non-Nuclear Principles," and "Examples of Japan's Announcement on the Three Non-Nuclear Principles," all available online at http://www.mofa.go.jp (accessed 30 June 2008); Ministry of Defense (MOD), "Japan's BMD," February 2009, available online at www.mod.go.jp (accessed 30 March 2009); Tetsuo Tamama, "Japanese Space Policy Revision and Future of National Security," *DRC Annual Report 2002*, online version available at www.drc-jpn.org (accessed 5 November 2009); Andrew L. Oros, *Normalizing Japan: Politics, Identity and the Evolution of Security Practice* (Stanford, CA: Stanford University Press, 2008), Appendices 5–7; U.S. Department of Commerce International Trade Administration's Trade Compliance Center (TCC), Japan Satellite Procurement Agreement and Japan Science and Technology Agreement, both available online at www.tcc.export.gov (accessed 30 June 2008); U.S. Government, National Executive Committee for Space-Based Positioning, Navigation, and Timing (PNT), GPS Agreement, available online at http://pnt.gov (accessed 30 June 2008); Government of Japan, Basic Science and Technology Plans, all available from the official site of the Council of Science and Technology Policy (CSTP) at http://www8.cao.go.jp/cstp (accessed 3 September 2008); Japan Science and Technology Agency (JST), Directory Database of Research and Development Activities (ReaD), available online at http://read.jst.go.jp (accessed 1 June 2009); Geographical Survey Institute (GSI), "Policy Planning" for the text of the Basic Act on the Advancement of Utilizing Geospatial Information (AUGI), available online at www.gsi.go.jp (accessed 28 July 2009); United Nations, General Assembly Resolutions, available online at www.un.org (accessed 16 April 2009); United Nations, *International Agreements and Other Available Legal Documents Relevant to Space* (Vienna: United Nations, 1999); United Nations, *United Nations Treaties and Principles on Outer Space*, ST/SPACE/11 (New York: United Nations, 2002); and United Nations, *United Nations Treaties and Principles on Outer Space and Other Related General Assembly Resolutions—Addendum: Status of International Agreements Relating to Activities in Outer Space as at 1 January 2009* (Austria: United Nations, 2009), p. 11. Of the five major international legal instruments on outer space, the Moon Agreement is the only one to which Japan is not a party.

1. With the exception of a few instruments sponsored by institutions, the focus here is primarily on government-related laws, resolutions, documents, and agreements. Where possible, some of the actors behind the instruments are identified as indicated in parentheses.
2. Between 1958 and 2008, close to ninety UN General Assembly Resolutions Relating to Outer Space have been passed (in such areas as cooperation in peaceful uses, transparency and confidence-building measures, and prevention of arms races). With some exceptions, Japan has voted in favor of such resolutions. The UN General Assembly has identified five declarations and legal principles as central to space law, and these are the only ones indicated here.
3. This was enacted in Japan in 1992.

ful Purposes Resolution (PPR) by the Diet, which was urged on by the prevailing elite and political sentiment that for a country like Japan the peaceful uses of outer space were to be one and the same as peaceful uses of nuclear power (as exemplified in the three non-nuclear principles earlier in 1967). Decrying both, the Diet resolved to severely circumscribe Japan's future space activities to exclusively peaceful purposes.[2] This took Japan's stance well beyond the provisions of the 1967 Outer Space Treaty, that offers no clear-cut statement as to whether the practical use of outer space for exclusively peaceful purposes signifies nonaggressive (which was accepted internationally) or nonmilitary uses (which Japan expanded domestically).[3] The other event was the establishment of the National Space Development Agency (NASDA) of Japan, which was designed to bring some coherence to the implementation of Japan's space policy. As the establishment law of NASDA lacked any such statement, a supplementary provision by the Diet stipulated that its activities were also to be circumscribed to peaceful purposes.[4]

Other bans and resolutions governing Japan's activities in outer space have also drawn attention. In 1967 Japan had a reasonably cast-iron policy on weapons exports.[5] A resolution on arms export controls—the Three Principles on Arms Export (TPAE)—committed the government to prohibiting exports to communist countries, to countries under weapons embargo by the United Nations, and to states or countries presently at war or likely to prosecute war. In 1976 there was an even stricter resolution expressly forbidding not only weapons sales abroad, but even dual-use machine tools and plants. The export of arms to areas designated in the TPAE was prohibited, the export of arms to other areas not subject to TPAE were to be restrained in line with the spirit of the Constitution and the Foreign Exchange and Foreign Trade Act, and, finally, equipment related to arms production was to be treated in the same category as arms.

The 1976 stricter application of TPAE made arms exports practically impossible, which was of grave concern to defense-related corporations. The resolution, however, was subsequently given an escape clause with respect to the United States.[6] A 1983 agreement specified that transfers of military technologies to the United States would be exempt from Japan's ban on arms exports. It also confirmed that commercial technologies with defense applications, so-called dual-use technologies, would be available. Perhaps most important, the notes indicated that if Japan improved on or modified technologies of U.S. origin, referred to as *derived technologies*, these would flow back to the United

States free of charge. Japanese-developed defense technologies, or *nonderived technologies*, would be available to the United States for a fee.

These constraints have also interacted with a set of international agreements and United Nations (UN) treaties that are relevant to Japan's space development and utilization. These include, for example, the 1963 Partial Test Ban Treaty prohibiting nuclear testing in outer space and the 1977 convention that generally prohibits parties from using military techniques that change the dynamics or structure of outer space. More directly on point, the UN has identified five treaties and agreements as central to space law—the 1967 Outer Space Treaty, the 1968 Rescue Agreement, the 1972 Liability Convention, the 1976 Registration Convention, and the 1984 Moon Agreement, which is the only one not signed by Japan.[7]

Of particular significance is Article IV of the 1967 Outer Space Treaty, which is a pragmatic blend of the national (especially that of the United States's and Soviet Union's) defense concerns during the Cold War and the multilateral interest in preserving space from militarized appropriations during that time and beyond.[8] The first clause of Article IV stipulates that states do not place in orbit or install on the moon any kind of weapons of mass destruction (WMD), including nuclear weapons (but leaving aside other kinds of weapons, such as American and Soviet intercontinental ballistic missiles, or ICBMs, that actually had to pass through outer space). The second clause of Article IV stipulates that the moon and other celestial bodies (but leaving aside outer space more generally) are to be used "exclusively for peaceful purposes" but that military personnel and any (presumably also military) equipment or facility are not prohibited so long as they are used for peaceful purposes or exploration.

The interpretation of the Outer Space Treaty's "exclusively for peaceful purposes" provision is of considerable significance to understanding Japan's maneuvering in the context of existing international space law. Since the beginning of the space age, the official position of the United States has largely been that the use of the word "peaceful" allowed for the "nonaggressive" military exploitation of space—implying that military use of space was permitted and legal as long as those activities remained passive or "nonaggressive."[9] Obviously, then, there continue to be two competing definitions of "peaceful purposes"—one is "nonaggressive" and the other is "nonmilitary." As no state has ever formally challenged the U.S. position, the consensus within the UN has also been that "peaceful" specifically equates to "nonaggressive."

As we noted earlier, Japan's 1969 resolution did not even admit to military uses, the common consensus in international society. Thus, from that historical benchmark in 1969, each and every step of Japan's official space policy until the first decade of the twenty-first century has been directed at finding nonmilitary justification for the nation's space development, whether it was to improve the life of the citizenry or to unravel the mysteries of the universe. It has been an unnatural struggle. As in other countries, the practical realities of the military use of space in Japan began to chip away at the domestic PPR edifice, beginning visibly just over a decade later with the controversies about the use of commercial satellites by the Self-Defense Force (SDF). This was the struggle that the Japanese space policy community had in mind when attempting to bring Japan's legal framework for space activities on par with international space law norms—certainly, "nonaggressive/nonoffensive" but not "nonmilitary" uses of outer space.

CHANGES AT THE OFFICIAL LEVEL

Although Japan was heavily criticized from within for not having a coherent national space strategy for most of the postwar period, we believe that a careful survey of the language of Japan's official space-related documents reveals a coherent realist progression in space policy that stemmed from security concerns.[10] More specifically, these documents indicate the Japanese government's increasing acceptance of the militarization of space technologies despite the constraints. For this reason, we begin by focusing on the major shifts in official texts that have taken place from the beginning of Japan's space program to the present.

Textual and Policy Shifts

From the 1970s through 2002, Japan's space policy has always emphasized the country's technological aspirations for space development, while giving only a vague agenda for their practical usage. The Space Activities Commission (SAC) released its first full-blown *Fundamental Policy of Japan's Space Activities* in 1978.[11] The *Fundamental Policy* was revised twice, in February 1984 and June 1989. SAC authored the final version in 1996, before the commission became an implementing authority under the Council for Science and Technology Policy (CSTP).

Over the course of roughly two decades in which there were enormous structural changes in global politics, Japan's official space policy continued to focus on lofty justifications for a national space program—the exploration of space and the solar system leading to different thinking about the universe, the consequent creation of a new philosophy and culture in an intellectually mature society, and so on. By and large, the *Fundamental Policy* was concerned with generalities. For example, it mentioned the essential nature of satellites in daily lives and the necessity of establishing a global Earth observation system in harmony with other nations. It also focused on how space development involves sophisticated generic technology, which can integrate various fields of science and technology and which can spill over creatively and propel developments in other fields, such as materials, computers, robotics, electronics, communications, and information processing. In addition, it urged that Japan use space development for environmental research and monitoring. None of this is controversial in the least, much less militarized in overtone. Overall, the basic policy at this early stage of Japan's space development seemed permeated with two important themes—civilian use and international cooperation.

However, the *Fundamental Policy* emphasized increasing the level of sophistication of key space technologies. For satellites, it was particularly clear about their utility for Earth observation, communications, broadcasting, and navigation. For launch vehicles, it was emphatic about ensuring Japan's independent access to space via developing a range of solid and liquid rockets. SAC's emphasis needs to be understood also in the context of the competitive military realities of space technologies, which had been impressed on the Japanese by the Americans. Even in the late 1960s, the United States feared that Japan's autonomous development of a technological capability to launch satellites could be converted to a means for launching ballistic missiles and, as such, would pose a threat to its own supremacy and leadership.[12] This led the Johnson administration to force Japan to accept the transfer of U.S. rocket technologies (based on Thor-Delta launch vehicles) under black box conditions. Large and important sections of Japan's space development community opposed this as a move by the United States to destroy Japanese indigenous technology. Despite setbacks, as we show in Chapter 4, Japan's indigenous rocket program did indeed go on to yield solid-rocket systems that were deemed by the United States to be suitable for conversion to ballistic missiles.

Starting in 2000, Japan began a marked shift toward the militarization of space assets that resulted from the shakeup in the science and technology policymaking structure (see Chapter 3). For now, we focus on the then newly created CSTP directly under the Cabinet Office (CO, the prime minister's office), which was pivotal in moving the space industry into the area of national security. With the Second Science and Technology Basic Plan (2001–2005), the CSTP effectively took strategic space planning out of the Ministry of Education, Culture, Sports, Science and Technology (MEXT), which was primarily concerned with securing budget for technological development and scientific research, and moved it into a wider community of experts linked to the prime minister's office.

In September 2001, the CSTP released the first post-realignment statement on space policy, focusing on a national promotion strategy across eight prioritized science and technology areas, including "frontier" areas such as space.[13] The significance of the *Promotion Strategy* report lies in the fact that it was the first official and visible statement to acknowledge the role of space in Japan's national security. Although pitched at a very broad aspirational level, its specifics on how safety was to be ensured left little doubt about the importance of the space sector for the future of Japan: information-gathering satellites and associated space transportation capabilities, advanced positioning and exploration technologies, and next-generation satellite technologies.

The CSTP followed that in June 2002 with a more comprehensive but essentially reinforcing statement on space policy.[14] As in the past, the *Basics of Future Space* report set out the objectives and priorities of Japan's space policy, such as the development of knowledge, and the advancement of economy and society. Other goals, however, such as ensuring safety, linked it to the future of Japan's national security. The priority for satellites, for instance, emphasized security and crisis management before listing communication and positioning, and Earth observation. There was, again, a distinct emphasis on maintaining independent access to outer space by developing and supporting launch vehicles. Overall, these policy moves in the early 2000s were subtle but important steps in highlighting the relationship between space planning and national security issues for Japan.

The tilt toward the militarization of space policy was cemented in 2004, when CSTP released the definitive version of a detailed space plan for the next ten years.[15] As the new *Basic Strategy* pointed out, with very limited budgets and personnel Japan had already reached a technological sophistication on par with the leading space powers of the day. That in itself was quite a feat. With the premise that space technology, as identified in Table 2.2, was strategic and would have ripple effects throughout the economy and society, the *Basic*

Table 2.2. Japanese Core Space Technologies and Their Potential Military Uses

Core Technology	*Component*	*Field*	*Examples of Potential Military Uses*
Observation Sensor Technology	High-resolution optical sensor technology	Earth observation, security, space science	Information, surveillance, reconnaissance Monitoring, data collection, intelligence Navigation, position location Environmental and meteorological services Precision targeting Early-warning system Nuclear test detection and surveillance
	Microwave sensor technology (including phased array antenna technology)	Earth observation, security, space science	
	Multiband optical sensor technology	Earth observation, space science	
	Satellite and sensor operation technology for imaging	Security, space science	
Communications Infrastructure Technology	Transmission and receiving technology (high-gain antennas, high-performance amplifiers)	Information communications	Command, control, communication Telemetry, tracking Navigation Precision targeting Troop movement, coordination
	Satellite data relay technology	Information communications	
	Technology required for high-speed communications (data relay, switching and transfer systems on satellite, high-speed IP technology	Information communications	
Positioning Infrastructure Technology	High precision time control technology (basis for improving precision of distance measuring)	Positioning	Precision targeting Troop movement, coordination
	High precision orbital estimation / determination technology (basis for positioning)	Positioning	

Rocket Technology	Liquid rocket propulsion engine Technology	Transportation systems	Ballistic missiles Antiballistic missiles Antisatellite systems Orbital reaction control systems Interceptor propulsion
	Rocket guidance technology	Transportation systems	
	Solid-propellant rocket system technology	Transportation systems	
Satellite System Technology	Satellite bus technology (including constellation technology)	Satellite systems, space science	Antisatellite systems Defensive/offensive counterspace applications
	Space robotic technology	Satellite systems, International Space Station, space science	

SOURCES: Columns 1–3 taken from Council on Science and Technology Policy (CSTP), *The Basic Strategy for Space Development and Utilization* (Tokyo: CSTP, 9 September 2004), p. 20. In addition to those mentioned in the text, some potential military uses listed here are from Joan Johnson-Freese, *Space as a Strategic Asset* (New York: Columbia University Press, 2007), pp. 31–32, table 2.1; and also Charles V. Pena and Edward L. Hudgins, "Should the United States 'Weaponize' Space? Military and Commercial Implications," *Policy Analysis* 427, 18 March 2002, appendix tables 1–3.

Strategy stressed the importance of autonomy and independence across the board. It left little doubt about the importance of space for national security or public safety in Japan. Although the CSTP did not address each identified core technology explicitly, we can extract the way that they can be used in military operations.

Overall, continuing the themes from its earlier reports, the CSTP deemed it absolutely essential that Japan maintain the ability to independently launch satellites and space transportation systems into space when needed. Both the government and, especially, the private sector were critical in advancing the state of space technologies: the development of information-gathering satellites (can be used for spying), maintenance of solid-propellant technology (can be used for ballistic missiles), and even the establishment of a Global Positioning System (GPS)-compatible regional satellite positioning system (considered an indispensable space-based force-multiplier technology) were now officially matters of state policy and in, various ways, critical to national security and crisis management. The distance Japan's space policy traveled under the CSTP was, by Japanese standards, dramatic. From there, the new national security rubric was also reinforced in the Kawamura initiative, which is detailed next.

The Kawamura Initiative

If the CSTP reports indicated only a market-to-military trend that challenged the principles governing defense policies, the ever more concrete proposals that came thereafter left little doubt about the future of Japan's space development. By 2006, there were concrete plans under way to militarize Japan's space development, both to bring it in line with international norms and to answer to pressing governmental and corporate interests that had long been constrained by institutional and legislative structures in Japan.[16] We dub this the Kawamura initiative, as these plans emerged from successive deliberations in 2005 within the National Space Strategy Planning Group (NSSPG) led by Takeo Kawamura.[17]

Set against the legal structure both domestically and internationally that had long attempted to constrain the militarization of space assets, the Kawamura initiative in Japan was striking. A combination of factors that form the subsequent narrative in this book—among them, the push for technological development without commercial payoffs, the failure to see and deal with space development as a coherent national strategic priority, the battles over the *Quasi-Zenith Satellite System* (*QZSS/Michibiki*), lack of funding for the

GX rocket program, for example — were brought home with clarity to concerned Japanese actors in the days after the November 2003 failure of the *H-IIA* rocket carrying the second pair of Japan's spy satellites that had drawn much attention. At that point, various players engaged in unsavory maneuvers to avoid responsibility for the debacle.[18]

Seeking a coherent strategy and institutional responsibility in such a context, the NSSPG focused its agenda on *using* space assets to respond to a range of pressing security concerns for Japan—nuclear weapons and missiles in North Korea, piracy in the Malacca Straits, natural resources in the East China Sea (the Senkaku Islands), the presence of unidentified ships, the proliferation of WMD, and terrorism. Within these geopolitical realities the Kawamura initiative was timely, and its emphasis on utilizing space for national security stemmed from more than a decade's worth of activity by a set of public and private Japanese actors to normalize Japan's defense posture in the face of an evolving series of threats.[19]

Institutionally, in an effort to eliminate crossed and tangled lines of command resulting from administrative reforms, the Kawamura initiative included a recommendation that control over space development be taken away from bureaucratic charge for the first time in postwar history and placed directly under the state minister for space utilization within the prime minister's CO. This would not only establish the strategic significance of space within the highest echelons of political decision making in the country, but it would also better equip Japan financially and mechanically in dealing with the United States as it moved into more complex space-related technologies.

On the legal front, in the Kawamura initiative, the NSSPG chided Japan's space policy insiders for being wedded to an unnecessarily narrow interpretation of the peaceful uses of outer space; and asserted that Japan's interpretation of peaceful as "nonmilitary" uses of outer space was out of line with the international (and especially the United States') interpretation of "nonaggressive" or "nonoffensive" uses. In addition, as the NSSPG pointed out, Article IV of the Outer Space Treaty does not prohibit the use of military personnel in outer space so long as the personnel were in place for scientific research and for a peaceful purpose. In short, the NSSPG used the Kawamura initiative to advocate the overturn of Japan's 1969-era PPR that had prevented Japan's space development from overtly militarizing and to replace it with a law echoing the norms of the Outer Space Treaty—one that allowed defensive military applications (such as those for gathering information) to be launched into orbit.

We believe that with the Kawamura initiative, the NSSPG sought to find ways to dispel long-standing frustrations voiced by some in the LDP, then JDA, and defense-related corporations. At the political level, the initiative went on to affect the wider political thinking on the subject and was crystallized in LDP proposals that spoke more cohesively to a national security paradigm for Japan in much the same way as the NSSPG report.[20] These included, for example, moves to replace the PPR with new legal foundations for the military use of space, steps to wrest control away from a plethora of bureaucratic actors, and consolidate strategic space policymaking directly in the CO. These efforts were actually then enshrined in a space bill that was submitted to the Diet in summer 2007 by Fukushiro Nukaga, head of the Special Committee on Space Development in the LDP.[21] The passage of the Basic Space Law in 2008 signified that Japan had entered a historic new era in its national space policy in which the market-to-military trend is unmistakable.

On the corporate side, the Kawamura initiative also spoke to concerns long voiced by corporations, most notably the Mitsubishi Group [with Mitsubishi Electric Corporation (Melco) and Mitsubishi Heavy Industries (MHI) in the vanguard] and Nippon Keidanren (Japan's formidable Business Federation lobby). Any overturning of the peaceful purposes paradigm would mean economic benefits for a specific set of Japanese industries, relieving them of the stress of having to compete for commercial systems in which they had, thus far, not been able to establish a foothold against Western competitors. It would also allow them to export and import next-generation technologies (such as those related to BMD), which would strengthen profitable relationships between Japanese and U.S. industry. This brings us to another critical strand in the evolution of Japan's space saga—Japanese defense-related corporations themselves.

THE ROLE OF CORPORATE INTERESTS

Japan's defense-related corporations have not exactly been disinterested in the changing structures and directions of Japan's space program. As early as 1961, Nippon Keidanren set up its Space Activities Promotion Committee, marking the start of its interest in the contents and directions of Japan's space industry.[22] The express purpose of this committee, which was formed well before the birth of NASDA in 1969, was and still is to conduct research on both industrial space developments and applications. Every year, on behalf of about fifty-six or so manufacturers, it puts forward policy proposals for the government's consideration from Nippon Keidanren.

Positing that industry is willing to work in partnership with the government, space agencies, and business enterprises, Nippon Keidanren's rationale for the continued focus on aerospace technologies of interest to Japanese corporations also echoes the market-to-military trend.[23] Nippon Keidanren has not been shy about demanding consistency between international space law norms and Japan's principle of peaceful utilization of space as described above. Such demands have a long lineage. Japanese industry has, in fact, been constantly and implacably opposed to the peaceful purposes rule from its inception, with open calls as early as 1968 for the promotion of scientific utilization of space in order to modernize national defense.[24] As discussed in the previous chapter, corporate demands have found sanctuary in the geostrategic environment surrounding Japan at present. The recent activities of Japanese industry, particularly related to technological application and legislative/policy changes as discussed below, are thus good indicators of the market-to-military trend.

Sectoral Background

An understanding of Japanese corporations' incentives for militarizing space assets requires a brief historical detour into the economic realities of Japan's space sector in the postwar period. First, it is helpful to briefly dispel the widespread and rather simplistic evaluations that generally fail to account for the full historical and integrated technological progress of Japan's space program.

As in other countries, Japan's space program has seen historic highs and lows.[25] In line with Japan's industrial rise that had caused much trade friction, many evaluations suggest that in the 1980s Japan's space program was a well-oiled machine that was a growing commercial challenge to U.S. and European interests across a range of technologies. As the 1980s drew to a close, the increase in Japanese content of major space launch vehicles (SLVs) and satellites boosted the confidence of key actors—the *H-II* had been commissioned as all-Japanese and the technology content of the *CS-3a* communication satellite was reportedly 80 percent Japanese. Major Japanese aerospace firms such as MHI and Nissan Motors involved in the production or design of advanced propulsion research considered SLVs as a new commercial frontier for profits in contrast to older industrial businesses and invested accordingly. Similarly, with the successful launch of a series of advanced communication and broadcasting satellites, Melco, Nippon Electric Corporation (NEC), and Toshiba were all keen on commercial satellite market entry in the 1990s and moved toward investment in plant and production facilities. For example, Melco was reported to be eyeing annual sales growth of 10 to 20 percent in the space sector.

Construction companies began to look to lunar bases as possible future expansion opportunities.

During this period, government actors were no less immune to the lure of space.[26] Japan's then NASDA was busy drawing up plans for new medium and heavy-lift capabilities for launching numerous satellites in development or under study. It was also planning large-scale space projects, such as a reusable orbital plane called the *H-II Orbiting Plane–Experimental* (*HOPE-X*) and a contribution to the International Space Station (ISS) with the *Japanese Experiment Module* (*JEM* also called *Kibō*). Even a manned space era did not seem out of the question for Japan. In retrospect, the mid-1990s were among the last years in which the Science and Technology Agency (STA), the Institute of Space and Astronautical Science (ISAS), and the SAC also pursued aggressive space development programs. They had ambitious programs for Japan spanning more than fifteen years—developing a series of geosynchronous (GEO) and low-Earth orbit (LEO) satellites as part of a global Earth observation system called *GEOS*, launching planetary and lunar missions, implementing the *JEM* for the *ISS*, and also developing a fully reusable shuttle.

From the mid-1990s, however, Japan's space program appeared to be on somewhat of a slide with some technical issues, and the program was repeatedly battered by the domestic press in particular. In reality, along with technical troubles, budget problems loomed ominously. The ambitious program planned in the late 1980s ran into trouble, largely because its founders had based it on the expectation of continuous increased funding in the era of seemingly unstoppable Japanese growth and had not anticipated the economic fallout from the lost decade of the 1990s. The stagnant growth and funding directly affected the economic livelihood of key corporations involved in the space game. The string of technical troubles began with the *ETS-VI/Kiku-6* satellite in 1994, the hypersonic shuttle prototype (*HYFLEX*) in 1996, the *Advanced Earth Observing Satellite* (*ADEOS/Midori*) in 1997, and even the successive failures of the *H-II* in 1998 and 1999. It seemed that Japanese space development had reached its technical limits. But we believe that from an evolutionary perspective, the lack of financial backing and markets was perhaps more critical to understanding the substantive new directions necessitated for the corporations' space divisions.

In early 1996, senior Japanese space officials were still confident (at least publicly) that NASDA could at least average 10 percent per annum increases in the budget to execute the plans.[27] Around that time, Keidanren requested huge budgetary increases for space development, suggesting about ¥7 trillion in funding for space-related activities to cover space projects over the next fifteen years.[28] In light of previous budgets—such as the one in fiscal 1994, which represented more than a 7 percent gain over the previous year—their requests were not out of line.[29] However, the budgetary situation in the late 1990s was dire. In 1996 the total space activities budgets began to be curtailed, just as increased investments were needed to complete the new ambitious programs. The space budget for fiscal year 1996 grew just 1.2 percent over that of 1995, marking a slowdown in space spending. For the fiscal year beginning April 1997, leading space-related institutions requested a 6.5 percent boost in the space budget. But they found themselves again under severe budgetary constraints, which forced a 14 percent reduction in Japan's long-term space programs. NASDA's budget allocation, for example, received a paltry 1.7 percent increase; ISAS, with a 0.36 percent increase, did even worse. In fiscal 1998, the space activities budget rose only about 1.4 percent over the previous year.

There were two upshots of the progressive budgetary tightening. First, space development projects suffered. Some, such as the *Engineering Test Satellite-VIII/Kiku-8*, the *Solar X-ray Observing Satellite* (*Solar-B*), the *Advanced Land Observing Satellite* (*ALOS/Daichi*), and the *SELENE* (*Selenological and Engineering Explore/Kaguya*) moon mission were either scaled back or delayed; and the *HOPE* unmanned shuttle program was cancelled outright. Second, as a result, the very same primary contractors who had been enthusiastic about their commercial space prospects now found that very little money actually reached their space divisions. In addition, they were hampered by the 1990 U.S.–Japan agreement that affected their ability to compete for satellite manufacturing projects even in the domestic market, with the notable exception of those satellites designated for science or research.[30]

Compounding the budget problems was the market situation confronting corporations. According to one estimate, Japan's space business had shrunk by nearly 40 percent, and the number of the workers in the industry was about a third less than it was at the turn of the century.[31] In this context, Japan's attempts to become a commercial SLV provider and commercial satellite builder were widely seen to have "failed," partly because of a dramatic plunge in these markets worldwide. Japan's industrial base for space development was scarcely big enough to provide the subsidies for Japan's satellite builders to achieve the economies of scale needed to compete globally.

The national space program was seen by many to be in crisis, lost, or at a crossroads.[32] Failures involving coolers for sensors, apogee kick motors, solar arrays, spacecraft communication, cryogenic first-stage and second-stage engines, and solid rocket motors were indicative of pervasive problems with rigorous testing and quality control and assurance, rather than Japan reaching its technological limits. Such problems continued because of a combination of features, such as underfunding and understaffing at the public level and inadequate incentives for investing in improved manufacturing and test facilities at the private level.[33]

By the close of the 1990s, Japanese industry was concerned enough by the economic situation that then Melco president Takashi Kitaoka asked publicly for funding, such as that for R&D projects, that contractors had expected to allow them to compete commercially.[34] While certain strategic programs were funded, the return to increased and increasing budgets for the space program overall did not arrive. Japanese companies' space divisions, geared to receiving increased work and contracts on a revolving carousel year after year, found themselves in dire economic straits. They were increasingly forced to consolidate and, as we discuss in Chapter 3, even Japan's major contractors could not buck the trend. Faced with high uncertainty about the future and dim prospects for commercial profits, Japanese defense contractors thus believed that they had little choice: militarize or see their investments in space technologies wither and die.

As it happened, the budgetary and commercial problems of Japanese corporations dovetailed nicely with U.S. moves toward an ambitious plan to harness the U.S.–Japan technology partnership for defense purposes.[35] As a precursor to the ballistic missile defense (BMD) system, Japan had been participating in the U.S.-led Strategic Defense Initiative (SDI) since the mid-1980s (see Chapter 6). Its leading defense contractors, such as MHI, Kawasaki Heavy Industries (KHI), Melco, NEC, and Fujitsu, had cooperated with U.S. contractors in the Western Pacific (WESTPAC) Missile Defense Architecture Study. Remarkably, given the strict arms export ban in place and long before any governmental level decision had even been made to pursue theater missile defense (TMD) or BMD, they had competed to submit an architecture study proposal for an estimated contract value of $3 million. There is little doubt that they saw such technologies as a potential avenue for bolstering their economic livelihood under government patronage.

But which technologies? With the North Korean missile tests in both 1993 and 1998, the geopolitical timing only legitimated Japanese contractors' interest in militarizing a wide range of space assets under the patronage of both the U.S. and Japanese governments—and it was quite a range. As formal U.S. government overtures to the Japanese government made clear in 1993, a complete missile defense system would include a satellite and surveillance system, a data-gathering-and-analysis controlling system, and a missile system. All this could be potentially lucrative for defense contractors. In 1994 the Higuchi Panel—the advisory group charged with drafting security policy for the twenty-first century that submitted a formal report to then Prime Minister Tomiichi Murayama—also emphasized the importance of Japanese acquisitions in military communications, including use of reconnaissance satellites and mid-air refueling and long-range transport aircraft, as well as indigenous capabilities in BMD.[36] These would be important, the report emphasized, not just for Japan's autonomous defense but also to give operational aid to the United States.

Other military-oriented space projects, such as the *Information Gathering Satellites* (*IGS*), did, in fact, prove to be economically important for Japanese corporations. Their saga also marked the beginning of a distinct trend in Japan's space policy—flat funding for many parts of the civilian space program but increasingly visible support for military space development. The counterpoint to the story of declining budgets over most of the 1990s is the fact that after 1998 a sudden gush of money went to the nation's first military satellite program, and there were moves to reverse, at least slightly, the downward spiral in military spending.[37] Why and how this money and program suddenly, as it were, appeared in the midst of a recessionary-budget space program was a windfall from a corporate point of view. Estimated to cost about $1.75 billion, the constellation of spy satellites constituted the single largest satellite procurement in Japan's history at that stage. And because they were originally labeled as research projects, the spy satellites were also able to bypass the strictures of the long-resented U.S.–Japan 1990 agreement.

Thus, overall, the first signs of Japan moving to military space development came with overt moves to develop spy satellites and, later in the decade, the conversion of missile and space technologies diverted to BMD research and development. Stepping back, these developments represent concrete moves by Japanese contractors to develop three crucial planks of a military space infrastructure—military communications, spy satellites, and missiles. This same infrastructure of course mirrored the official pillars of the Japanese space program—SLVs, satellites, and associated technologies—then being developed by the very same defense contractors.

With the pursuit of profits stymied, Japanese corporations have been left with little choice but to seek more military business, entering a symbiotic nature with government and policy makers who have taken space technologies to heart as a way to bolster national security. This, then, is the critical causality embedded in our market-to-military trend, which we maintain has been a driving force in Japan's space technology development for some time and is accelerating pace in the current geopolitical realities surrounding Japan.

Industry Moves

That defense contractors would seek better business and more profits is not controversial in Japan or elsewhere. As is now evident, corporate moves to militarize space development have their roots as far back as the late 1980s with the formation of the WESTPAC Missile Architecture Study (see Chapter 6, Table 6.1). At that point corporate players had already identified the constraining effects of Japan's arms export ban as the principal impediment to greater bilateral defense technology collaboration with the United States.[38] But there has been a significant change in just over a decade: in the 1990s few corporate players advocated an outright end to the ban; by the close of the 2000s, they had banded together to get it revised outright. As we discuss below, such institutional and legislative changes in Japan's space policy are continuing to catch up with the real-world demands inherent in the *business* of Japan defending itself.

When it comes to specific demands regarding space development, the public got its first taste of the real militarization issue in Japan in May 2004, when the Society of Japanese Aerospace Companies (SJAC) requested that the government fund four major space programs to give direction to Japan's space industry in the next two decades.[39] These were the development of a two-stage reusable launch vehicle based in part on existing Japanese designs, the launching of a series of huge dirigibles (floating balloons) to act as substitute satellites for communication applications, the construction of space robotics aimed at in-orbit satellite repair and service capabilities, and the planning of a $100 billion project to erect a giant space solar power station in orbit.[40] SJAC believed that the four high-technology projects could help double the nation's share of the global commercial space market to 20 percent by 2020. Because SJAC and Nippon Keidanren share the same business groups and their shared strategic goal is increased taxpayer funding, it would stand to reason that if SJAC then asked Nippon Keidanren for an endorsement, the four-pronged program would be received warmly.

It did not turn out to be so. Nippon Keidanren apparently had no interest in any one of these proposals, which were brushed aside by key officials. Instead, the focus fell on a crisis for Nippon Keidanren and two of its main clients.[41] The two clients were MHI, the main systems integrator for the *H-IIA* rocket, and Melco, the main builder of Japan's present and future constellations of *IGS* spy satellites, as well as the country's regional *Quasi-Zenith Satellite System* (*QZSS/Michibiki*) GPS system.

The crisis was the future of the *H-IIA*—Japan's launch vehicle that was supposed to have put Japan on the map on the global commercial space launch industry. For Nippon Keidanren and its clients, the most important issue was not the additional SJAC programs; it was the restart of the launch of the *H-IIA*, which had been halted because of a launch failure that had destroyed two of the nation's Melco-built *IGS*. In addition, at that point the *QZSS/Michibiki* faced deadlock, and there were concerted efforts under way to push politicians and lawmakers, especially in the LDP, to deal with the problem.[42] A key element of Nippon Keidanren's strategy to support Japanese space spending was the overturning of the nation's ban on arms exports and also the overturning of the PPR. As far as we are able to ascertain, in 2003 Nippon Keidanren formally began active lobbying within the LDP to overthrow both restrictions. As a corporate player put it only too well:

> In Japan, the space has been not related to or separated from the defense. But in order to enhance Japanese defense capability, the space is getting more important and necessary in terms of information and communication. Thus the change of resolution is hoped so that the space can be fully utilized in the national defense capability enhancement. Another reason the change is asked for is that even high performance space products or systems are developed in Japan, one example of which is the LE-5 series second stage rocket engine that has world-class performance, [but] it is not allowed exporting such products or systems. This export restriction that comes from the narrow interpretation of the "peaceful purpose only" policy is impeding promotion of Japanese aerospace industry so that the change is hoped. Emergence of new defense oriented space programs will activate space development in Japan. Such activation as well as removal of export restriction will help promotion of Japanese industry. Also it will support the nation for the reason described above.[43]

Nippon Keidanren, and specifically the heavyweights within it that represent the Japanese space and defense industry, was thus challenging the conceptual basis of Article 9 of Japan's Constitution. In 2005, parsing Article 9, Nippon Keidanren stated that while the philosophy of peace in the first paragraph should be retained, the second paragraph diverged from reality in various ways. In its view, without going through formal constitutional amendments (which were cumbersome), immediate measures were required to legalize "necessary activities" for security.[44] We believe this to be one of the major turning points in the tide to move Japan's space development from the market-to-the-military—a trend that was pushed from within the LDP itself with the Kawamura initiative.

In 2004 at the behest of industry, there were other landmark events that pushed Japan onto a more visible path in terms of the militarization of its space program and that underscore our theme that legislation and policy were only catching up with the real-world technology developments of interest to Japan's corporate actors. The first significant event was the announcement in January 2004 by then JDA Director General Shigeru Ishiba (who was later appointed minister of defense of his willingness to review the ban on weapons exports, which was based on notions prevalent during the Cold War.[45] Because this was obviously of concern to Japan's neighbors, Prime Minister Junichiro Koizumi then moved to clarify the announcement, saying the projected move was to enable Japan to proceed with procurement and development of BMD with the United States. But the announcement can also be seen as Ishiba publicly floating the opinions of Japan's military and especially space contractors who saw BMD as an opportunity to expand their business. Both Yasuo Fukuda, then chief cabinet secretary, and Fukushiro Nukaga, former chief of the JDA and chairman of LDP's Policy Affairs Research Council (PARC), publicly supported Ishiba's idea, with Nukuga making clear that such a legal change would not only allow Japan to maintain its technological capacity but also allow it to become a big arms exporter. The revision of the TPAE was first publicly floated, in fact, by Chief Cabinet Secretary Hiroyuki Hosoda following December 2003 remarks made by Ishiba that Japan should revise the articles. These moves were closely linked to successful lobbying efforts by Japan's defense industry and represent a clear victory for *kokusanka* (indigenization), which served corporate interests.

Beginning roughly in mid-2003, industry broached the subject directly in a series of reports and proposals.[46] In one, there was a distinct emphasis on strengthening national security and crisis management as the number-one priority of Japan's future space direction. This was followed by improving Japan's GPS positioning capabilities in space, which was a direct reference to the *QZSS/Michibiki* program, Japan's super-accurate version of GPS. By building these priorities into Japan's space development, Japan could secure its direction and industry could recover more efficiently. In July 2004 two Nippon Keidanren proposals were also floated in Nagatacho (home to Japanese politicians) and Kasumigaseki (home to Japanese bureaucrats).[47] The target of these proposals was change in the existing framework governing the peaceful uses of space. Specifically, the goal was to review the peaceful purposes of space policy and to allow the defense establishment to utilize space assets. Doing so would allow Japan to use space for defensive purposes, bringing its position in line with international norms.[48]

More visibly, Nippon Keidanren's position, which was lobbied for by Nippon Keidanren's Space Activities Promotion Committee (headed by Melco's Ichiro Taniguchi) and which had significant support in the ruling LDP, was that the Japanese government should commit to "space development in (the) national security area."[49] The individual and cumulative purpose of these proposals was an attempt by Nippon Keidanren's powerful business lobby to radically shift the legal, conceptual, and strategic framework of Japan's space activities. We believe that Nippon Keidanren's two proposals are as relevant to the debate about Japan's apparently off-on quest to become a "normal" country as any of the heated exchanges on the Diet floor about the dispatch of the SDF to Iraq. For this reason, they deserve close attention.

Nippon Keidanren's first proposal contained four requests. The first two formally asked the Japanese government in polite but blunt language to expediently fix the nation's *H-IIA* rocket, which had blundered catastrophically in November 2003, when it blew up Japan's second pair of *IGS* satellites. The third clause essentially asked the government to make a concrete plan for space development that would help preserve the nation's technological base (or at least not waste the approximately $20 billion spent on space activities over the previous decade), and to "make clear" exactly what the government intended to do and what it wanted further from space activities. While the third clause reflected the subtle and inexorable shift to the designation of Japan's space development as a basic strategic national technology, the fourth clause is more directly on point. It not only sought government assurance on the industrialization of space activities to achieve improvements in and launches of satellites for communications, position determination, information gathering, and observation purposes. It went further and bluntly stated that with the increasing importance of space utilization for national security and crisis management, the government needed to reconsider its stance on the peaceful uses of outer space and to bring it into conformity with international interpretations.

These positions represented the official view of Nippon Keidanren, whose Space Activities Promotion Committee, in particular, has been an indispensable player in Japan's market-to-military trend. In short, Japan's leading business group was asking legislators to consider relaxing the principle built into space development that strictly prohibited the use of space for any military or defense-related purposes. The subcommittee, which also represents another sixty or so space, aerospace, and related companies, distributed the document to politicians. It proposed that Japanese companies, and Melco in particular, be allowed to develop and use defensive national security systems and technologies in space, in line with international norms and, specifically, the Outer Space Treaty of 1967. As we discussed earlier, international norms meant moving closer to the official U.S. position and the UN general consensus on "peaceful purposes" such that it was not to be interpreted as "nonmilitary"; rather, it was to be interpreted as allowing military use of space as long as such use was "nonaggressive." The primary reason for asking for a new, broader interpretation of the PPR was to remove the increasingly ridiculous inconsistencies that first became apparent in the 1980s, when the JDA began using satellite communications, and then more specifically in 1998, when the nation started its *IGS* reconnaissance program. The request in July 2004 was therefore only catching up with the reality of the pace of Japan's steady militarization of space for defense purposes.

The first proposal was only half the issue in considering the corporate role in the market-to-military trend. Another Keidanren committee—the Defense Production Committee—submitted a draft of a second proposal to legislators after the Lower House elections in July 2004, causing an uproar in the Diet. Essentially, this proposal asked the Diet to consider returning to a 1967-era resolution that allows Japan to export technologies that could be used both for peaceful and defensive applications—a resolution that was explicitly tightened in 1976 to forbid Japanese industry from exporting dual-use technology under any circumstances. Nippon Keidanren's proposal, which also challenged the 1976 resolution banning nearly all arms exports, would allow directly for more open militarization of Japanese industry's space technologies.

Japan's defense contractors have long chafed against such restrictions. There has been considerable public debate on the changing role of arms exports and technology sharing regimes, dramatically spurred by the FSX (Fighter Support Experimental) row, which played a significant role in helping ensure Japan's place at the core of BMD.[50] Since the mid-1990s, industry has been working hard to change the restrictions, most specifically, perhaps, with the establishment of the U.S.–Japan Industry Forum for Security Cooperation (IFSEC) in

1996.[51] IFSEC members include some of the world's largest defense contractors located in the United States (for example, Boeing, Lockheed Martin, Raytheon, Northrop-Grumman) and Japan (for example, Mitsubishi, Ishikawajima-Harima, Melco, NEC). Taking as its departure point a combination of changes in external security threats, mechanisms of U.S.–Japan cooperation, defense acquisitions, and industry trends, the revised statement in 2003 built on a 1998 report that called for the promotion of effective defense equipment and technology programs between the two countries. Given both geopolitical and global industry trends, the IFSEC noted that future U.S.–Japan cooperation on defense programs would mean, among other things, increased interaction among defense industries.[52] The same report called for a more "flexible application" of Japanese arms export control policies and continued that very theme both generally and in the specific context of BMD.

The revised IFSEC statement makes it clear that the organization's goal was not just to seek "enlightened" U.S.–Japan cooperation on defense programs but also to combat reduced military spending. More directly, while the joint statement was dubious on the possibility of a general relaxation of Japan's principles on arms export, it proclaimed that it was the *major* obstacle to more effective defense equipment and technology cooperation and sought at the very least a more "flexible interpretation" of it. This combined lobbying effort is not unimportant. It makes evident that Nippon Keidanren and Japanese industry heavyweights in space were—and still are, in general terms—supported by their U.S. counterpart in terms of pushing for changes that support the market-to-military trend in Japan along several defense-related dimensions, including space-related ones.

Nippon Keidanren and the backers of the change for the arms export resolution change have been quite clear about their motivations and quite clear about the changes. The 1976 arms export resolution classified rocket stages and engines, as well as key components like valves, as arms integrated subsystems.[53] MHI, for example, wants to be allowed to export tanks and valves for the U.S. Delta space launch vehicle and to market abroad its LE-5B second-stage liquid-fuel engine, which powers Japan's *H-IIA* rocket. Extending, clarifying, or changing the 1976 weapons export definition, or at least expanding the parts and systems lists and target countries to which sales can be made, will help bolster Japanese industry and contribute to the U.S. missile defense program. As MHI officials put it,

> There are two barriers that restrict the promotion of the industry: the separation of space from the national defense and the export restriction. Changing them

> will bring us national defense enhancement by full utilization of space, and activation of industry by both will lead to the emergence of new defense-oriented space programs. Allowing participation of Japanese industry in U.S. defense programs will bring merit to both countries. The U.S. can utilize Japan's excellent technologies, such as propulsion, sensing, and so on, into their program. Participation in the U.S. programs, of course, encourages Japanese industry.

Japan's desire to participate in the U.S. missile defense program has not been dampened by the current resolutions because of a series of factors: weapons technologies are exempted for BMD technology transfer from Japan to the United States, MHI and some other electronic companies are key partners in the system development, the United States can avail itself of Japanese engineering and semiconductor electronics and other know-how, and U.S. industry will accrue huge profits from BMD sales to Japan.

The same logic extends to other space-related ventures. Nippon Keidanren, for example, is interested in the construction of a satellite network contributing to safety and security, as well as support for the *H-IIA* and the *H-IIB* heavy launch vehicles. One specific request that shows concern with corporate profitability is that the government commit to a minimum of three *H-IIA* launches (with their concomitant satellite payloads) in order to maintain a decent revenue stream for Japan's space industrial base.[54] The message is clear, through the contractors, through their powerful lobby Nippon Keidanren, and through direct contact with politicians that a governmental response is necessary. This is not just to stop the flat-lining of the national space budget that corporate Japan strongly believes has long been damaging the ability of Japanese industry to respond to the needs of the country in the future. It is also, more fundamentally, necessary to stabilize the fluctuations in revenues suffered by companies due to the start-stop nature of development since the late 1990s under a fractured government policymaking structure for space ventures.

CONCLUDING ASSESSMENT

In the late 1960s, Japan apparently committed itself to the peaceful development of space activities, or at least avoided the appearance of overtly developing increasingly sophisticated dual-use technologies. However, as Nippon Keidanren's gathering momentum for legal changes shows, the ideal of keeping space development away from the development of dual-use and military technologies turned out to be an impossible dream. All space endeavors are built on military technology, and the market-to-military trend in Japan is well on its way, as recent changes in space-related laws and policies make clear. While the domestic Japanese defense industry has been demanding that weapons exports to all nations be allowed in principle, given the huge economic stakes, the government has responded by lifting the arms-export ban. For the present, however, arms exports will be related to missile-defense products developed with the United States, which will be the only recipient.[55]

Our more basic point throughout this chapter has been simple: when assessing the concrete direction of Japan's military capabilities, whether in space or otherwise, the economic interests of corporations must be taken into account. They are certainly not the only actors shaping Japan's space policy, but by virtue of making the actual technology, they have a stake in government policies like no other set of actors. Looking across the range of its lobbying activities, Nippon Keidanren has been very effective in leveraging the burgeoning national security concerns in Japan in the interests of its own defense contractors, who are involved in both civilian and military space-related projects. We have concentrated not on the money or contributions they may have expended in influencing policymakers (on which information is unclear) but, rather, on the changing structure of the legislative and policy outcomes that can be linked to their economic interests. These efforts are not likely to dissipate anytime soon, especially given the upgraded status of the MOD in Japan's policymaking apparatus and the by now acknowledged centrality of space-related ventures to Japan's defense. With an eye on the future of Japan's national space policy and the economic impact on some of its most influential member corporations, Nippon Keidanren has lobbied the government for the designation of the space sector as a "critical technology" in the nation's industrial base.[56] Above all, like other space powers, it has stressed the importance of a coherent national space strategy for Japan.

Working through the Space Activities Promotion Committee, its proposals to this end have also been specific.[57] In June 2006 Nippon Keidanren let the government know that it strongly supported the enactment of a basic space law because it was indispensible to moving forward from development to three new pillars for the space industry, namely, national security, industrialization, and research and development. It further lobbied for the setup of a space development strategy headquarters with the prime minister at its head. In July 2007 it again urged the necessity of a basic space law, calling for a unified space promotion system that would smooth the path of the Japanese space industry and take it to a new stage. Slowly but surely, as we show, the Japanese space industry has made its mark felt on the contents and directions of Japan's space technology, policy, and laws.

3 THE PLAYERS

FOR MOST OF THE POSTWAR PERIOD, the succession of administrative changes governing the space sector in postwar Japan left insiders and outsiders struggling to find the locus of decision-making power. As indicated in the previous chapter, this became a major impetus behind the Kawamura initiative and the subsequent Basic Space Law of 2008. In this chapter we focus on the main public and private players involved in Japan's space program since its inception. We do not shine as much light on the actual technologies they produce, such as rockets and satellites, which are taken up more specifically in the next three chapters. Rather, here, we focus on the makeup of the institutions and players in Japan's space saga: Who are the players? How did they come to be doing what they do today? What does the sum total of their activities and ambitions say about the militarization debate in Japan at the specific level of space technologies?

This chapter has two main parts. First, it lays out the changes in the organization of the nation's principal space agencies and institutions, with due consideration for their historical evolution and their current role in space policymaking. It examines the government ministries and associated organizations, their various roles, and the key programs identified with the construction of a military space infrastructure. Second, it turns to the crucial industry players that have not garnered as much attention in shaping outcomes in Japan's space policy. With a focus on the economic fortunes of the various corporations, it examines their historical, current, and projected space-related activities, with an eye on those involving military ones.

GOVERNMENT PLAYERS

Japan's space policy has been disbursed over several government institutions in the postwar period.[1] This was precisely the situation that the Kawamura initiative attempted to alter.[2] Figure 3.1 lists the main public players as of 2009. The new Strategic Headquarters for Space Policy (SHSP), whose existence was mandated by the new Basic Space bill as it came up for passage, held its first meeting in September 2008.[3] At that time the SHSP became responsible not only for drafting the Basic Space Plan but also for coordinating it across ministries and bringing it into effect. Its space projects and plans will similarly have to be strategized in the context of existing institutional structures.[4] Figure 3.1 reveals that the governance of space activities in Japan, while now presumably dominated by the SHSP, is still functionally characterized by long-standing divisions across ministries and agencies that will need considerable work to cohere under an overarching national space policy.

The SHSP is a new and high-profile player in an institutional setting that is characterized, as discussed below, by older, established, and less well known but very influential players in Japan's space saga. Although its legal status signals far more coherence than ever before in Japan's space policy establishment, it remains to be seen whether or not the SHSP will play a dominant and overarching role in shaping space policy in the national interest in the future.[5] Apart from budgetary afflictions, in the short term it too will have to tread carefully among the existing actors that have been critical in bringing up Japan's strategic space technologies.

Within the fractured organization of Japan's space-related policymaking structure in the postwar period, not all ministries and agencies have wielded equal power as they have pursued their various space projects, visions, and plans. Below we lay out some of the key facts regarding the main public players in this period, paying attention to two elements: their weight in shaping space-related directions, and their oversight specifically of those systems that might feed into military space capabilities. Table 3.1 provides information on the key space-related projects of the main public players, almost all of which will be discussed in the book.

Figure 3.1. Japan's New Space-Related Establishment, August 2009[1]

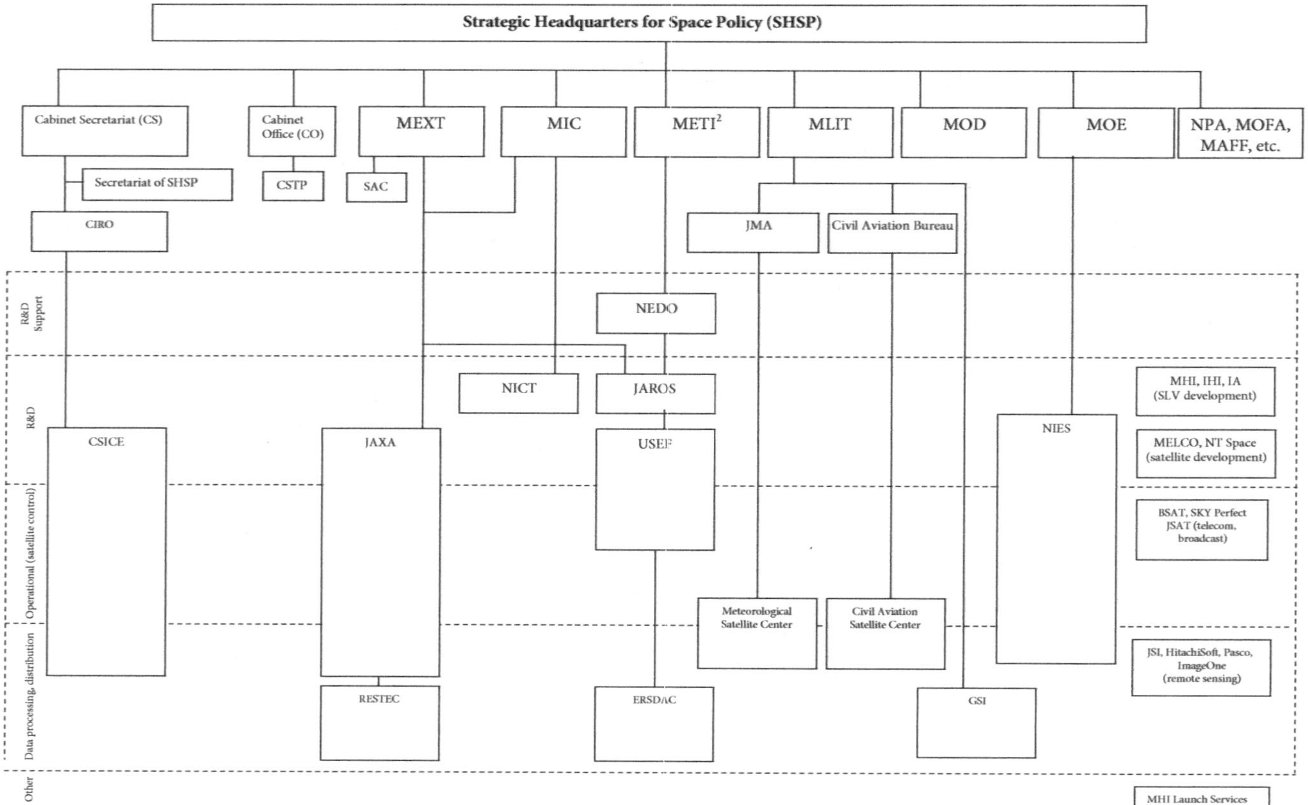

SOURCE: Condensed from the Secretariat of Strategic Headquarters for Space Policy, "Wagakuni no Uchū Kaihatsu Riyō Taisei ni Tsuite" [About the Organization of Japan's Space Development and Utilization], Reference No. 4, Tokyo, Japan, October 2008, p. 1. Some additions based on official e-mail correspondence, secretariat of SHSP, Tokyo, Japan, 30 August 2009. Based on current trends, there are likely to be further changes in Japan's space-related establishment in the near future.

NOTE: Key to acronyms: Cabinet Secretariat (CS); Cabinet Office (CO); Ministry of Education, Culture, Sports, Science and Technology (MEXT); Ministry of Internal Affairs and Communication (MIC); Ministry of Economy, Trade and Industry (METI); Ministry of Land, Infrastructure and Transport (MLIT); Ministry of Defense (MOD); Ministry of the Environment (MOE); National Police Agency (NPA); Ministry of Foreign Affairs (MOFA); Ministry of Agriculture, Forestry and Fisheries (MAFF); Council on Science and Technology Policy (CSTP), Space Activities Commission (SAC); Cabinet Intelligence and Research Office (CIRO); Japan Meteorological Agency (JMA); New Energy and Industrial Technology Development Organization (NEDO); Cabinet Satellite Intelligence Center (CSICE); Japan Aerospace Exploration Agency (JAXA); National Institute of Information and Communications Technology (NICT); Japan Resources Observation System and Space Utilization Organization (JAROS); Institute of Unmanned Space Experiment Free Flyer (USEF); National Institute for Environmental Studies (NIES); Remote Sensing Technology Center (RESTEC), Earth Remote Sensing Data Analysis Center (ERSDAC), Geographical Survey Institute (GSI); Mitsubishi Heavy Industries (MHI); Ishikawajima-Harima Heavy Industries (IHI); IA (IHI Aerospace); Mitsubishi Electric (MELCO); Nippon Electric (NEC); NEC-Toshiba Space Systems (NT Space); Broadcasting Satellite System Corporation (BSAT); and Japan Space Imaging Corporation (JSI).

1. Apart from ministries, includes all public-interest legal persons (JAROS, USEF, RESTEC, ERSDAC), independent administrative agencies (JAXA, NICT, NEDO, NIES), and private enterprises (MHI, IHI, IA, Melco, NEC, BSAT, SKY Perfect JSAT, JSI, HitachiSoft, Pasco, ImageOne, MHI Launch Services). Flows between ministries and such institutions, agencies, and enterprises are not always indicative of direct jurisdictional control but, rather, of close associations.

2. METI has direct jurisdictional control over the manufacturing range and long-standing institutionalized relations with the space-related corporations.

Table 3.1. Projects of Interest to Main Public Actors

Actors	*Key Projects*
Cabinet Office (CO)	*IGS* program management; overseeing ground station network; image and intelligence analysis and distribution; guiding development of second and third generation satellite constellations
Council on Science and Technology Policy (CSTP)	Overall strategic planning for Japan's space policy, superseding SAC in October 2001; development of satellites and technologies, especially sensor technologies, for national security; articulation of major change in national policy guidelines
Space Activities Commission (SAC)	Implements CSTP policy and closely manages programs; authorizes launches; manages technology development and implementation
Strategic Headquarters for Space Policy (SHSP)	Mandated by Basic Space Law to coordinate space policy, draft and implement Basic Space Plan, and carry out general research and consultation on all national space projects
Ministry of Education, Sports, Science and Technology (MEXT)—Japan Aerospace Exploration Agency (JAXA)	Overall design and program control of Satellite programs, equipment, environmental and advanced weather sensors, global warming gas sensor technology, and so on; NASDA legacy projects, *ALOS, GOSAT, GPM/DPR, ETS-VII, WINDS, OICETS*; ISAS legacy projects, *LUNAR-A, SELENE, PLANET-C, BepiColumbo, SOLAR-B, ASTRO-F*; *H-IIA* and *H-IIA* heavy version; *HTV*; International Space Station (*JEM/Kibō*); management for project *QZSS/Michibiki* positioning system technology and infrastructure, *Mu-5* and possible solid booster derivates, *Galaxy Express* (*GX*, now cancelled) Medium Launch Vehicle LNG engine; *USERS* reentry pod technology, *Advanced Solid Rocket* (*ASR / Epsilon*)
Ministry of Land, Infrastructure and Transport (MLIT)	*MT-SAT* program; Positioning system and weather satellite research; *QZSS/Michibiki* basic technology development; advanced GPS-type positioning infrastructure
Ministry of Economy Trade and Industry (METI)	*USERS* and *SERVIS* commercial satellite prototype programs; natural resources monitoring technology program; commercialization promotion for *GX* (now cancelled); promotion of Space on Demand (SOD); *USERS* reentry pod technology, *QZSS/Michibiki* basic technology development; *ASNARO/Sasuke* satellite technology development
Ministry of Internal Affairs and Communications (MIC)	Space communications technology research; space utilization research; gigabit bandwidth technology research; *QZSS/Michibiki* communications systems research and development
National Institute of Communications and Information Technology (NICT)	Satellite communications technology R&D, including broadband, optical, and millimeter wave; Smart Satellite Technology Group (SSTG, now disbanded); satellite rendezvous and repair technology; highly advanced atomic clocks.
Ministry of Defense (MOD)	Seeking for national defense: ballistic missile early warning; signals intelligence, dedicated communications; military reconnaissance (optical and radar), Operationally Responsive Space (ORS) systems, Space Situational Awareness (SSA) systems, satellite protection technology

NOTE: Table represents a summary of findings/projects as covered in Chapters 3, 4, 5, and 6.

The Space Activities Commission (SAC) and the Council for Science and Technology Policy (CSTP)

Although not well known, Japan had set up a Space Development Council in May 1960 and a Space Development Promotion Headquarters in the Science and Technology Agency in 1964.[6] When the Space Activities Commission (SAC) was established within the prime minister's office in 1968, it abolished the Space Development Promotion Headquarters, seeking to replace it with the National Space Development Agency (NASDA). From then on SAC became responsible for formulating the fundamental policies of Japan's space program at the broadest levels, publishing Japan's first ever *Fundamental Policy* for space in 1978, which it continued to revise until 1996.

In 2001 the Council for Science and Technology Policy (CSTP) was established within the Cabinet Office (CO). Focused at the broadest planning level, one of its seven expert panels was designated as the Special Research Committee on Space Development and Utilization. The CSTP is officially headed by the prime minister and presided over by a minister of state for science and technology policy, also located in the CO. Its objective is to articulate a basic long-term national science and technology policy, allowing for coordination across Japan's ministries. Its primary purpose is to support the prime minister and cabinet in scientific matters—space activities included. The establishment of the CSTP within the CO downgraded SAC from its political perch. In the wake of administrative reshuffling in the early 2000s, SAC was then appended to the one ministry, discussed next, that had long been dominant in space policy.

Ministry of Education, Culture, Sports, Science and Technology (MEXT)

In 2001 the former Ministry of Education merged with the Science and Technology Agency (STA) to become the present Ministry of Education, Culture, Sports, Science and Technology (MEXT).[7] This single ministry was historically a key player in Japan's space policy, and it was only at the end of 2009 that its star began to decline in the wholesale recalibration of Japan's space policy to focus on a national security role. Because the relative power of MEXT will remain strong until either the Ministry of Defense (MOD) or the Ministry of Economy, Trade, and Industry (METI), or a combination of both, assume more power, it is helpful to start here. Even as budget increases have been flat, the fact stands out that in 2005, MEXT accounted for about 65 percent of all science and technology-related expenses by the government, and assumed prime responsibility for the promotion of the space field.[8] As Figure 3.2 indicates more specifically, MEXT's heavyweight status in the overall Japanese space program is evident in the budgetary allocations across principal ministries. There are other ways in which MEXT has been important. On the implementation side, historically MEXT (and its predecessors) oversaw or coordinated the work of three principal government agencies that came to be central players in Japan's early space policy—the Institute for Space and Astronautical Science (ISAS), the National Aerospace Laboratory of Japan (NAL), and NASDA.[9] It is helpful to know something of their background as these three space agencies were critical to advancing Japanese space technologies over the postwar period.

Several committed engineers and scientists began launch vehicle research principally at the University of Tokyo from April 1955 onward under the auspices of the Institute of Industrial Science. In 1964 the University of Tokyo also became home to the Institute of Space and Aeronautical Science (ISAS),

Figure 3.2. National Space Budget of Japan by Ministry, 2009

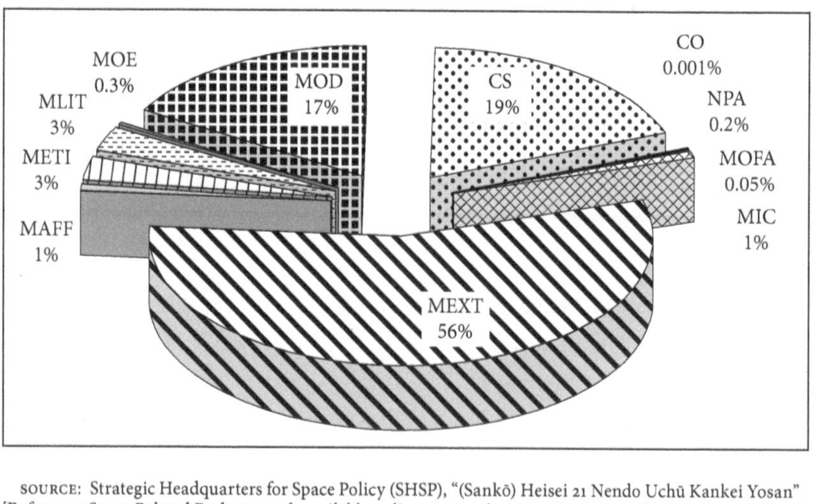

SOURCE: Strategic Headquarters for Space Policy (SHSP), "(Sankō) Heisei 21 Nendo Uchū Kankei Yosan" [Reference: Space Related Budget 2009], available online at www.kantei.go.jp (accessed 30 June 2009). Total space-related budget at approximately $3.5 billion.

NOTE: Key to acronyms: Ministry of Education, Culture, Sports, Science and Technology (MEXT), Cabinet Secretariat (CS), Cabinet Office (CO), Ministry of Defense (MOD), Ministry of Economy, Trade and Industry (METI), Ministry of Land, Infrastructure and Transport (MLIT), Ministry of Internal Affairs and Communication (MIC), Ministry of Agriculture, Forestry and Fisheries (MAFF), National Police Agency (NPA), Ministry of Foreign Affairs (MOFA), and Ministry of the Environment (MOE).

which was even more focused on rocket development. In 1981 this institute was reorganized under the name of the Institute of Space and Astronautical Science (ISAS) as a joint research organization among Japanese universities. In January 2001 it was placed under MEXT's jurisdiction. NAL was established in July 1955 as an auxiliary body to the prime minister's office, and placed under the jurisdiction of the new STA in 1956. It opened an aerospace division in 1963 and was renamed the National Aerospace Laboratory. It developed large-scale test facilities, including the Kakuda Research Center, which was built in 1965 to allow research on a wider scope. Since the late 1960s, much of NAL's work has focused on the research of key technologies for space and aerospace transportation systems. In 2001, after being placed under MEXT, its status was changed to an independent administrative institution and it transformed into the Kakuda Space Propulsion Laboratory.[10] With respect to launch vehicles, the focus in the early stages was on the development of small-scale, solid-fuel rockets that could launch scientific satellites, the first of which was launched in 1970. It was the general emphasis on getting into space that led to the creation of NASDA in October 1969. NASDA was able to rely on the technology transfer provisions in the 1969 U.S.–Japan Space Technology Agreement to specifically develop and launch rockets and application satellites. In January 2001, it too was placed under MEXT's jurisdiction.

NASDA and ISAS had a healthy rivalry, which was not alleviated despite the fact that by 2001 both agencies were under MEXT's umbrella. NASDA was always the bigger of the two agencies, both in terms of personnel and budgets. Around 2003, NASDA had a staff of about 1,000 personnel, whereas ISAS had only about 400, and NAL's staff was just shy of 300. On budgets some international perspective is necessary.[11] In the United States, the National Aeronautics and Space Administration (NASA) spends on an average between $15 and $16 billion a year for its space projects. In comparison, Japan's combined official budget for space programs in 1994 was approved at $3.2 billion, a figure that left very little room for operational and developmental mistakes across the board. Of this amount, NASDA garnered $2.3 billion, and ISAS about $300 million.

There was a sharp bifurcation between NASDA, with its focus on building liquid-fuel rockets and satellites for strategic technology acquisition, and ISAS, with its focus on solid rocket technology and scientific research. Irrespective of their competencies and achievements, a succession of failures in both satellites and launch vehicles, even after multiple rounds of enquiries and blue-ribbon accident investigation committees, led to widespread domestic criticisms of the Japanese space program as a whole—especially its jurisdictional splits and its ineffectiveness in establishing Japan's credibility as a player in the global commercial space arena.[12] There were thus many initiatives and action plans focused on quality control, program management, contractor relationship with manufacturers, and peer reviews. While consolidated under MEXT and waving the cooperation flag, all this suggested that ISAS and NASDA cooperated minimally and kept their various fields of competence intact.

Japan Aerospace Exploration Agency (JAXA)

These operational afflictions led to further efforts to streamline the policymaking structure—to create a continuous and systematic approach to space development from basic research and development (R&D) to practical application. A bill on the merger of the three agencies was therefore submitted to the Diet in October 2002. At that point the idea was for MEXT to have looser oversight over the proposed merged agency, which was to become an independent administrative agency or semi-privatized entity with more power to decide its own direction.[13] As an independent agency, the organization's management was supposed to be able to make decisions without the need for program-by-program oversight by MEXT and SAC, to draft members in from industry, and to outsource more work to industry contractors.

The thinking no doubt was that both MEXT and SAC would be taking a "hands-off" approach to the new organization that, in theory, would have more autonomy for its management and responsibility for its decisions. Institutionally, the expectation was that there would be synergistic effects from merging different technical competencies. The new agency was also to have responsibility for both aeronautics and space development that would circumvent some of the legal problems related to the militarization of space assets. Although the aerospace portion of the new agency's mandate would involve developing jet fighters and other military-related technologies, those programs would not be subject to the "peaceful purposes" clause that had applied to NASDA.

Structurally the effort bore results, even though NASDA and ISAS officials resisted the merger because of their different competencies (technological development and basic science, respectively).[14] On 1 October 2003, ISAS, NAL, and NASDA were merged into one independent administrative institution, the Japan Aerospace Exploration Agency (JAXA)—the month and year in which China succeeded in its first manned spaceflight.[15] The goal was to make it a more efficient and vigorous organization, and to work in close collaboration with industry to forge a new role for Japan in space activities.[16]

As Figure 3.3 reveals, JAXA ended up with consolidated powers in terms of Japan's space development ambitions. Of the three original agencies, only ISAS appears to have survived with its focus on space sciences and research, whereas the functions of the other two are split among the various directorates. ISAS reportedly only accepted the merger after assurances that it would be allowed to continue much of its science missions without interference within the larger agency.[17] Just as crucially, mission selection has been kept within familiar ISAS internal mission selection and other crucial decision-making processes. In practice, overall administrative planning is carried out by the Strategic Planning and Management Department. NASDA's functions seem to have been incorporated in the Space Transportation Mission Directorate, which oversees key projects involving space access, such as the *H-IIA* and *H-IIB*, and the *Advanced Solid Rocket* (*ASR/Epsilon*). Satellites, ranging from Earth and environmental satellites to the *Quasi-Zenith Satellite System* (*QZSS/Michibiki*), are generally controlled by the Space Application Mission Directorate. NAL's functions seem to be split between the Aerospace Research and Development Directorate and the Aviation Program Group. Human Space (the *Japanese Experiment Module*, or *JEM/Kibō*) seems separated from the rest of the organization and, as noted above, ISAS is in its own kingdom.

What is important is that, along with the other conventional space agency programs, the spy satellite program is also housed in JAXA. As there is no public information available on the issue, it is difficult to say what this means for the government's earlier decision to circumvent the turf wars by placing *Information Gathering Satellite* (*IGS)* control within the CO and creating an independent body, the CSTP with its own space-related committee.[18] At the time of writing, the CO remains in charge of overseeing the development and operation of the spy satellite fleet (originally this was to avoid the issue of the then Japanese Defense Agency (JDA) and now MOD directly controlling the satellites). While it is also not apparent from official organizational charts for either MEXT or JAXA, if the current JAXA law continues to stand it will mean that MEXT has the dominant supervisory role over JAXA through legal provisions (for example, concerning the appointment of the agency's executives or its own designation as the major competent ministry).[19] Here too it remains to be seen how jurisdictional issues will be resolved involving JAXA, especially with the arrival of the SHSP, and its efforts to put its own stamp on

Figure 3.3. Organization of JAXA by Principal Competencies, April 2009

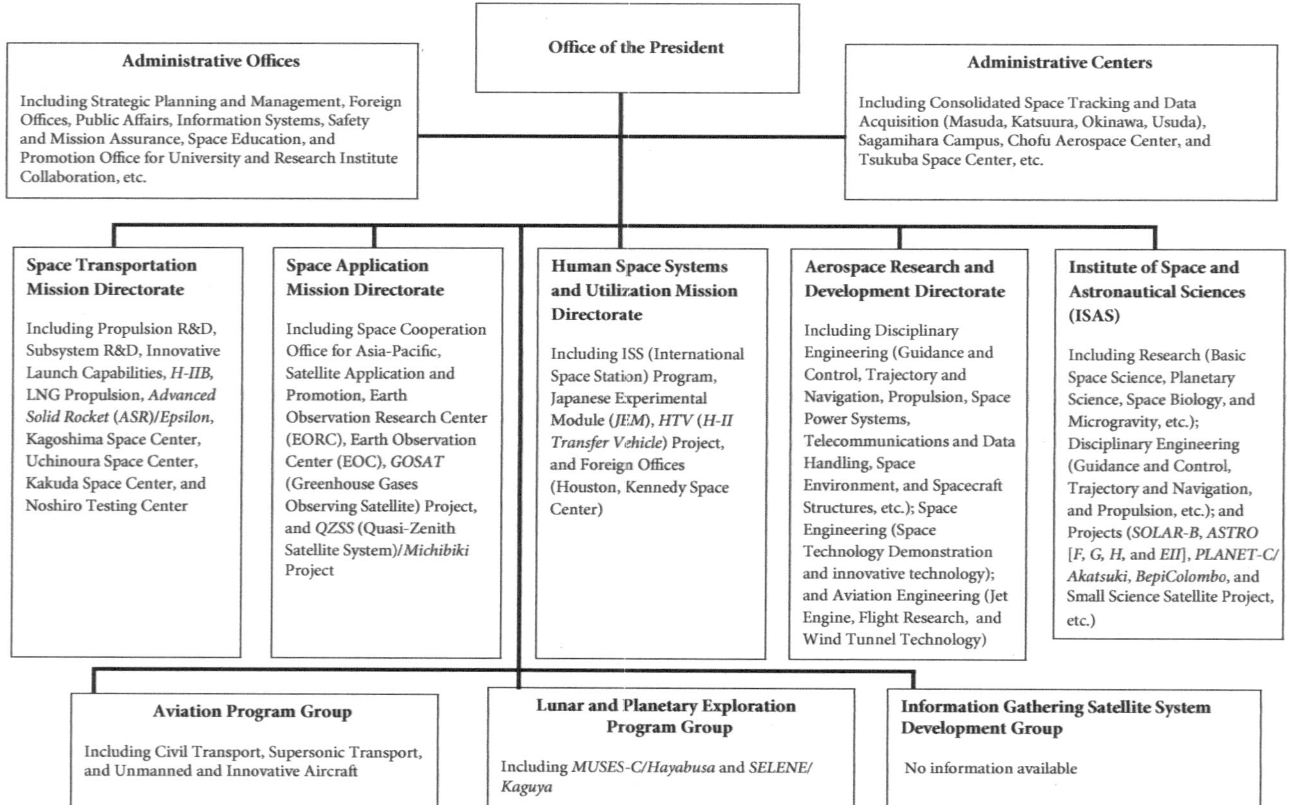

SOURCE: Japan Aerospace Organization Agency (JAXA), Organization Chart, available online from www.jaxa.jp (accessed 30 June 2009). Based on an interview, JAXA official, 6 August 2009, as well as on information from the Basic Space Law, Supplementary Provisions Article 3 (see Appendix II in this book), JAXA's organizational structure and its institutional relationship to Japan's entire space policymaking infrastructure may undergo further changes in the near future.

the space policymaking structure in Japan. Perhaps most important of all with respect to JAXA's future, it is helpful to keep in mind that the new Basic Space Law mandates a review of its administrative, jurisdictional, and organizational structure.[20]

Ministry of Internal Affairs and Communication (MIC)

The Ministry of Internal Affairs and Communication (MIC) was formerly the Ministry of Posts and Telecommunications (MPT) before it was transformed into the Ministry of Public Management, Home Affairs, Posts and Telecommunications (MPHPT) and then renamed MIC in mid-2004. MIC has long had an interest in space activities related to radio waves, and it oversaw other interested players, such as the Communications Research Laboratory (CRL), Nippon Telegraph and Telephone (NTT), Kokusai Denshin Denwa (KDD), and the Japan Broadcasting Corporation (Nippon Hōsō Kyōkai, the NHK).In its previous incarnation as MPT, it was the national governmental agency that defined and funded R&D projects in the space communications area.

In particular, its CRL research arm was responsible for developing a large portfolio of satellite communications technologies that were designed to enable Japan to become a major commercial space power.[21] The CRL was also closely tied to the development of communications technologies with NASDA and the former big three of Japan's satellite building companies: Mitsubishi Electric (Melco), NEC Corporation, and Toshiba Corporation. It turned out to be the lead government agency in conjunction with NASDA/Melco, for Japan's *CS* (*Sakura*), *CS-2*, and *CS-3* communications satellites. The CRL was responsible for developing the highly regarded radar technologies, such as the precipitation radar on the joint-NASA *Tropical Rainfall Measuring Mission* (*TRMM*) environmental satellite. Other technologies have included sophisticated inter-satellite and inter-orbit communications, broadcasting, and spot beams. In the 1990s CRL was praised for its fundamental and advanced research in satellite communications, primary focus on mobile and high data rate multimedia satellites, optical communications technology, and satellite/terrestrial communications systems experiments. Its integration in global satellite research networks and especially close work with Japanese industry allowed it to optimize the technology transfer process. On the other hand, the CRL also appeared cursed, with its involvement in the highly advanced *ETS-VI/Kiku-6* and *COMETS/Kakehashi* satellites, both of which were placed in incorrect and unsuitable orbits.[22]

CRL's institutional fortunes evolved, and, as with other agencies, it was spun off into an administrative agency in 2001. In April 2004 the CRL was merged with another quasi-independent communications research agency, the Telecommunications Advancement Organization of Japan, to form the National Institute of Information and Communications Technology (NICT) at present.[23] As noted above, the then-CRL was responsible for developing the communications technologies for a series of satellite programs over the decades that significantly advanced Japan's abilities, such as satellite constellations, laser inter-satellite technologies, and extremely high-bandwidth-to-mobile devices technologies. Continuing this line of work, the NICT's budget for the fiscal year 2005–2006 was about ¥3 billion. Initially the NICT's space-related programs were run in six space technology and communications departments—namely, Broadband Satellite Network, Optical Space Communications, Space Data Transmission, Millimeter-Wave Devices, Mobile Satellite Communications Group, and the Smart Satellite Technology Group (SSTG). Its related organizations and facilities include Kashima Space Research Center (KSRC), which succeeded in broadcasting the Tokyo Olympic Games to the world in 1964. KSRC, which presently studies formation and cluster-flying satellite control and plate tectonics, includes the Space Cybernetics Group, which also looks into multiple satellite control and orbit dynamics. High-bandwidth, narrow-beam laser communications technologies, as well as the development of microsatellites by SSTG for on-orbit rendezvous, repair, and refueling technologies are discussed in Chapter 6 with other cutting-edge technology development programs that are controversial because of their potential dual-use.[24]

In late 2005, NICT and MIC launched a bid to be major players in Japan's next generation satellite communications infrastructure.[25] Under the rubric of ensuring security and safety, their highly ambitious proposal involved an infrastructure stretching over twelve fields utilizing space communications in the service of a network society. While the MIC did not receive a budget for the project, many of the ideas are closely echoed in the new space policy of 2009. However, in 2007, NICT's ambitious space technology programs were restructured and consolidated into one research group under the New Generation Wireless Communications Research Center. While the Center is officially in the NICT's Yokosuka laboratories, space technology research actually continues in Kogane.[26] In our view, this was primarily because the laboratory's major work, namely building Japan's space communications technology base, was seen to be accomplished.

There is also evidence of a backlash against the CRL/NICT in today's environment, where the laboratory's broad-based approach to research is no longer seen as needed. Some have protested, saying that Japan needs to continue to reinvest in satellite communications research and development for a network-centric defense system.[27] But we believe that with Melco and NEC in the lead, Japan's contractors are quite capable of conducting the research and development they need for next-generation technologies under the rubric of funding for national security.

Ministry of Economy, Trade, and Industry (METI)

The Ministry of International Trade and Industry (MITI, the predecessor of METI) first established its Space Industry Office in 1979, upgrading it to a Space Industry Division in 1987.[28] In July 1997 two separate divisions at MITI, namely aircraft and space, both under the Machinery and Information Industries Bureau, formally merged into the Aircraft, Ordnance and Space Division. At present, METI's space policy is focused within the Manufacturing Industries Bureau—specifically, what is now the Aerospace and Defense Industry Division.[29] Despite its prominence in Japan's industrial catch-up, METI appears at first blush to have been out of the loop for much of Japan's space policy development. For the most part, it also appears that METI's main and most visible activities were focused on the promotion of the industrial utilization of space, with a specific focus on remote sensing and uses of microgravity for applied science and space components experiments. Linked to these have been programs designed to develop standardized systems for building satellites and lowering costs to make them commercially competitive.

However, METI- and METI-related institutes have carved out a critical role for themselves that we believe is tailor-made for today's environment. Over the course of the 1990s, they have forged or used their close links with the private sector to come to the fore in programs with dual-use technology that might prove to be extremely useful in pushing development of systems of military potential. Of particular interest in this respect is the Institute for Unmanned Space Experiment Free Flyer (USEF), the New Energy and Industrial Technology Development Organization (NEDO), and the Japan Resources Observation System and Space Utilization Organization (JAROS).

USEF, which was established in 1986, is METI's main space organization for its non-rocket-related programs and has publicly stressed themes of cost-cutting and commercialization of the Japanese space industry.[30] As we discuss below, USEF's ties to industry are as direct as is its prominent role in developing dual-use technologies, both of which lend weight to the market-to-military thesis. In the last decade alone, three of USEF's heads have been former CEOs of Melco, namely, Takashi Kitaoka, Ichiro Taniguchi and Setsuhiro Shimomura. As Melco CEO in 1998, Kitaoka foreshadowed his later moves by publicly calling for the government to infuse more money into satellite development, particularly the *Unmanned Space Experiment Recovery System* (*USERS*) project. Shortly after this, he resigned from Melco in disgrace over a financial scandal and the company's first losses in fifty-two years. This did not stop him from assuming the chairmanship of USEF, where he oversaw the *USERS* project and several other next-generation satellite systems for Melco. He was replaced at Melco in 1998 by Taniguchi, the former head of the company's defense and aerospace section. After becoming chairman of Melco in 2002, Taniguchi went on to serve as chairman of the Space Activities Promotion Committee at Nippon Keidanren, working assiduously to change Japan's space policy to its present national security role. As of 2010, his replacement as chairman of USEF is yet another former CEO and president of Melco, Setsuhiro Shimomura. All of the major corporations in Japan's corporate space infrastructure, the most prominent of which are discussed below, are members of USEF and likely to remain so as long as it continues to exist.

USEF has been involved in technologies that are suitable for warhead targeting and reentry from orbit, particularly in two projects: the *EXPeriment RE-entry Space System* (*EXPRESS*) and *USERS*.[31] *EXPRESS*, a joint project between Japan and Germany, was intended to test autonomous reentry capability in 1995, but the capsule was not injected into orbit correctly and ended up landing in Ghana. The *USERS* project was aimed at establishing capability in long-duration, self-return space experiments. The *USERS* space capsule successfully splashed down in the Pacific Ocean near the Ogasawara Islands. The experiment—which basically tested the ability of the capsule (the shape of which was easily mistakable with the nose cone of a warhead) to perform automatic maneuverable pod reentry—incorporated ablative shielding technologies that ensured the capsule's accurate targeting. As of 2010, USEF is involved in the *Advanced Satellite with New System Architecture for Observation* (*ASNARO/Sasuke*), which can potentially be integrated in a military space infrastructure.

METI was also involved in direct support of rocket development, known as the *GX* program, that was to have used the first stage of a U.S. Atlas for a medium-sized rocket and was supposed to be commercially competitive, which we discuss later in the book. Looking ahead, METI and USEF have played a prominent role in developing radar and sensor technologies that are technological building blocks of what is to be Japan's next-generation Operationally Responsive Space (ORS) and spy satellite constellation. These technologies include, for example, the optical sensor and synthetic aperture radar for the *Japanese Earth Resources Satellite* (*JERS-1/Fuyō-1*) Earth observation satellite (Melco), the *Phased Array type L-band Synthetic Aperture Radar* (*PALSAR*) for the *Advanced Land Observing Satellite* (*ALOS/Daichi*) satellite (Melco) and the Hyper Spectral Sensor (NEC).

NEDO, which was established as a semi-governmental organization in 1980, became an independent administrative agency in 2003.[32] In promoting and coordinating nationwide R&D on industrial technology, its main ministerial coordination is with METI. With respect to current space ventures, NEDO is focusing on advancing fundamental technologies for the next-generation of small-scale satellites, as well as the *QZSS/Michibiki* system (which will be discussed in Chapter 6). Through the *Panel Extension Satellite* (*PETSAT*) concept, it is promoting R&D in modular satellites with unfolding multipurpose panels. Echoing ORS concepts, such small-scale technologies are relevant as building blocks for Japan's upcoming METI-led "Space on Demand" (or SOD) programs. The new SOD concept by METI targets consumer-oriented, efficient, innovative, low-cost, short-term delivery technologies across satellite systems, launch systems, technology verification, and ground systems.[33] As the emphasis on these types of small technologies suggests, METI's star may well now be rising as of 2009 just as the role of other established players begins to wane. In fact, METI may be uniquely positioned for a more prominent role in Japan's space policy on two fronts: user-friendly commercial operations (telecommunication, broadcasting, car navigation, weather, and so on) which still constitute the bulk of profits for space-related enterprises; and, as described above, dual-use and now small-scale technologies that are in line with the new emphasis on defense.[34]

Finally, JAROS had its origins in corporate sponsorship in 1986, when a range of satellite makers—among them, Toshiba, NEC, and Melco—sought then MITI's as well as the prime minister's approval to conduct R&D on resource observation technologies for use on the *JERS-1*.[35] *JERS-1*, launched in 1992, continues to observe and collect data related to land and coasts, forestry, environment, disasters, and so on, with both a Synthetic Aperture Radar (SAR) and an Optical Sensor (OPS). Research to advance its SAR technology, in particular, was started in 1993, and it was instrumental in advancing Japan's Earth observation (EO) and, subsequently, the spy satellite program. Presently, JAROS receives its sponsorship and commission from METI as well as MEXT and is charged primarily with developing space-borne remote sensors to better observe Earth resources and environmental changes. It merged with the Japan Space Utilization Promotion Center (JSUP) in April 2006. At present, JAROS continues to work in cooperation with JAXA, the Remote Sensing Technology Center of Japan (RESTEC, connections with MEXT), the Earth Remote Sensing Data Analysis Center (ERSDAC, connections with METI), as well as with prominent space-related corporate players.

Ministry of Land, Infrastructure, and Transport (MLIT)

MLIT has long been investigating the establishment of multipurpose satellite systems that would essentially cover the telecommunication needs of the ministry as a whole (but the multifunctionality of which would allow military-related work as well).[36] One of MLIT's major programs, which focused on improving air traffic control over the Northern Pacific route between the United States and Japan, suffered a setback in November 1999 when a faulty turbine on the *H-II* rocket led to the blowing up of the *Multi-functional Transport Satellite-1* (*MTSAT-1*) satellite. The replacement *MTSAT-1R* and backup *MTSAT-2* were launched by *H-IIA*s in February 2005 and February 2006, respectively. The MLIT is also a major player in the *QZSS/Michibiki* system, as are other ministries, such as MEXT, METI, and MIC.

Ministry of Defense (MOD)

Another rising player in the space policy saga is MOD, which was formed in January 2007.[37] Setting aside MEXT, one indication of the importance MOD has gained in such a short time is the space budgetary allocations as set out in Figure 3.2. Citing the dependence of many nations on space assets, MOD is clearly going to be an even more important consumer of BMD-based systems as well as satellite-based communications, imagery, and meteorological information in the future. It has not wasted much time in attempting to put its stamp on the directions of space policy, having already established a Committee for the Promotion of Outer Space Development and Utilization (CPSDU) in 2008. As discussed in Chapter 5, in January 2009 this committee came up with MOD's own basic policy guidelines, which speak to protecting satellites and supporting small satellite programs that have strong implications for highly militarized space assets. Although the research is being carried out openly by universities across Japan, it is our view that at some point military communications and counterspace technologies—especially those based on small satellites, if the trend continues—will probably be aggregated in Melco and NEC, Japan's two primary satellite makers, with direct contracts through MOD and METI's USEF. Given our emphasis on the market-to-military trend in Japan's legal and policy rubric for space at present, we also believe that MOD's overall weight in the space establishment is likely to grow.

CORPORATE PLAYERS

METI's jurisdictional oversight over Japan's manufacturing base is an important segue to considering the key role corporations play in the technical development of space assets, and now we turn more fully to the private actors themselves. As Figure 3.1 reveals, private corporations and enterprises are now acknowledged as an integral part of Japan's space policy establishment.

The basic transition toward a more visible national security infrastructure in Japan's space policymaking is what distinguishes the past from the present. For most of the postwar period, it was within the fractured nature of Japan's space policymaking apparatus that government and industry attempted to move elements of the nation's space program from a technology acquisition phase to an industrialized phase to the applications phase. Because of the risks and costs involved in "commercial" space projects, industry has been and remains of course unwilling to pursue investment unless it is either underwritten or heavily subsidized by the government. Government, on the other hand, has often proved unwilling to play the roles that industry requires in the PPPs (public private partnerships) that have been formed. This has been evident in cases of the PPPs for both the *QZSS/Michibiki* system and the *GX* medium rocket that were originally proposed by industry to promote "applications" or "commercial" space activities.[38] Such issues have had a profound impact on the ability of future products to become *commercially* successful in any meaningful sense and have played an important role in pushing toward government procurement based on building the nation's national security infrastructure.

Central to Japan's space policy is a formidable set of Japanese corporations, some with manufacturing experiences stretching back to Japan's earliest efforts at industrialization and militarization of industry assets. Until a wave of consolidation hit Japan at the end of the 1990s—itself an echo of the massive consolidation that took place in the United States and Europe when the global commercial satellite and launch market bubble burst—Japan's space development industry was controlled by six contractors that, between them, accounted for the lion's share of government funding and that accumulated the vital systems integration technologies. These six contractors can be split into two camps: the rocket and launch facility builders and the satellite makers. The three space launch vehicle (SLV) makers were Mitsubishi Heavy Industries (MHI), Ishikawajima-Harima Heavy Industries (IHI), and Nissan, while the three satellite makers were Melco, NEC Corporation, and Toshiba.

Table 3.2 sets out the principal private players in the Japanese space saga with a focus on those that have helped shape the nature, contents, and directions of Japan's space technology. As is also indicated, almost none of these players is a novice in the production of military-related products. Perhaps the most important thing to note is that the industrial group that sits atop all of the defense contractors in Japan is the very same that is also presently at the apex of Japan's space-related production—namely, the Mitsubishi Group. At this stage, the Mitsubishi companies figure in the top rankings of aggregate defense revenues around the world. Although, relative to other global players, their reported revenues from defense are a small part of their overall makeup, a focus on aggregate defense revenues still means that MHI is among the top twenty players in the world, and Melco is within the top fifty. While ordinal rankings should not be taken too seriously because the currency's swing value could make a company fall in the list while the acquisition of a new program could make it go up, these placements nevertheless show that these are quite important firms. As the next section suggests, more important than rankings is the actual capabilities and present technical knowledge of the key corporate players across the stated pillars of Japan's space program, rockets and satellites. Looking at these two sets of programs more closely, there is little doubt that it is these very players that stand to benefit economically from any expansion in the business of militarized space products over the next few decades.

Rocket Makers

Until consolidation, Japan's three major contractors involved in rocket vehicle design, integration and production were Nissan Motor Company's Aerospace Division, MHI, and IHI. They are discussed in turn below, with attention to the principal historical turns and twists that affected Japan's rocket development.[39]

Nissan Nissan was established in 1933 and acquired its present name in 1934.[40] It has rightly billed itself as a major contributor to Japan's space pro-

Table 3.2. Estimated Rankings of Japan's Major Space-Related Contractors (US$, in millions)

Contractor	*Space Sales*[1]	*Defense Revenues*[2]
Mitsubishi Heavy Industries (MHI)	$216 (rank 29)	$2,516.7 (rank 19)
	Major projects: SLVs (*H-IIA*; design and integration technologies for multi-stage launch systems from *N-I, N-II, H-I, H-II, H-IIB*); rocket engines (LE-5A, LE-5B, LE-7A, MB-XX with Boeing Rocketdyne (now Pratt & Whitney Rocketdyne); reaction control systems; *JEM/Kibō, HTV*; infrastructure (launch and test facilities for SLVs, rocket engine combustion test facility); rocket motor cases	Major projects: F-2, SH-60K, UH-60J, Type 90 battle tank, Patriot, Type 04 AAM, Type 88 SSM, Submarine, Destroyer
Melco	$1,018 (rank 11)	$959.8 (rank 48)[3]
	Major projects: satellite programs (*MTSAT-2, OPTUS-C1, DRTS/Kodama, SUPERBIRD-C2, ALOS/Daichi, ADEOS-II/Midori-II, GOSAT/Ibuki, SERVIS 1 &2, ETS-VIII/Kiku-8, USERS, SOLAR-B/Hinode, HTV, JEM/Kibō*); control systems (JAXA New Ground Network, JAXA Tracking Control System for SLVs); satellite communications Earth control systems; optical and radio telescopes (VERA, RAINBOW, SUBARU); *IGS satellites*; *QZSS/Michibiki*	Major projects: Type 03 SAM, Type 99/AIM-7M AAMs, air defense radar, NOLQ-3 electronic jammer
Nissan	—	—
	Major Projects: (see IHI Aerospace below)	
IHI	$190 (rank 31)	$458.5 (rank 83)
	Major projects: (see IHI Aerospace below)	Major projects: Engines for C-X/P-X, F-2, SH-60K, AH-64D, UH-60A, CH-47J
IHI Aerospace (merger of Nissan and IHI space business)	—	—
	Major projects: SLVs (*GX* [now cancelled], *ASR/Epsilon* [new solid-fuel rocket];[4] rocket propulsion systems (solid rocket boosters (SRB-As), R&D of reusable rocket propulsion system); guidance and control systems (second-stage reaction control systems for *H-IIA*; liquid oxygen turbopump for *H-IIA* rocket's first-stage LE-7A and second-stage LE-5B; movable nozzle TVC system); satellite propulsion systems (*HTV* propulsion), *JEM/Kibō* (extravehicular experimental platforms); Moon Explorer (Penetrator); utilization equipment for space environment; ground test and support facilities	

(*continued*)

Table 3.2. *(continued)*

Contractor	*Space Sales*[1]	*Defense Revenues*[2]
NEC	—	$842.6 (rank 56)
	Major projects: (see NT Space below)	Major projects: Base air defense ground equipment system, communication systems, sonor/sonobuoy
Toshiba	—	$385.9 (rank 91)
	Major projects: (see NT Space below)	Major projects: Type 93, Type 81 SAMS, Type 91 portable SAM, JFPS-4 air defense radar
NT Space (merger of NEC and Toshiba space business)	—	(see NEC and Toshiba separately, above)
	Major projects: satellite systems (involved from 1970s onward in series of engineering test, communication/broadcast, Earth observation, scientific, and small satellite; e.g., *Ohsumi, GMS/Himawari, BS/Yuri, MS-T5/Sakigake, MOS/Momo, ETS-VII/Kiku-7 (Orihime/Hikoboshi), ASTRO-D/ASCA, MUSES-B/Halca, COMETS/Kakehashi, OICETS/Kirari, GCOM, PLANET-C/Akatsuki, Planet-B/Nozomi, ALOS/Daichi, MUSES-C/Hayabusa, SELENE/Kaguya, etc.*); satellite communications subsystems (transponder equipment, receivers, convertors, solid-state power amplifiers, etc.); satellite on-board antenna; Earth observation sensors (optical, radar); space robotics (*JEM* remote manipulator system (JEMRMS), manipulator flight demonstration (MFD), on-orbit serving; ground systems (telemetry, tracking and command (TT&C), data processing); remote–sensing geospatial infrastructure system (RSGIS); ORS/SOD spy satellite constellation	

1. Taken from *Space News*, "Space News Top 50: 2004," available online at www.space.com (accessed 23 September 2008), based on 2003 space sales. Figures in brackets reflect global ranking. Depending on the product/component space, sales may or may not be reflected in total defense revenues. Major space-related projects are as cited in text.
2. Taken from *Defense News*, "2005 Defense News Top 100," available online at www.defensenews.com (accessed 23 September 2006), based on 2004 defense revenues. Some revenues reflect different ending dates for fiscal years. For Japanese companies highlighted above, the fiscal year ends on March 31, and all defense revenues reflect then Japanese Defense Agency (JDA) contract awards. Figures in brackets reflect global ranking. Major military projects are taken from Eiichio Sekigawa, "MHI Still on Top," *AWST*, 20 June 2005, p. 58.
3. Based on estimates of defense revenues from *Defense News* and Sekigawa, the total value of Melco's 2004 contracts ranges from $959.8 and $1,032, respectively.
4. As of July 2009, although IHI Aerospace only states that it is now engaged in developing a new solid-fuel rocket as a successor to the *M-V* launch vehicle, it is likely that it will be the central player in the development of the *Advanced Solid Rocket* (*ASR/Epsilon*), given Nissan's historical track-record in solid-fuel rockets and also the fact of ongoing research in which IHI Aerospace is involved. See, for example, Yashuhiro Morita, Takayuki Imoto, Hirohito Ohtsuka, and Advanced Solid Rocket Research Team, "Research on an Advanced Solid Rocket Launcher in Japan," 2008-g-02, paper presented at the 26th International Symposium on Space Technology and Science, Hamamatsu City, Japan, 2–8 June 2008, available online through Scientific Journal Editing System (SciEd) at www.senkyo.co.jp (accessed 19 July 2009).

— Indicates data unavailable.

gram with a focus on rockets. The company was able to enlist some of the very same key engineers and physicists who had also played a significant role in the development of Japan's warplane industries in the 1930s, such as Ryoichi Nakagawa (chief of the team of Nakajima Aircraft Company that developed the Zero's sakae-21 engine) and Yasuakira Toda (who helped another Nakajima engineer, Hideo Itokawa, known as the father of Japan's space program, to launch the first *Pencil* rockets). At Nissan, both Nakagawa and Toda were part of the intellectual team that powered a series of Japanese rockets, such as the *Pencil*, *Kappa*, *Lambda*, and *Mu*.

Nissan's postwar role in Japan's rocket development dates specifically back to April 1953, when it started R&D on rocket motors. By 1986 all solid rocket motors used by NASDA and almost all solid rocket stages used by ISAS were produced by Nissan, and the company was confidently building on successive sets of technologies to develop the first *H-II* solid rocket boosters (SRBs). This technology will feature prominently later in the book, as it is the basis of Japan's SOD space access program, the new *Advanced Solid Rocket* (*ASR/Epsilon*).

Looking back over the early postwar period, there is little question that Nissan's technology was behind a series of solid launches for Japan. In February 1970, when Japan successfully launched its first satellite, *Ohsumi*, it was Nissan that had developed the rocket engine and launch vehicle, *Lambda 4S-5*. Nissan also went on to design and produce the fixed apogee motor for the *H-I* rocket that successfully launched the engineering test satellite, *ETS-V/Kiku-5*, in August 1987. The company also developed and manufactured a series of sounding rockets under contract from then ISAS, all of which were also successfully launched between January to February 1998. Subsequently, under contract with NASDA, it developed and manufactured one *J-I* rocket, *TR-1A* rockets, and the SRBs for the *H-II* launch vehicle. Despite its historical prominence and solid performance, however, Nissan faced immense restructuring pressures in the late 1990s. As part of its revival program to concentrate on its core automotive operations, it subsequently sold off its aerospace division to IHI, discussed later below, in early 2000.[41]

Mitsubishi Heavy Industries (MHI) Although Mitsubishi was established in 1950, its origins can be traced back to 1884 in the Nagasaki Shipyard and Machinery Works.[42] Through a series of integrated corporate moves across shipbuilding, electrical machinery, machine tools, and aircraft, MHI has come to stand at the apex of Japan's aerospace industries. At present, in fact, MHI bills itself as the leading company in Japan's aerospace industries—and with good reason, given the range of its products across the defense sector (such as jet fighters, anti-submarine helicopters, aero-engines, missiles, torpedoes, and BMD program technologies) and the civil aircraft sector (such as airframe components, including fuselage panels for Boeing 777, composite-material wing boxes for Boeing 787, and known ambitions to produce an entire fixed-wing aircraft). More recently, MHI's complete systems integration prowess is being demonstrated by its decision to launch the Mitsubishi Regional Jet (MRJ), a domestically produced plane that remains the long-cherished dream of the Japanese aircraft industry.[43]

In the space systems sector, MHI has also loomed large.[44] It has historically been a major developer in a series of liquid engine (LE) technologies, beginning with then NASDA's *N-I* launch vehicles that were developed and launched between 1970 to 1982 and also the *N-II* launch vehicles developed from 1976 to 1987. It was the *N-I* that actually launched Japan's first geostationary satellite, *ETS-II/Kiku-2*, in 1977. The *N-II* had a 100-percent success rate, with the launch of eight satellites over its lifetime. The successively improved LE technologies also led to greater efforts to break away from reliance on American technologies. This began with the *H-I* rocket in 1981, which had an exclusively Japanese-built upper stage. An estimated 84 percent of the *H-I* was built with equipment designed by Japanese producers like MHI, and it had a faultless (nine-satellite) launch record until its phase-out in 1992.

MHI became primus inter pares when it secured the lion's share of the work for the *H-II's* development and systems integration from the STA and NASDA, including the advanced cryogenic first- and second-stage LE-5, LE-5A, LE-5B, LE-7, and LE-7A engines. Begun in 1986, the *H-II* was designed with the goal of technological autonomy, and considerable strides were made. At present the LE-7A rocket engine, initially developed for the first stage of the *H-II* rocket, is one of the main space products manufactured by MHI and is used in the first stage of the subsequent rocket, *H-IIA*. In addition, MHI pioneered the LE-5B for the second stage of the *H-IIA,* which it asserts is the first rocket engine in the world that uses hydrogen for cooling the thrust chamber of the turbine gas.

The *H-IIA* launch vehicle, which is a low-cost version of the *H-II*, now forms the backbone of Japan's independent access to space. MHI has loomed so large in the *H-II* and *H-IIA* development program that it was the Japanese government's only choice when it decided to privatize the rocket. It was not just MHI's manufacturing skills but also its commitment to marketing the

H-II rocket family through a Mitsubishi-led consortium of thirty companies in the form of Rocket System Corporation (RSC) that endeared it even more to the government.[45] RSC received orders for some thirty launches, including several tens of commercial orders, but the commercial orders were canceled due to the problems with the *H-II*. MHI subsequently took over RSC's activities, and the latter was closed down in March 2006. Given the tremendous spread and depth of its technologies and integration abilities, MHI unsurprisingly remains the main system integrator as well as the prime contractor for launch services of the *H-IIA*. MHI has now developed an advanced, heavy-lift version of the *H-IIA* version, the *H-IIB*, which has successfully lofted a huge autonomously guided cargo supply ship, the *H-II Transfer Vehicle* (*HTV*), to the International Space Station (ISS).[46]

Indeed, if Japan wants to ever build its own complete space station, it must work with MHI, which owns the core technologies: the *JEM*, also called *Kibō*, which is a manned space facility designed for long periods of stay as part of Japan's role in the ISS; and the joint development of the space rocket engine MB-XX, which is a hybrid liquid oxygen/liquid hydrogen engine family for superior and low-risk upper-stage propulsion. It is MHI's mastery of cryogenic engine technology and propulsion development that has brought it to the attention of U.S. and European players. In the early 1990s, for example, Boeing approached MHI to co-develop or provide engines for *Delta* rockets, but the move had to be abandoned because of the PPR.[47] However, MHI did go on to develop the MB-XX rocket engine with Boeing (now Pratt & Whitney Rocketdyne), which the companies have billed as being critical to positioning for the next-generation upper-stage engines.[48] MHI has also agreed to team up with Arianespace and Boeing to give commercial customers options to use each other's rockets as backup launchers for on-time delivery to orbit.

Ishikawajima-Harima Industries (IHI) and IHI Aerospace (IA) Through a series of mergers and expansions across business lines dating back to 1853, Ishikawajima-Harima Heavy Industries (renamed IHI Corporation in 2007) has become another historic heavy-machinery player across Japan's shipbuilding, aircraft, and automotive industries.[49] Its reputation for being a Japanese leader in the manufacturing of jet engines is well known, as it holds about 60 to 70 percent of the market and is also the primary contractor for aircraft engines used by the Japanese defense establishment. These technological strengths have spilled over into other industries. Even as IHI competed with MHI to become Japan's main integrator for liquid-fuel launchers, it has played a major role in Japanese space development efforts over the years and, more specifically, in producing key components of rocket engines and other technologies. Its key products for the *H-IIA* rocket include turbopumps (for the LE-7A engine, the heart of the rocket), the SRBs (from Nissan), and the second-stage reaction control systems. It is also engaged in the HTV propulsion system, as well as other satellite propulsion systems. For *JEM/Kibō*, billed as Japan's first manned facility in space, IHI has developed extravehicular platforms and pallets, as well as experimental racks and devices installed in onboard labs.

But IHI's efforts to establish itself even more solidly in space ventures came with the creation of IHI Aerospace (IA) in 2000, a wholly owned subsidiary of IHI and now one of the IHI group companies.[50] IA was established primarily to take over the aerospace business of Nissan Motors for an estimated ¥40 billion. For IHI the rationale for the deal was that it would allow two complementary businesses to come together and thereby help enhance its own aerospace and defense divisions both in terms of technology and government contracts: Nissan's rockets were based primarily on solid-fuel propellants, whereas IHI focused on liquid-fuel ones; in addition, Nissan's defense contracts were with the Ground Self-Defense Force (GSDF) whereas IHI's main contracts came from the Air Self-Defense Force (ASDF) and the Maritime Self-Defense Force (MSDF). IA was also responsible for the dual-use *Reentry Module* (*REM*) for *USERS*, which means the company has become familiar with both ballistic missile and payload (warhead) reentry technologies.

Both IHI and IA threw their weight behind the *Galaxy Express* (dubbed *GX*) launch vehicle (see Chapter 6), which was ultimately unsuccessful.[51] With the *GX*, IHI attempted to become a major launch services provider and thus to also play a major role as a prime contractor, instead of a subcontractor, in Japan's space activities. Toward this goal, IHI and IA formed Galaxy Express Corporation (GALEX) in March 2001 to manufacture the *GX* rocket in cooperation with Lockheed Martin. The *GX* represented a joint public-private development between GALEX, METI, and JAXA focused on an advanced second-stage engine. GALEX was supposed to be oriented toward commercial business, with launches that were projected primarily for small and medium payloads into LEO or Sun Synchronous Orbit (SSO). However, the Japanese government noted the potential use of the *GX* rocket for the launch of national satellites even as the program came under intense scrutiny.

As discussed later, the origin, politics, and development of the *GX* rocket were tortuous, and its future often appeared somewhat uncertain.[52] In March 2003, SAC finally reauthorized the *GX* program after many years of attempts to kill it because of doubts about its development costs and performance. At that point, SAC finally committed to a three-year program to finish development of the second-stage engine for the rocket. The engine, a new design fueled by liquid natural gas (LNG) and liquid oxygen (LOX), was to be provided by IA and JAXA with money from METI. This painful saga does not necessarily signal the end of IHI or IA in the SLV arena, as they may well continue to be the major players in Japan's emerging new *Advanced Solid Rocket* (*ASR/Epsilon*) that is also taken up in Chapter 6.

Satellites

Before looking at the individual Japanese players in the satellite market, it is helpful to have an understanding of their rivalries to determine the dominance of any one of them in particular. Until a series of changes in the late 1990s, Japanese satellites were developed by a revolving carousel of government contracts between Melco, NEC, and Toshiba. Each of these companies took turns to be the prime integrator on a NASDA contract, while the other two companies would receive important subsystems contracts, so that each satellite program could maximize the spread of skills.

During this time, externally there were requests from the United States Trade Representative (USTR) to open the Japanese communications satellite market.[53] Under the Super 301 provisions of U.S. Trade Law, a U.S.–Japan satellite agreement was signed on 15 June 1990. Its goal was to bind the Japanese government to open non-R&D satellite procurement to foreign producers and to enable foreign suppliers to compete in the procurement of broadcast satellites by NHK and NTT. This effectively sabotaged Japan's domestic commercial satellite development for the rest of the decade.

Domestically, a defense and space contracts bidding scandal that unraveled in October 1998 had a negative effect on NEC's space business, which had been vying with Melco to be Japan's biggest space contractor.[54] NEC, which had amassed losses of ¥8 billion, was implicated in overcharging NASDA (as well as then JDA) contracts to cover the deficit, and the company was barred temporarily from bidding for NASDA programs. This in effect killed NEC's chance to beat Melco for the then ¥170 billion *IGS* spy satellite program, which Melco subsequently won in November 1998. The upshot of this scandal, along with a string of engineering concerns, was that it changed the nature of strategic competition in the satellite business. While NEC went on to merge its space division with its equivalent in Toshiba, a combination of bad luck, poor design, and a number of specific technical and integration problems for NASDA satellites (such as *ETS-VI/Kiku-6* and even *ETS-VII/Kiku-7*) cast a considerable pall over both NEC and Toshiba. The die was cast when Toshiba failed to win the prime contract for the *ETS-VIII/Kiku-8*, which went to Melco. At that stage, NASDA ended up with four satellite contracts with Melco, three with NEC, but only one with Toshiba.

In small twists, the damage to NEC, Toshiba, and their joint offspring proved to be deeply wounding as Melco, in the meantime, successfully lobbied the CO with its own proposal and maneuvered for itself a prime position in the spy satellite business. As was widely known, Melco had long been agitating to overturn the Japanese government limitations on building satellites largely for civil use. As we will now show, the history of Japan's satellite makers in the second half of the 1990s clearly underscores our market-to-military trend, and the history of Melco, especially, is most instructive. By 1998 NEC and Melco were, by their own accounts, poised to enter the commercial satellite market, leveraging considerable experience with components and competitive space-related technologies. But as the global commercial satellite market also got stuck in low gear, both Toshiba and NEC found themselves with fewer venues.[55] Melco, meanwhile, after considerably enhancing its satellite manufacturing processes across virtually all types of systems, was devoting most of its satellite production and know-how to building Japan's spy satellites by adapting work done by itself, NEC, and Toshiba for NASDA over the preceding five years. The dramatic switch to military space production, which occurred in a few short years and in which Melco was named lead contractor, reveals how it is not so much commercial but rather military production that now underpins Japan's satellite industry.

Mitsubishi Electric Corporation (Melco) From its origins in 1921 when it was spun off from then Mitsubishi Shipbuilding Company (now MHI) to its manufacturing experiences across a diversified set of innovative electrical equipment and home appliances over the years, Melco has emerged as one of the foremost high-tech producers in Japan and around the world.[56] Melco has also turned out to be the undisputed leader in Japan's non-rocket space program, and it is presently Japan's most advanced and accomplished satellite builder. It rightly claims that it has the most significant track record of any Japanese company in completing satellite projects. In addition, its extensive manufacturing capabilities in control and ground station system technologies gives the company a distinct advantage, not just in designing and building satellites, but also in launching and controlling them. Melco's goal is to deliver all aspects of space infrastructure information technologies—satellites, ground-based equipment and terminals, key subsystems (such as solar array panels, communications equipments and antennas, structural panels, power amplifiers, transmitters and transponders)—as well as software and other services necessary to make these elements work in coordination.

Melco has long been involved in the development of space systems, especially on the commercial side.[57] This is evidenced in a series of projects for governments, communication concerns, and other large-scale clients, such as Intelsat (International Telecommunications Satellite Consortium), Eutelsat (European Telecommunication Satellite Organization), and Singtel Optus.[58] In 1966 it started technological cooperation on space technology with the United States' TRW (then a leading developer of military and civil space systems and satellite payloads, acquired by Northrop Grumman in 2002). A year later, it had delivered an Earth-station antenna to Mexico for use with an international satellite. In 1969, Melco was chosen as the prime contractor for Japan's first working satellite for ionosphere sounding. From these beginnings, the company went on to attain the status as the world's eleventh largest satellite operator in 2004; during that time several of its achievements are worth noting as they relate both to satellite and ground systems.

From the late 1970s to the present, Melco has won an impressive number of prime contractorships from NASDA. These include the production of some high-profile satellites, such as Japan's first domestically produced communications satellite (*CS-2a/b/ Sakura-2a/b*) in 1983 and the first large-scale Earth resources satellite (*JERS-1/Fuyō-1*) in 1985. The latest trajectories reveal that Melco has gained considerable know-how and confidence as a player and since 2000 has increasingly become the prime contractor in all types of satellite systems—communications (including *MTSAT-2*, Optus C1, *DRTS/Kodama*, *Superbird-C2*[7], ST-2); observation (including *MTSAT-2*, *ADEOS I, II/Midori I, II, GOSAT/Ibuki*), Engineering Tests (including *SERVIS*, *ETS-VIII/Kiku-VIII*, *USERS*); and science (*SOLAR-B/Hinode*). In addition, Melco (along with MHI) is in charge of the electronic modules and system design for the *HTV*, as well as electrical components for the *JEM/Kibō*.[59]

By 2002, Melco claimed to have been involved, as a prime contractor or subcontractor, in the development of more than 280 satellites, with 30 of those for international customers. It has also scored a number of notable firsts. Melco has increasingly sought to expand its manufacturing and marketing for communications satellites on a global basis, with a specific focus on the Asia-Pacific region, where it anticipates the most growth for space-based infrastructures given geographical barriers. There are indications that it is becoming successful in the continued search to break into the international broadcast and telecommunications market. In 1998, Melco was designated the prime contractor by its first client outside Japan, SingTel Optus, which meant it had responsibility for the manufacture of all related communications systems.[60] In 2003 it successfully launched Optus-C1, among the largest and most sophisticated commercial satellites covering Australia, New Zealand, Southeast Asia, and Hawaii. Although this was hailed largely as a commercial venture in press releases, it was known that half of the Optus C1's transponders were for use by the Australian Defense Department.

Another unprecedented victory came in November 2005. At that point Melco became the first Japanese company to enter the domestic commercial communications satellite market, which was entirely dominated by U.S. manufacturers who, up until that point, had provided all eighteen of Japan's broadcast and commercial communication satellites. After an international bid for the contract, Melco received a "delivery-in-orbit" contract for the *Superbird 7* from Space Communications Corporation (SCC), which means that it was responsible for managing the project from start to space.[61] SCC itself was formed in 1985 as a fixed satellite service operator and was part of the Mitsubishi group, with its top shareholders being Mitsubishi Corporation, Melco, and MHI. It had been launching its *Superbird* fleet of satellites since 1992, and also had an extensive ground network infrastructure in place.

In March 2008, SCC subsequently merged with its competitor, SKY Perfect JSAT Corporation (holding company integrating JSAT and SKY Perfect in spring 2007)—a move which brought all of Japan's private satellite operators and satellite broadcasting platforms in one place.[62] Company officials had stated that they saw future technological opportunities in integrated satellite, applications, and media systems (from broadcasting to mobile services, communications, and remote sensing, and onto positioning and land-, sea-, and air-based measurement systems). Importantly for our angle, along with commercial areas, they also foresaw new business opportunities in the "security" realm for government agencies, local governments, and other users. The government and security market angle, which has been key to Melco's corporate strategy, was also important because the satellite communications industry in Japan continues to face tough competition from terrestrial commercial services even in 2009. Sky Perfect JSAT Group's first satellite (and SCC's last) was the Melco-built *Superbird-7*, and was deemed critical for advancing Japanese manufacturers, such as Melco, in a market dominated by Americans and Europeans.

The *Superbird 7*, with its focus on the Asia-Pacific region, thus signaled a potential shift in the competitive structure for American players in the domestic commercial satellite market in Japan.[63] This is because the satellite bus (its basic satellite frame, propulsion, and avionics system) for *Superbird 7* is produced by Melco, which claims to have adapted a design originally created for the *DRTS/Kodama* and *ETS-VIII/Kiku-VIII* platforms (and some say also the Boeing 601 bus) into what it presently calls the DS2000 standard satellite platform. Thinking ahead to other domestic and international clients, Melco is banking on the cost competitiveness and performance of the DS2000 platform. This was satisfactorily demonstrated in 2006 in the successful launch of the *MTSAT-2*, and also confirmed by the 2008 announcement that Melco would be building the ST-2 communications satellite for a joint venture between Singapore Telecommunications and Taiwan's Chunghwa Telecom Company.

Melco has also advanced in another niche. From virtually its inception in the space business, Melco began to spread its manufacturing capabilities to ground systems and software, which are not only critical to space ventures but are also a highly lucrative market.[64] Since the late 1960s, Melco has established a substantial track record in supplying a number of various satellite-communications Earth station systems, including antennas, transmitters, and receivers.[65] Between 1968 and 1992, it supplied about 60 stations to Intelsat, whose satellite systems serve as the basis for its technology across international and regional networks worldwide. Between 1985 and 2001, it supplied more than 400 Earth stations to KDD, NTT, and then the Ministry of Home Affairs. During roughly the same time, it supplied more than 30 telemetry tracking and control stations to a variety of government agencies, such as NASDA and ISAS. In 1995 it also began to supply mobile Earth terminals for North American Mobile Satellite (MSAT) systems, which was set up to provide an unprecedented range of innovative and satellite services.[66] In 1998, Melco was chosen as the prime contractor for upgrading the satellite ground network system in Japan by what was then NASDA. About four years later in 2002, JAXA's new ground network went operational, spanning both domestic tracking stations (Katsuura, Usuda, Uchinoura, Masuda, and Okinawa, with Tsukuba as the control center) and overseas ones (Perth, Santiago, Maspalomas, and Kiruna). Melco also contributed to JAXA's tracking control for launch vehicles, with deliveries of precision radar and telemeter command antennae.

Over the years Melco has, of course, had its share of setbacks on satellite projects for which it was the prime contractor, such as *ADEOS-2/Midori-2* (which was lost due to a malfunction in the solar-array electrical circuitry in November 2003), and one of the radar spy satellites (which was abandoned due to electrical problems in March 2007).[67] However, its integrated experiences, know-how, and incremental improvements across space technologies has undeniably secured its role in the competitive satellite market. In part this can be attributed not just to its competitive prowess but also to the way it has skillfully exploited the geopolitical uncertainties surrounding Japan at present. As noted above, Melco's considerable manufacturing expertise in satellite and ground systems proved highly useful when it began to chafe against exclusively civilian and commercial uses of satellites in Japan. From the viewpoint of its corporate profitability, both commercial and military satellite markets offered windows of opportunities. These were, after all, in some ways mirror markets, yielding technological and, more importantly, cross-over institutional advantages that thus far had been denied to Japanese corporations like Melco, unlike their U.S counterparts. Melco has undeniably been the prime mover behind a military-based space infrastructure for Japan, a development that has reinforced its own competitive advantages, nurtured painstakingly over decades across commercial and military markets.[68] Melco's win over NEC and Toshiba to design and build Japan's spy satellites proved a tremendous economic victory for the company. As Japan's satellite story is covered more fully later, we merely note the following here.

To position itself in the burgeoning satellite opportunities under the exigencies of an increasingly national-security bound paradigm at home and the possibilities of commercial-defense contracts abroad (like *Optus CI*), Melco moved to establish a full-scale satellite assembly and testing plant in 1998.[69] The facility, the Kamakura Works, completed the following year, is the largest in Japan to date and has been designed explicitly to allow Melco to compete with European and U.S. manufacturers' clear superiority in cost and delivery time. We believe that Melco's major contribution to the militarization of Japan's space development is as the prime contractor for the spy satellite program, for which it rebuilt and redesigned Kamakura Works.[70] The investments made by the company were actually aimed at boosting its ability to fulfill its long-term contracts to the military satellite-building program. After the company was awarded contracts to build the initial constellation of Japan's four *IGS* satellites, including spares and replacements, Melco moved to redevelop its plant and, more importantly, completely overhaul its satellite-building capabilities to reach cutting-edge levels that would allow it to target all possible markets. Of course, the new facility has also allowed it to position itself as the prime contractor for the *Quasi-Zenith Satellite System* (*QZSS/Michibiki*), which originated with Melco and which the company is keen to see extended as a regional system.

Even with the improvements the company has made to its Kamakura Works, Melco probably does not have the capacity to build more than a handful of satellites per year in addition to its domestic procurement duties for the spy satellites. The latter, we believe, will take up the lion's share of Melco's capacity and keep it occupied with building national security infrastructure. In this sense, Melco's own journey echoes the market-to-military trend quite clearly, but this is hardly the end point. Some of Melco's global-cum-regional ambitions took an additional step forward in 2001.[71] It announced a strategic alliance with Boeing in order to respectively boost their satellite and space businesses across a range of fields—satellite communications, air traffic control technologies, space technologies, a block agreement for Melco to launch its satellites on up to six Delta-IV missions, and possible provision of subsystems or components by Boeing for Melco's new satellite bus. The deal was in fact part of a three-way package between Melco, MHI, and Boeing, and it cemented the Mitsubishi group's dominance in Japanese space development with one of the world's leading aerospace companies.

Nippon Electric Corporation (NEC) NEC was established in 1899, and it is presently one of the world's giant electronic firms with a wide range of products, such as computers, semiconductors, telecommunication equipment, software, and home appliances.[72] Its space business began with the delivery of a rocket telemetry transmitter-receiver system to a lab at the University of Tokyo in 1956. In 1969 the company built a Yokohama plant for its space business, a facility that was designed to enhance the prowess of its expansive space business that was spread across not just a formal Space Systems Division but also various business technology units, such as aerospace, defense, and electro-optics. From the early 1970s until late 1998, when NEC apologized formally for its scandalous overcharging in JDA and NASDA projects, there appeared to be little doubt about the company's viability as a player.[73] It was well known, of course, that NEC had been vying with Melco to be Japan's primary satellite builder. As Melco pushed to develop large satellite bus technology, NEC focused on smaller satellites, was the primary developer of ISAS missions, and was given responsibility for developing many of Japan's communications technologies. In the late 1990s, NEC had set its sights on becoming a top-notch satellite maker and seemed well positioned to compete with Melco in the space business.[74]

The company had, after all, been central to some of the early successes in Japan's space program and had begun to reach around the world with its satellite and ground systems. At home, NEC was the renowned supplier of Japan's first engineering test satellite, *Ohsumi*, launched in 1970. From that time until 2007, it was involved in the development of more than sixty-one satellite development projects—engineering tests, communications/broadcast/GPS, EO, and scientific ones. Whatever their subsequent successes and failures, the important point is that these projects allowed NEC to be involved in Japan's acquisition of a number of pioneering space technologies:[75] Japan's first geostationary weather satellite, *GMS/Himawari*, in 1977; Japan's first interplanetary spacecraft, MS-T5/*Sakigake*, in 1985; Japan's first full-fledged Earth (specifically marine phenomenon) observation satellite (*MOS-1/Momo-1*), in 1987; Japan's first successful experiments for lunar swingbys on the *MUSES-A/Hiten* in 1990; the world's first precision spectrosocopy and photography (allowing observation and study of X rays in stars and galaxies) on *ASTRO-D/ASCA* in 1993; the world's first space VLBI observation (with resolution 300 times that of the Hubble space telescope) on *MUSES-B/Halca* in 1997; Japan's first efforts to utilize rendezvous docking technologies on the *ETS-VII/Kiku-7* in 1997; Japan's first space probe, *PLANET-B/Nozomi* in 1998, which embarked on the country's first Mars surveyor mission; Japan's *COMETS/Kakehashi* communications engineering satellite that performed a maximum range of communications in 1998; the world's first space probe, *MUSES-C/Hayabusa*, launched in 2003, which succeeded in landing on and taking off from an asteroid (Itokawa); the *OICETS/Kirari* engineering test satellite in 2005, which allowed for the world's first optical communications between a LEO satellite and a ground station using a laser beam (and whose laser intersatellite technology is considered un-jammable and thus the next frontier for secure military communications); and the *ALOS/Daichi* land observation satellite in 2006, which allowed for the world's first satellite equipped with different types of simultaneously usable sensing instruments, radar and optical (and which was central to the construction of the *IGS* satellites).

Abroad, NEC had also built on its reputation to expand its space business to other government and commercial clients in the late 1990s.[76] It had, for example, been designated a subcontractor by Matra Macroni Space, UK Ltd., for a complete communications repeater system for *ORION-2* because of its success in providing one earlier for the *ORION-1* satellite. It was also involved in the provision of a satellite ground station from Paraguay's public telecommunication company and had provided Tracking, Telemetry, and Command (TT&C) equipment for ground stations in Saudi Arabia and Sweden. Perhaps most important, NEC had banked on demand for satellites from abroad and had even unveiled an advanced satellite bus design suitable for LEO and GEO applications.

Certainly there was plenty to spur on the company's wild expectations. First, in 1995 there was Intermediate Circular Orbit (ICO) Global Communications, a commercial spinoff of the London-based INMARSAT (now International Mobile Satellite Organization) consortium, which was meant to compete with the Motorola-led Iridium, Loral-led Globalstar ventures, and TRW-led Odyssey.[77] NEC won a $600 to $700 million contract as the main supplier of equipment and associated services for the ground infrastructure—such as satellite access nodes in Korea, Indonesia, Mexico, the United Arab Emirates, China, and Brazil, as well as network management systems and systems integration services.[78] It was also given a supply order for 100,000 dual-mode satellite/cellular handsets for of ICO and charged with the completion of twelve TT&C Radio Frequency Terminals (RFT) in the United States, India, Australia, South Africa, Chile, and Germany. NEC had also looked to satellite demand from Teledesic, which had begun with an ambitious "Internet in the sky" scheme in the early 1990s.[79] Teledesic sought to put 840 data-relay satellites in LEO in order to provide seamless global services without the signal delays from geostationary satellites.

Both external businesses ran into trouble. ICO Global Communications had a troubled beginning in a crowded global market, with some competitors immediately charging patent infringement at the time of its startup.[80] Moreover, analysts had seriously warned that the demand for non-geosynchronous mobile satellite services seemed to be sufficient to support just one (mobile satellite phone) operator, and investors were already greatly troubled by the problems with Iridium. By the end of 1998, ICO had already begun to post losses, and its follow-on stock offering the following year remained undersubscribed ($1.7 billion shy of the needed $4.7 billion) as confidence in the market continued to erode. By 1999 it had filed for bankruptcy, although it was rescued by investors and emerged as New ICO in 2000.[81] Teledesic also ran into trouble. By late 2002, the company, which had scaled back the number of satellites, first to 240 and then down to 30, pulled the project altogether, citing uncertain financial markets and commercial prospects.

The late 1990s were to prove the zenith of NEC's fortunes in terms of its space business, which at the time made it Japan's largest space contractor by revenue. However, neither its size nor its considerable first-mover technological advantages in satellite programs helped the company in the highly competitive market. As the satellite business abroad began drying up and its overbilling scandal at home exploded as noted earlier, NEC was irreversibly damaged on both fronts. In the aftermath of the scandal, both the then JDA and NASDA moved to suspend NEC from contracting activity. The timing of the scandal was especially disastrous because the Japanese government was moving toward considering bids for an estimated $1.3 to $1.7 billion *IGS* program to develop two optical-imaging and two radar-imaging satellites.[82] NEC was not even considered a likely bidder. NEC then moved forward through a merger with Toshiba, which led to the establishment of a new space business company. At this point, though, NEC has not recovered its former stature as a dominant player in Japan's space business.

Toshiba Corporation Although Toshiba did not acquire its present name until 1978, it was formed in 1939 through the merger of two manufacturers (one, established in 1875, with a focus on heavy electric equipment; the other, established in 1890, focused over time on consumer products).[83] At present, Toshiba is unquestionably one of the world's premier high-technology firms, with a manufacturing range across some of the most advanced electronics and electrical products. It holds the distinction of completing Japan's first radar systems in 1942, and is an important defense contractor today.

Certainly, the company has always had ambitions in the space field. Toshiba has built an impressive array of space-related technologies, and even as late as 1997 was infected with the same enthusiasm about commercial satellite prospects as NEC and Melco were. Toshiba launched Japan's first medium-sized experimental broadcast satellite, *Yuri*. It was subsequently responsible for the integration of several of Japan's satellite programs, such as the 385 kilogram *ETS-III*, *BS-2A/Yuri-2a*, *BS-2B/Yuri-2b*, *ETS-VI/Kiku-6*, and *ETS-VII/Kiku-7*. In late 1997, still enthusiastic about expansion, it went on to invest in the Alcatel-led skybridge venture, which sought to place sixty-four satellites in LEO for broadband communications.[84] It was also moving forward with satellite components. Since 1982, when it delivered its first solar array panel, it positioned itself as a leader in this area of space technologies. In late 1997 it entered into an agreement with U.S.-based SS/L for manufacturing and supplying solar panels for three years.[85] To continue to expand its space business as aggressively as it was doing, Toshiba invested more than ¥1 billion in updating its facilities, such as at its Komukai Works, for design, manufacture, and testing of space equipment and systems.

In the 1990s Toshiba was so proud of its space technology that it hung a model of the *ETS-VI/Kiku-6* satellite, for which it was the prime contractor, in the gigantic lobby of its Tokyo headquarters.[86] This satellite was a particularly advanced design for its era. However, because of a basic fault in its apogee kick-motor, it was condemned to a highly damaging orbit that swung it repeatedly through the Van Allen radiation belt, quickly degrading it. The satellite's veering off course was subsequently reflected in the fate of its parent, Toshiba, in the space businesses. The satellite was later removed, which also seemed to presage the company's impending decline in space activities.

Toshiba had built an early record of failure with its *BS* satellites series in the 1980s, which were constructed with the help of General Electric.[87] The principal customer for these satellites was NHK (Japan's public broadcasting corporation). A variety of malfunctions plagued the *BS-2A*, and although Toshiba and General Electric were favored by NASDA to develop the next generation *BS-3* series, NHK vehemently opposed the move, and the follow-on contract then went to NEC. Toshiba also began to lose to its competitors with the *ETS* series.[88] Following the malfunction of *ETS-VI/Kiku-6*, Toshiba suffered embarrassment with the *ETS-VII/Kiku-7*, for which the company was bitterly criticized for installing a robotic arm in an incorrect position. The *ETS-VII/Kiku-7* was the last satellite for which Toshiba served as prime contractor for NASDA. The *ETS-VIII/Kiku-8* contract went to Melco and signaled a shift away from the revolving carousel. Toshiba began to lose other contracts, such as those from the Ministry of Transport to supply computers in a ground station for Japan's GPS-based air navigation system. Given dying prospects in government and commercial ventures, it was no surprise that Toshiba began to consider moves toward a merger, first with Melco (which failed as the firms had overlapping product lines) and then with NEC.

NEC-Toshiba Space Systems Ltd. (NT Space) Formed in April 2001, NT Space represented the consolidation of NEC and Toshiba's space divisions, with NEC holding 60 percent and Toshiba 40 percent of the initial capitalization of about ¥7 billion.[89] The justification for its formation was the well-known story—the domestic market for launch vehicles, ground systems, spacecraft and the international space station valued at about ¥200 to ¥300 billion was essentially flat, the government was the only source of orders, and little commercial growth was expected; additionally, the external market was valued at ¥3 to ¥4 trillion and was transforming with numerous large-scale mergers under intensifying competition. NT Space—a modest initial enterprise seeking to expand its space business in the design and integration of satellites, satellite subsystems and components (transponders, sensors, solar array panels, large deployable reflectors), and ground systems to about ¥100 billion in the first five years—was pooling its resources in order to more effectively position its products and services in the tight commercial market both at home and abroad.

As a business strategy under those conditions, it certainly made sense. NT Space could well boast that it had successfully integrated 70 percent of all Japanese satellites. But NT Space had a troubled start and, from there on, a muddled existence. In early 2002 it was revealed, and subsequently acknowledged by the company, that one of its employees had illegally accessed computer data at NASDA on the ultrahigh-speed internet service due to be launched through the work of both NT Space and Melco.[90] The NT Space employee had accessed documents, meant for NASDA and Melco eyes only, that evaluated satellite component parts by Melco. Although NASDA did not press formal charges, the agency removed NT Space from a list of designated bidders for a month. NT Space also had to pay a penalty of about $3 million to ISAS (which became part of JAXA) when it incorrectly copied an assembly drawing for an electrical connection that caused the *Demonstrator of Atmospheric Reentry System with Hyper Velocity* (*DASH*) experimental satellite to fail in separating from the *H-IIA*.[91]

The company has made investments that continue to suggest its interest in the space business even though the projects are as yet tenuous.[92] Theses include the Advanced Space Business Corporation (ASBC), which was set up to help fund and operate the *QZSS/Michibiki*, and GALEX. NT Space was also involved in the successful launch of an asteroid sample-return mission on the *MUSES-C/Hayabusa* spacecraft, which it had designed and built. Perhaps most significant, it was also responsible for the optical sensor called Prism for the *ALOS/Daichi*, one of the largest EO satellites launched by JAXA in early 2006. The instrument could simultaneously see forward (70 kilometers) and backward (35 kilometers), which allowed it to develop three-dimensional images of terrain. The capability of this satellite's sensors forms an important part of our analysis of Japan's military reconnaissance capabilities later in the book.

After only a few years of NT Space operation, its parts began to be restructured as NEC moved to boost the quality and competitiveness of its own divisions under the larger umbrella of the company's Aerospace and Defense Operation Unit. In April 2007, NEC brought NT Space's satellite design and development to its own Space Systems Division and NT Space's satellite equipment sales function to its own Commercial Satellite Sales Group. Although NT Space faced an uncertain future outside its specialties of communications, EO equipment, and scientific satellite contracts, the company has held its ground somewhat. In terms of scientific missions, it has a manifest of three important new satellites- the *ASTRO-G/VSOP-2* radio telescope successor to *Halca*, *PLANET-C/Akatsuki* Venus Climate Orbiter, and the *Mercury Magnetosphere Orbiter* (*MMO/BepiColombo)* missions—all highly challenging and major interplanetary missions that only the United States and Russia also have the technology to mount with confidence, underscoring NEC's long legacy of accomplishment. In addition, for JAXA, NT Space is building both *Global Change Observation Mission* (*GCOM*) satellites. So while Melco has emerged as the dominant player in a de facto government satellite duopoly, it would be unfair to say that NEC/NT Space has been left completely in the cold. In line with our market-to-military thesis, NT, with its heritage of small satellite bus technologies, was chosen by METI/USEF to build Japan's upcoming ORS/SOD spy satellite constellation, *Advanced Satellite with New System*

Architecture for Observation (*ASNARO/Sasuke*). The procurement is for, initially, up to four small, high-resolution (50-cm) optical spy satellites.[93]

EVOLUTION OF PUBLIC AND PRIVATE PLAYERS

This chapter provides an overview of the main players in Japan's space policy as of the end of 2009—from government agencies that have amalgamated in the interest of streamlining the space policymaking infrastructure to the key private companies whose economic fortunes have risen and fallen in the highly competitive space business. It shows how the origins of and relationships between sets of players across the public-private domain shape Japan's space policy.

In tracing their activities, it is clear that the tactics of Japan's space-related players have waxed and waned, fallen apart and come together, under a set of historical conditions: technical glitches and failures across rockets and satellite technologies, pressures from the United States for market access, the small size of the domestic and global space-business market, brutal internal competition and rivalries among especially the private players, and, increasingly, national-security bound paradigms for Japan as a whole. Although no outcome was predetermined, Japan's present institutional and corporate landscape is getting more pruned: MEXT, METI, MOD, JAXA, and the Mitsubishi group companies, such as MHI and Melco, are presently the dominant players, and their space efforts are to be coordinated in the national interest through SHSP.

On the government side, it remains to be seen how and in what ways SHSP will fundamentally put its stamp on Japan's space policy establishment which has long lacked a center of gravity. Even as MEXT's role is being challenged, and JAXA's operational and administrative structure in which it has a dominant place comes under review, the ministry continues to be a sizable player. MEXT, like other players, has to come to grips with two other ministries who are henceforth likely to be visibly significant players. The first of these is METI which has successfully managed to cover the gamut of space industry and policy—the former with its jurisdiction over the manufacturing range, the latter through the activities of independent administrative agencies and public-interest institutions with which it has close associations such as NEDO, JAROS, ERSDAC, and especially USEF. In painstaking bits and pieces, METI has thus been a much less visible (at least compared to MEXT) but perhaps far more critical player in bridging both the market and the military aspects of space policy. As the MOD rises in importance, both as a consumer and defender of space assets, such as BMD and space-based communications and security, there may continue to be further changes in the institutional structure. MOD has already attempted to put its own visible stamp on the directions and contents of space policy in Japan, more specifically with the formation of the CPSDU.

On the corporate side, Japanese corporations on both the SLV and satellite side have long been stymied by the lack of commercial prospects—domestically, they have been restricted by technology-transfer and procurement agreements with the United States; externally, they have faced volatile business scenarios which have been dominated by foreign rivals. Their search for domestic and foreign markets, and their already considerable investments in space products and technologies, has also slowly but surely pushed them toward the militarization of space assets. At this stage, after all the mergers and restructuring, MHI (on the rocket side) and Melco (on the satellite front) have emerged as the top space corporations. But despite the track record of MHI and Melco, government institutions now or in the future may not want to rely on them as sole suppliers. Calls for the pairing of NEC and Toshiba into NT Space, and from there toward a mega-merger with Melco into an All Japan Space Corporation, must be understood in this context.[94] Already as the *IGS* continue to be dominated by Melco, the *ASNARO/Sasuka* have brought NT Space into the fold; and even as MHI has been pivotal to the development of the *H-IIA* and beyond, IHI and IA are likely to be the main developers for the emerging *ASR/Epsilon*.[95]

Thus, rather than concerns with the dominance of any one player or another, the more important point is that the sum total of all the public, and especially the market players,' activities over the postwar period has allowed Japan to acquire an impressive set of space-related technologies—those that double as being central to the militarization debate. We find that there are common threads that weave through all the players' stories as they set about making a name for themselves in competitive global markets. We find it is too much to claim that there has been some grand national security strategy at work as these players went about their respective work and businesses in the space field over the postwar period. Rather, in retrospect, the more mundane reality turned out to be the accumulation of a vast array of technologies of world-class, sometimes cutting-edge, quality and nowhere to sell them in the commercial sphere but with increasing possibilities to do so in the military one. But what kind of products exactly, and with what implications for Japan's national security? To shed light on these issues, we next turn to showing how the activities of these players across key components of Japan's space policy—launch systems, satellites, and other emerging technologies—lend more specific support to our market-to-military thesis.

4 LAUNCH VEHICLES

IN DECEMBER 2004, Japan dropped a proposal to develop the nation's first long-range surface-to-surface missile. The was largely due to protests by the Liberal Democratic Party's (LDP) junior partner, the New Komeito Party, which is backed by Japan's largest lay Buddhist group, the Soka Gakkai.[1] While most LDP members on the government security panel supported the plan, the much-highlighted concerns of the junior partner in the coalition were that such offensive capabilities would run counter to Japan's postwar defense-only policy and, not unimportantly, the pacifist ideology of the Soka Gakkai. The proposed missile research was to have been incorporated into the nation's 2005–2009 defense buildup plan, which was approved by the Cabinet together with the new National Defense Program Outline (NDPO).

If implemented, the expected specific result would have been that Japan could launch preemptive strikes against foreign enemy bases at a maximum range of up to 300 kilometers; the more consequential general result would have been an effective (and visible) end to the self-imposed ban on offensive weapons. The then Japan Defense Agency (JDA) had been assessing the possibility of such a missile buildup due, no doubt, to growing apprehensions about North Korean and Chinese incursions in the region and the fact that such a missile buildup would have allowed Japan to defend its remote islands and repel invasion on its own soil. The missile development program had surfaced after an October 2004 advisory panel to Prime Minister Junichiro Koizumi called for Japan to consider having offensive capabilities against enemy missile bases—in the absence of alternatives and only as a last resort.

Behind these concerns about whether or not Japan would suddenly move forward with its own full-fledged offensive missile program lay, to our minds at least, a far more critical fact: Japan already has the technological capability to make missiles—even ballistic missiles. After all, the nation's missile budget has exceeded the space budget and, more importantly, most of the very same firms that have built missiles for Japan's defense establishment (as shown in Table 3.2) also build rockets for the peaceful utilization of space.[2] As for the aversion to "offensive" missiles that appears to grip the popular Japanese imagination, that too needs to be reassessed in light of the fact that corporations such as Mitsubishi Heavy Industries (MHI) have been developing and producing missiles, classified as offensive weapons, since the early 1970s.[3] Why might all this be important as background? Thus far it has been the seamless role of one corporation, namely MHI, that is critical to the missile-space industry crossovers. As Chapter 3 made clear, there is little doubt that the Mitsubishi Group is, at this stage, at the top of Japan's space systems game. In short, MHI is an experienced missile and space launch vehicle (SLV) builder.

It is, therefore, simply not true, as press coverage at the time seemed to suggest, that the proposed missile research would be starting virtually from scratch. As we show in this chapter, Japan painstakingly acquired its ballistic missile-centric capability in the early postwar period, and later companies such as MHI became involved in Japan's space program. In particular, companies like Nissan, and now MHI and Ishikawajima-Harima Industries (IHI), have become accomplished SLV developers—that is, they possess the very same technological processes that potentially also make them competent ballistic-missile technology integrators. From the viewpoint of Japan's national security, since the early 2000s various aspects of this corporate competence have been fortified even further by the ballistic missile defense (BMD) technology cooperation with the United States (discussed in Chapter 6).

In this chapter, we start from the premise that Japan's long postwar trajectory to acquire SLV technology has been deeply interwoven with military technologies. Indeed, almost all space launch vehicles today, as in the United States, can trace their ancestry back to the sophisticated V-2, which was deliberately acquired from Germany after World War II.[4] The point is simple, and the Federation of American Scientists (FAS) puts it well: A rocket is called a launch vehicle when used to launch a payload (such as a satellite) into orbit or deep space; it becomes a missile when its intended use is as a weapon and its payload is a warhead.[5] Thus a rocket, and the underlying rocket technology, is one means of possessing "high-end" delivery systems such as ballistic missiles.[6]

Although Japan was careful to distance its space program from military purposes, a range of commentaries and intelligence estimates alike make clear that all civilian SLV programs, whether in Japan or elsewhere, inherently encompass intercontinental ballistic missile (ICBM) technologies.[7] This has long been known, and several reputable (some later declassified) sources in the United States have confirmed it more openly.[8] The U.S. Congress is well aware that much of the technology used in sounding rockets and SLVs can be directly applied to surface-to-surface missiles. The U.S. Central Intelligence Agency (CIA) has pointed out that almost all aspects of SLV technology are applicable to ballistic missiles: the staging, propellants, airframe, engines, thrust control, and exhaust nozzles of an SLV use the same technologies and function on the same principles as ballistic missiles; the reentry vehicles, separation, internal guidance and control, and strap-on booster SLV technologies may be adequate for ballistic missiles; and the only unique ballistic missile technology is the warhead. The National Defense Industry Association (NDIA) has similarly pronounced that there is virtually no distinction between a civilian space program and a military program in the development phase and that a range of ballistic missiles can be developed from indigenous SLV programs. In assessing strategic weapons worldwide, the much trusted Jane's Information Group provides details on SLVs, primarily because they share common technologies with ballistic missile programs.

The ease of such market-to-military conversions has taken on greater significance under conditions of increasing economic globalization, the latter in which Japan has participated both as an agent and a beneficiary over much of the postwar period. The writers of the so-called *Rumsfeld Commission Report*, who had sought to assess the ballistic missile threat to the United States, were especially alert to the ease with which scientific, technical, and industrial information could move across borders in the globalized commercial economy. This, then, was the dark side of globalization at play. These very same commercial exchanges and technology transfers also continue to serve as pathways for the dissemination of military technologies necessary for the construction of ballistic missiles and weapons of mass destruction (WMD)—themes that echo particularly well in the case of Japan, which has been characterized as having a mercantile realist bent that advances the country's technoeconomic security agenda.[9] As noted in the *Report*, there appears to be little distinction between the defense technological base and the commercial technological base in Japan, and the government takes for granted the "spin on" of commercial technologies to defense applications.[10] More directly on point, as the U.S. government recognized early on, countries with the demonstrated capability for developing SLVs should be considered capable of developing ballistic missiles as well.[11] As discussed later, it was this reality that no doubt alarmed the United States enough to try and constrain Japan's launch vehicle technology development and to give it a more peaceful orientation in use in 1969.

Taking a lead from such dual realities, this chapter looks at the development of Japan's SLV programs over the postwar period. Several questions guide us, in line with our theme of the market-to-military trend: First, what components has Japan developed that speak to its ballistic missile capability? Second, how did Japan acquire such components through the technical and institutional infrastructure in its civilian SLV program over the postwar period? And third, does Japan have the institutional, technical, and political infrastructure to put together and launch ballistic missiles if needed in the near future? Like the National Intelligence Council (NIC) in the United States, which sought to assess foreign ballistic missile threats and programs in 2001, we too would like to be clear that Japan's ability to convert its SLV program into future ballistic missile development is inexact.[12]

Nevertheless, there is considerable evidence that Japan has acquired a solid capability for ballistic missile development, which has surely not been lost on the U.S. government or military, or even other governments or militaries around the world. Of the open sources available in the United States, several have come to clear assertions with respect to Japan.[13] One suggested that Japan has accumulated enough industrial and technical knowledge in its space launch programs to support long-range ballistic missile operations. Another has claimed that Japan had developed rocket capabilities in its civilian space program that could quickly be converted into ICBMs. Other observers have declared that although Japan does not currently possess ballistic missiles, it could nevertheless develop them in a short period of time. Of course, Japan has never openly stated any grand ambition to develop ICBMs. Indeed, for half a century, there was no need for Japan to develop ICBM technology given that it was protected by the United States. But, the geostrategic concerns in areas surrounding Japan at least make consideration of a Japanese ballistic missile capability no doubt very important for Japan's defense planners.

This chapter will focus on the progression of Japan's ballistic missile-centric capabilities through an examination of the SLVs in the country's official space program over the postwar period. In looking chronologically at the historical progression of Japanese liquid- and especially solid-fuel SLVs, we assess the component technologies that are considered to be critical in the manufacture and launch of ballistic missiles from start to finish—propulsion systems, structures, staging ability or large boosters, sophisticated guidance and control systems, reentry vehicles, and flight operation skills, as well as the technical and institutional infrastructure necessary for integration.[14] We show that the sum total of these by-now-matured parts of the country's civilian SLV program leaves little doubt that Japan has the technical experience and capabilities to build ballistic missiles (particularly in terms of propulsion, guidance, and also reentry, which is discussed in Chapter 6), although it remains to be seen whether Japan will choose to do so *overtly*. We would also like to be clear that while it is certainly true that a country's missile capability, such as that in Japan, needs to be assessed in parallel with its capability to produce warheads (such as nuclear ones), that is not the central focus of this book and we only take it up tangentially at the end of the book.

Thus, here we concentrate only on the delivery systems in order to show just how advanced and experienced Japan has become with respect to ballistic missile technologies as the country's rocket-building infrastructure has moved from the interwar period to the present day. To make clear the successive phases that have brought Japan ever closer to ballistic missile technology and beyond, the remainder of this chapter has three parts. The first part focuses briefly on the interwar period, when Japan's efforts to build launch vehicles got under way, until the end of World War II, when those efforts were cut short. The second part, the bulk of the chapter, focuses on Japan's postwar developments from its first rocket, *Pencil*, to the largely indigenous *H*-series, which presently holds a privileged position in shaping the future of Japan's SLV capabilities. The background in this chapter paves the way for understanding the future of Japan's SLVs, such as those beyond the *H-IIA*, the *GX*, and the *Advanced Solid Rocket* (*ASR*/*Epsilon*) which are taken up in Chapter 6.

FROM THE INTERWAR PERIOD TO 1945

Rockets are not new to Japan; they are not even new to East and Southeast Asia.[15] They have occupied a colorful place at traditional and religious festivals in Japan at several shrines for centuries and continue to do so today in such places as the *Ryusei* festival (so called because of visual similarity to dragon power or a shooting star). Some attribute their introduction to Japan either near the end of the thirteenth century, when Mongolian and Korean armadas attacked the southern parts of Japan, or near the end of the sixteenth century, when the shogunate imported rocket arrows from China and fire arrows from the Netherlands. But such "technology transfers" remain a matter of historical speculation, and we remain unsure about their subsequent impact on Japan's more formal efforts to boost its military rocket capabilities.

By most accounts, Japan's contemporary rocket program is deemed to have started in 1955.[16] However, it benefited from the pioneering developments during the interwar period when both the Japanese Army and Navy began solid- and liquid-fuel rocket research and development in 1931.[17] Their efforts led from rocket-propelled cars in 1932 to rocket-bomb guns and onto a launcher-based system for winged rocket bombs. Perhaps best known is the *Ohka*, a rocket-propelled manned attack glider.[18] This Japanese suicide bomb, which was equipped with three solid-propellant rockets, entered mass production and use in 1945, when Americans subsequently christened it the "baka," or stupid bomb.

Japan had other rocket-based advances that are less well known and that were perhaps more significant from an evolutionary perspective. In 1943 the army deployed small-caliber versions of the rocket bomb and its launcher that went on to become an official set of weapons known respectively as "*Funshin-dan*" and "*Funshin-ho*." Both the army and the navy developed variants of the *Funshin-dan*, based on length, weight, velocity, and range; and they gained experience in advanced techniques such as tail wings and/or spin for flight stability and the use of uniform propellant quality and burning rate to ensure ballistic trajectories. The greatest advantage of the *Funshin-dan* was the ease of its transportation, installation, and launch, and it was used in battle in the South Pacific. The navy used them on Iwo Jima and continued to conduct rocket research for rocket bombs to destroy other targets such as enemy planes, landing craft, and B-29 bombers. Heavier versions of the *Funshin-dan* were also used at Iwo Jima and Okinawa.

In 1943 the Japanese Navy began more advanced rocket research; it succeeded in the flight test of what can best be described as a solid-propellant surface-to-air missile, the *Funryu-2*, guided by a neutral position radio guidance system. The *Funryu-3* and *Funryu-4* followed. The latter benefited from technical crossovers from liquid-fuel rocket technology developed for the *Shusui*, Japan's rocket fighter. The *Funryu-4* itself remained untested. The army also began research on developing liquid-propellant rockets, but the focus of Japan's rocketry was on solid propellants until the final stages of the war. The army's endeavor to develop radio-controlled missiles was known as the "*I-go Project*," consisting of large (*I-go-1A*) and small (*I-go-1B*) guided missiles. Importantly, it was MHI, today Japan's preeminent rocket maker, that had primary responsibility for the larger version. The *I-go-1A* was a small pilotless rocket with an 800-kilogram bomb in the nose; the rocket was designed to be attached under the fuselage of a mother plane and launched from there toward the target. It took from August to October 1944 to design and actually test it, by which point U.S. air power had made its use impractical. The *I-go-B* was designed, tested, and produced by Kawasaki Aircraft Company, but air raids in June and July 1945 destroyed the production facility and, consequently, the company's prospects.

Japan's liquid-fuel rocket base also took an alternate step forward in March and April 1944, when Germany gave two sets of technical data—on the body and rocket engine of the Me-163B *Komet*—to a Japanese naval attaché. Germany had developed the Me-163 rocket fighter during World War II. Part of one set of technical data made it to Japan, and it became the basis for further technical expertise in liquid-propellant engines. This expertise was used to develop the Tokuro rocket engine and its subsequent variations as in the *Tokuro-2* (or *KR 10*). The *Tokuro-2* went on to power the *Shusui*, Japan's first liquid-propellant rocket fighter, which targeted the U.S. B-29.[19] Once again, working with the army and the navy, MHI was heavily involved with *Shusui's* body, engine, and propellant technology. After appearing to function smoothly, the *Shusui* crashed and burned on its first test flight in July 1945. In total about five *Shusui* were completed or were nearly complete before the end of the war, but this was too late to make any impact. Whatever the eventual outcomes, from an incremental technological and experiential viewpoint the *Shusui* was critical to Japan's and MHI's rocket development efforts long after the war was over. The *Shusui* was not an ordinary tactical achievement; at that point it actually earned Japan the distinction of developing a manned rocket plane during World War II.

None of Japan's rocket fighters or missiles affected the trajectory or outcomes of World War II. But it is clear that Japan, through military and corporate research, showed early competence in propellant production and loading technologies, as well as some competence in guidance and flight stabilization technologies. It achieved all of this by the end of the war in 1945, leading to an authoritative claim that Japan was second only to Germany in terms of the technological level of rocket development at the time.[20]

THE POSTWAR CAMPAIGN: FROM ZERO TO *PENCIL* AND ONWARD

Appendix I at the end of this book lays out some of the principal launches (SLVs and satellites) by or involving the Japanese space program between 1945 to 2009, some of which have drawn general attention.[21] We have a more specific goal. Relying primarily on the materials made available through JAXA, we concentrate here only on the chronological development of Japan's ever more sophisticated and indigenous SLVs in the postwar period.[22] Using a range of other sources, the idea is then to see whether the experiential and technological SLV base acquired by Japan over fifty-plus years can also allow the country to construct ballistic missiles at present.[23] The following narrative shows how Japan has gradually acquired and tested all the component technologies necessary to construct especially solid-propellant ICBMs under its civilian space program—if it so chooses.

Why Solids?

In order to focus in on Japan's SLV trajectory more closely, it is helpful to begin with some distinctions between solid-propellant rockets over liquid-propellant ones.[24] Most contemporary analysts agree that solid-propellant motors are more suitable for tactical missiles (air-to-air, air-to-surface, surface-to-air, or short-range surface-to-surface) and ballistic missiles (short- and long-range surface-to-surface) than liquid-propellant rockets. Liquid-fuel rockets do have their own advantages including that they have the highest specific impulse (a higher figure indicates higher merit of performance); they can be throttled, stopped, and restarted; they can be tested at full thrust on the ground or the launch pad prior to flight; and they can be designed for reuse after field services or checkout. However, their relatively complex design, with many parts and components, means a greater probability of things going wrong. Propellant loading occurs at the launch pad, and special design and storage facilities are needed for cryogenic propellants if they are to be stored for a long time or even at time of launch. Cryogenic propellants also require a start delay to cool the system's flow passage hardware to cryogenic temperatures.

However solid-propellant rockets have critical advantages for military applications: These include simple design (few or no moving parts mean fewer things go wrong), ease of operation, compactness, instant readiness, lack of leakage of hazardous materials, and long storage periods (five to twenty-five years).[25] Today, U.S. missiles almost exclusively use solid-propellant rocket motors for strategic missiles (such as long-range ballistic missiles) aimed at rival military targets and also tactical missiles designed to support or defend air, land, and sea forces in a military theater. In short, if the United States is a guide, there is presently a distinct emphasis on solid-propellant missiles across all militaries. Although we provide a discussion of all of Japan's SLVs—solids, liquids, and hybrids—it is with the above background in mind that we are particularly interested in starting off with Japan's postwar solid-fuel campaign in the narrative below.

Starting from Zero

In 1945 the United States banned all research and development pertaining to aviation in a defeated Japan. This still could not destroy the knowledge or keen interest of a critical group of people who were to come together and set Japan on the path to rocket development. In this regard, the crossovers from Nakajima Aircraft Industries, established formally in 1919 and a key player in the imperial Japanese aircraft industry, are especially instructive.[26] Like other large companies deemed to have played key roles in Japan's military industrial infrastructure, Nakajima was broken up into fifteen companies (which subsequently morphed into Fuji Heavy Industries and Nissan) but retained major connections to postwar Japan's space-related ventures. From the start, Nakajima stood out from other manufacturers with an emphasis on its own engineers and designers rather than a reliance on foreign licenses. Its engineers were centrally involved in the Zero fighter, which was of Japanese design and which became a symbol of Japan's technological advances.

The direct ancestry of postwar Japan's solid-fuel rockets reaches back to the era of those ingenious engineers at Nakajima who developed the radial engines for the Zero fighter. Among them were Ryoichi Nakagawa, the principal designer for the Zero's radial engine, who became Nissan Motor Company's senior managing director; physicist Yasuakira Toda, who is primarily attributed with devising igniters for rocket fuel, developing the material for as well as the shape of a rocket nozzle for emitting burned fuel, and leading the team of engineers at Nissan's aerospace division that developed the earliest series of Japanese rockets; and the legendary Hideo Itokawa, who designed the Hayabusa and Shoki fighters and who is widely credited as the force behind the launch of Japan's first rocket, the *Pencil*. It was this critical group of men whose experience, knowledge, and ambitions moved from one era to the next and who helped institutionalize solid-fuel rocketry in Japan.

There is little question about Itokawa's ambition in particular.[27] In fact, he was once forlornly heard to say that, minus aircraft, he himself was zero. It was not until 1952, following the end of the U.S. occupation and the San Francisco Peace Treaty, that Japan's solid rocket development began to take shape, with Itokawa figuring prominently. In an age where the majority of Japan's aeronautical engineers and business interests were focusing on the exciting new field of jet engines and research into plane manufacturing, Itokawa, following a six-month sojourn in the United States in 1953, is widely credited for single-handedly turning Japan's attention toward rockets. He accomplished this in October 1953 in a seminal lecture about rockets and guided missiles he delivered to an audience of forty engineers from representative companies and members of Japan's National Safety Agency (Hoanchō, predecessor to the JDA/MOD) at Keidanren. From this, Itokawa was able to build the human and capital support needed to begin Japan's postwar rocket program.

He also helped create the Avionics and Supersonic Aerodynamics (AVSA) research group at the University of Tokyo, which rapidly attracted funding from both industry and the government and which subsequently evolved into the Institute of Space and Astronautical Science (ISAS). AVSA took the first steps toward the construction of hypersonic shock tunnels and experimentation with rocket telemetry. The sole company that reacted positively to AVSA was none other than Fuji Seimitsu Company, which had evolved from Nakajima Aircraft and which subsequently morphed into Nissan Motors and later IHI Aerospace. This core nucleus of actors and companies (along with the expertise in solid-fuel propellants by one individual, Tsutomu Murata at Nippon Oil & Fats), was critical to Japan's subsequent efforts. After its first official meeting in 1954, AVSA and its supporting companies, with contributions from then Ministry of Education (MOE) and the Ministry of International Trade and Industry (MITI), started operation with an estimated ¥3.3 million with the goal of testing tiny solid-propellant rockets. This intellectual team developed the engine systems that powered the earliest series of Japanese rockets, named the *Pencil*, *Kappa*, *Lambda*, and *Mu*.

The Rockets

From that point on, Japan's rocket series have moved in stages, showing a country clearly in quest of an indigenous and independent launch service capability.[28] We begin with a discussion of these early rockets, in which the work of pioneering individuals and corporate players was critical, primarily as a gateway to showing how Japan has arrived at its present rocket capabilities.[29]

Pencil Though tiny, the *Pencil* rocket—23 centimeters long, 1.8 centimeters in diameter, and 200 grams in weight—was Japan's first concrete postwar step toward building solid-propellant rockets and, from there, missile capabilities.[30] Its psychological and historical significance cannot be overstated. Although it did not start that way, AVSA's research became inextricably linked with the goals of the International Geophysical Year (IGY) program. In showcasing postwar technological advances, IGY enlisted cooperation from scientists around the world to provide a comprehensive mapping of the earth's surface, divided into observation of Antarctica and the earth's upper atmosphere. The Americans had offered to let Japan's observation equipment ride on their rockets, but the Japanese had other plans. AVSA's *Pencil* rocket would become a sounding (observation) rocket that would allow Japan to participate in the IGY activities related to observations of the upper atmosphere.

The minimum requirement was the ability to launch a rocket to an altitude of about 100 kilometers by 1958. Itokawa was convinced that the *Pencil* could help achieve that goal. Between March and April 1955, it was tested twenty-nine times in horizontal launches at the western Tokyo suburb of Ogikubo. In June 1955, a two-stage version failed when the main rocket was ignited before the booster. Subsequent technical tests and improvements continued at Ogikubo, which was to become Nissan's rocket production site until the company moved to a bigger facility north of Tokyo in May 1998.

Pencil's next major hurdle was a vertical launch, requiring a safer site, facing out to sea. Michikawa in the north (Akita) was selected, and until 1962 it remained central to rocket technology. In August 1955, it was the historic site for the vertical, or more appropriately diagonal, launch for *Pencil* 300. The rocket failed to launch properly the first time around. But after a patch-up with vinyl tape for support, the rocket then flew to an altitude of 600 meters, covered a distance of 700 meters, and had a flight time of about 16.8 seconds. Whatever the subsequent evaluations of the *Pencil* itself, at various stages from start to finish the project participants acknowledged the significance of the venture from a rocket learning and development point of view. They had, after all, helped launch Japan's first postwar indigenous rocket.

Baby After the *Pencil*, Itokawa's team moved on to the two-stage *Baby* rockets that were launched between August to November 1955.[31] Physically, the

Baby rockets were 8 centimeters in diameter, 105–120 centimeters in length, and about 10 kilograms in weight and were all able to reach about 6 kilometers in altitude. They were divided into three types: *Baby-S* (the first of the series to be launched with verifiable flight operations), *Baby-T* (the first postwar rocket to carry a telemetry system), and *Baby-R* (the first to carry onboard instruments, including a camera that was successfully retrieved in the air). As the successive experiments with the *Baby* rockets revealed, many problems with onboard equipment or rocket stages were overcome and from an evolutionary perspective, several important milestones are still worth noting. To start, armed with little equipment but much impromptu ingenuity on the part of a few pioneers, the project participants had every right to be heartened by the fact that all the experiments were carried out almost back to back in the same year. Additionally, the *Baby* rockets were deemed to have broken the sound barrier. As such the *Baby* rockets were instrumental in pushing Japan down the path to bigger and better rockets.

Kappa With participation in the IGY activities still a goal, a Japanese rocket had to come close to achieving a required minimum altitude of 100 kilometers.[32] The AVSA group had originally planned to develop *Alpha, Beta, Kappa,* and *Omega* series rockets to reach the goal to launch 20 kg instruments up to a 100 km altitude. To join IGY, however, the group was forced to speed up the pace and went directly to *Kappa*. Building on the *Baby* rocket series, the *Kappa* series went on to come close to that goal from its humble beginning in 1956. Its conception was hardly auspicious. Unlike the American or French liquid-fuel rockets that had cleared the required IGY-mandated altitude, the Tokyo group was focused more on the challenge of doing the same with a solid-fuel rocket—a focus for which it was widely criticized within Japan as, many sensibly argued, there were then no foreign models for such a venture. It was in prevailing against this commonsensical wisdom of the time that Itokawa showed his independent streak, and Japan moved down a more autonomous rocket path than it might have otherwise.

Contemporary observers consider Japanese *Kappa* rockets to be excellent, and these rockets involved a number of revolutionary innovations, including multi-staging capability. In June 1958, the *K-6* allowed the development and successful testing of composite propellant, the fuel used by U.S. ICBMs as well as the boosters for the U.S. space shuttle. The *K-6* was a two-stage, 5.4-meter-long, 255-kilogram vehicle that was able to reach the threshold altitude primarily because of the composite propellant. It was the data on winds and temperatures in the upper atmosphere collected through the *K-6* that finally allowed Japan to participate in the IGY in September 1958. Leaving aside the United States and the then Soviet Union, only Japan and the United Kingdom were able to achieve this feat during the IGY time limits.

The *K-8*, which had the *K-6* for its second stage, used more advanced fuel and improved nozzle and engine design and was much larger. In July 1960, a *Kappa-8* reached an altitude of more than 200 kilometers, entering the ionosphere, and detected cosmic rays. *Kappa* formed the basis for Japan's full-fledged solid-fuel sounding rocket program, moving through earlier rockets such as the *K-9M* (of which eighty two were launched) and the currently operational *S-310*, *S-520*, and *SS-520*.[33] The purpose of these sounding rockets is to observe and obtain information about the upper atmosphere and to conduct scientific experiments in suborbital flights. But as noted earlier and confirmed by ISAS, sounding rockets, including Japan's present ones, can also be used to test a host of other technologies that are also critical to ballistic missiles, most notably reentry technologies and propulsion systems.

The *Kappa* series also helped power a more cohesive national and governmental focus on Japan's rocket future. Over time, a new and more permanent site to accommodate the increasingly large and powerful rockets became of the utmost importance. Itokawa had a direct hand in choosing the site, and that site subsequently became ISAS's launch facilities at Uchinoura-machi in Kagoshima, which today is known as the Uchinoura Space Center (USC), and was previously called the Kagoshima Space Center (KSC). With the site eventually secured, planning and construction took place from 1961 onward, and the space center was established in 1962. Moving beyond the experience and technologies acquired through its sounding rockets—which at that point meant largely the *Kappa* series—Japan then began a concerted two-decade campaign to build satellite-launching capabilities that would bring it ever closer to ballistic missile technology.

Lambda Even as construction at Kagoshima continued, ISAS was working in conjunction with Nissan Motors to develop the *Lambda* series.[34] At that time, Itokawa had asked what it would take for Japan to put a 30-kilogram payload in space in the following five years or so. According to a draft plan, the envisioned rocket that would be able to do this was a three- or four-stage *M* rocket. En route to that goal, the *Lambda* series was the first response to

Itokawa's quest, and while it began as a general sounding rocket, it was, perhaps more than anything else, a technical and experimental tester for injecting a payload in space.

The maiden flight for *Lambda*, the *L-2-1*, occurred in 1963. The goal, achieved through successive *Lambda* rockets, was to surpass a 1,000-kilometer altitude and reach the inner Van Allen radiation belt. In January 1965 the *L-3-2* met the 1,000-kilometer goal, and in July 1966 the *L-3H-2* went to 1,800 kilometers. The team then focused on a four-stage solid-propellant *Lambda* rocket, the *L-4S*, which built on the *L-3H* and on which was pinned Japan's hope of placing a satellite into orbit. The initial research and development focused on ensuring successful staging—specifically, that the fourth-stage kick motor of the *L-4S* would work successfully with the three-stage *L-3H* rocket. At first things did not go well, and the failures of the first four rockets in the series from *L-4S-1* to *L-4S-4* (as well as a fifth test of the *L-4T-1*) were disheartening. Malfunctions occurred, the various stages did not operate in sequence as planned, and attitude-control went awry. But after these successive failures, in February 1970 Japan put the 24-kilogram *Ohsumi* satellite into space using the four-stage solid rocket *L-4S-5*. While *Ohsumi* was largely a test satellite, it did earn Japan the distinction as the fourth nation in space.

Here too it is important to look beyond the failures and successes. The ability to overcome the failures implied the acquisition of complex technologies to launch *Ohsumi*. Over the course of the four failures, third- and fourth-stage trajectory and guidance became assured, as did attitude-control, kick-motor technology, and stage separation so that the fifth launch put Japan's first satellite into a 335-kilometer-by-5,150-kilometer orbit. The engineering achievements that enabled Japan to reach this level of technology were also highly important. By their nature, solid rockets can deliver thrust for only a short time and optimal flight dynamics can be achieved only through staging technologies. Japan made impressive strides here. By August 1972, the country had successfully placed three scientific satellites in orbit using the *Lambda* as a rocket. The incremental acquisitions during the *Lambda* stages were invaluable lessons that boosted Japan's capabilities—all of which came fully to light during the launch of the next series of rockets.

Mu Flush with the hard-won victory in the *Lambda* series, Japan's space program shifted its focus decisively to bigger rockets that would put actual functioning payloads, such as scientific satellites, in space.[35] This is where the

Mu (*M*) series came into play, and from our perspective, it also allowed Japan to become far more familiar with missile technology in successive steps. In September 1970, building on the four-stage solid-fuel *L-4S*, the first *M-4S-1*, capable of putting 180 kilogram payloads into LEO, was stabilized using a large tail fin and spin. However, due to a malfunction in the fourth-stage ignition, it was not able to put a satellite into orbit. The *Mu*, however, did not suffer the *Lambda*'s early fate. The three successive launches after that between 1971 and 1972 confirmed Japan's ability to put satellites into orbit stably with the new series: *M-4S-2* with *Tansei*, M-4S-3 with *Shinsei*, which was Japan's first scientific satellite, and *M-4S-4* with *Denpa*.

Even as the trend toward larger rockets got under way, the *Mu* series also took a step back toward a three-stage rocket to launch satellites—one that would make it easier to integrate Thrust Vector Control (TVC) technologies on which research had been taking place since 1966. Chronologically, from the *M-4S* (still four-stage) onward, Japan began installing the TVC in the second stage of the transfigured *M-3C* (then three-stage) with the stated goal of improving the precision of payload injections. Between 1974 and 1979, the *M-3C-1*, *M-3C-2*, and *M-3C-4*, which allowed for further testing and refinement of ground and attitude guidance over that time, enabled Japan to put the *Tansei-2*, *Taiyo*, and *Hakucho*, respectively, in orbit. Although there were still technical glitches, such as when *M-3C-3* failed to put *Corsa* into orbit due to a control system failure in early 1976, it is safe to say that by the onset of the 1980s the trajectory of Japan's rocket development was clearly on an upward, and, thanks to TVC, a more precise, trend. The *M-3H* series (numbers 1 to 3) was an extended physical version of the previous *M-3C*. It did, however, give Japan the ability to put heavier payloads, up to about 300 kilograms in space between 1977 and 1978, namely, the scientific satellites *Tansei-3*, *Kyokko*, and *Jikiken*.[36] By this point, Japan had slowly but carefully acquired considerable control over stages of a solid-propellant rocket, but TVC integration continued. In 1980 the *M-3S* generation further integrated TVC in the first stage and successfully launched *Tansei-4* on the very first tested *M-3S-1*. Subsequent launches and payload injections by the *M-3S* series (numbers 2 to 4) between 1981 and 1984—the *Hinotori*, *Tenma*, and *Ohzora*—went on to demonstrate that, at last, Japan had gained rocket precision and control, thanks to the developments in the *Mu* series.

TVC must be put in context. Japanese rocket developers admit that until the late 1960s, although rocket performance was enhanced, associate technologies, including avionics, remained comparatively undeveloped. Within avionics, guidance and control systems were backward and remained so because of opposition to developing missile technologies. In the *Lambda* series, the final-stage attitude was controlled merely by the thrust of the hydrogen peroxide jet. The first model of the *M-4S* was equipped with an attitude control system only for the last stage in order to be able to inject the satellite to the local horizon. According to some, the then scientists consciously avoided developing missile-related technologies as the Japanese public was nervous about military technologies back in those days.[37]

The *Mu* series systematically broke down barriers to the use of missile technologies. Like most missiles, the *M-4S-1* was equipped with tail fins that help provide stability in flight; in fact, tail control is probably the most commonly used form of missile control, such as in longer-range air-to-air missiles (like the Advanced Medium-Range Air-to-Air Missile/AMRAAM) and surface-to-air missiles (like Patriot and Roland). More significantly, the *M-3S* generation (that is the fourth generation of the *Mu* series) also went on to deploy "full equipment" thrust vector control (TVC) systems to every stage of the rocket, as well as enhanced capabilities for each motor.

What does TVC mean exactly? The official word on this important development is that it allowed Japan to considerably improve the accuracy and performance of all of Japan's solid-propellant SLVs. TVC technology and its subsequent application at each stage of the *Mu* series is significant as TVC is also one of the crucial technologies for missiles.[38] Put simply, missile trajectory is controlled by thrust vectoring. TVC systems generally work two ways. They either change the direction of the thrust from within a fixed nozzle or change thrust direction via a moveable nozzle. TVC is one of the main steering technologies for solid-fuel missiles for ICBMs, such as the U.S. Navy's Trident and the U.S. Air Force's Minuteman.[39]

Japan's technology capabilities continued to advance especially through the M-3S series. It started with Japan's stated desire to emulate the achievement of the U.S. and Soviet space programs to go beyond low Earth orbit, launching a probe to study Halley's Comet, then set to appear in 1985. To do this, ISAS developed significantly more know-how: a rocket big enough to escape the gravitational pull and get close to the comet, a probe that would function in interplanetary space, secure communications between ground control and the probe over the huge distance through the construction of a giant new antenna, and new software to cope with communication lags. In 1981 it helped pioneer and test the technologies that would ultimately result in *M-3SII*. The *M-3SII* comprised the first stage of the previous fourth-generation *M-3S* series, combined with new second- and third-stage engines, as well as strap-on boosters to increase the payload capacity.

In 1985, the *M-3SII-1* and the *M-3SII-2*, fitted with an optional fourth stage, were used to launch Japan's first interplanetary probe (*MS-T5/Sakigake*) and also the first probe to study Halley's Comet (*PLANET-A/Suisei*). In the process, the *M-3SII series* became a solid-fuel rocket capable of steering a payload beyond LEO. Combining multiple large and small solid-rocket motors used for propulsion, stage separation, and attitude control, the *M-3SII* represented another stage in the *Mu* series march to excellence. The Freon TVC system and gimbaled nozzles showed Japan could combine complex technologies into an efficient launcher with three times the payload capacity of the U.S. Scout booster.[40] Over roughly the next decade, the *M-3SII* (numbers 3 to 7) then went on to successfully launch a series of scientific satellites: *ASTRO-C/Ginga*, *EXOS-D/Akebono*, *MUSES-A/Hiten*, *SOLAR-A/Yokoh*, and *ASTRO-D/Asca*.

In 1995 the last in the series, *M-3SII-8*, had a second-stage failure and could not place the *Experiment Reentry Space System (EXPRESS)* satellite into orbit; it subsequently fell to Earth. In conjunction with then MITI and Germany, the mission was to have conducted weightlessness and heat-resistant material tests and to have a recoverable capsule. In the event the capsule, which was briefly mistaken for a warhead, was recovered in Ghana; whatever it did not achieve, the capsule was still considered extremely successful in providing important data for Japan on reentry and recovery tests—the former of which is especially critical in the final phase of a ballistic missile. In the late 1990s, noting both the performance and guidance system of the *M-3SII*, the *Rumsfeld Commission Report* classified the rocket as a potential intermediate range ballistic missile (IRBM), which if repurposed as an IRBM could carry a 500-kilogram warhead approximately 4,000 kilometers; the report further judged that any transfer of this missile or its related technology would be in violation of the Missile Technology Control Regime (MTCR).[41]

ISAS's final class of vehicles was called the *M-V*, conceptualization of which had begun in 1985. The three-stage *M-V* (first and second stages with axis attitude control, third stage with spin stability) was able to eschew strap-on boosters and an optional kick motor fourth stage. Development on the *M-V* started in earnest in 1990, with a projected launch year in 1995. The goal was to launch astronomy satellites and planetary missions in the interest of space science. After delays, *M-V-1* launched the first satellite, *MUSES-B/Halca*, in 1997.[42] From there until 2006, when the series was retired, the *M-V* (numbers 2 to 8) was used to launch a number of significant scientific probes and satellites, including Japan's first Mars explorer, *PLANET-B/Nozomi*, the autonomously controlled sample-return spacecraft, *MUSES-C/Hayabusa*, that arrived successfully on the asteroid Itokawa, *ASTRO-EII/Suzaku* by *M-V-6*, *ASTRO-F/Akari* by *M-V-8*, and *SOLAR-B/Hinode* by *M-V-7*, which was the last of the series.[43]

In 2000 the *M-V* series suffered its only loss—that of *M-V-4*, which was launched to loft the fifth X-ray astronomy satellite (*ASTRO-E*)—due to a first-stage nozzle failure. Responding, ISAS and Nissan introduced the latest carbon-carbon technology in the nozzles of each of the *M-V's* stages and, all in all, the rocket received several sets of refits. An earlier refit replaced the second stage's steel alloy casing with carbon fiber and knocked 900 kilograms of weight off the rocket while allowing a 10–15 percent increase in power. The *M-V* also got a Movable Nozzle Thrust Vector Control system (MNTVC).[44] The post-*ASTRO-E* refit improved the first-stage nozzle with the introduction of advanced 3D-C/C (three-dimensional carbon-carbon) technologies.

There is little question that with the development of the *M-V* class, Japan mastered a range of extremely sophisticated technologies—such as MNTVC, and Liquid Injection TVC—in separate stages. MNTVC is now the principal TVC of solid missile systems. Thus, far from avoiding guidance issues as in an earlier time, it appears that Japan overcame its inhibitions when the need arose and the technology was ready. The *M-V* payload and thrust capabilities provoked the *Rumsfeld Commission Report* to conclude that if adapted as a missile, the resultant system would give Japan an ICBM comparable to the then U.S. *MX Peacekeeper* missiles.[45] The *M-V* rockets, widely acknowledged as the best solid-propellant rockets around, were formally discontinued in September 2006. This forced retirement came about due to much-highlighted official concerns with cost reductions, estimated at about ¥7 billion per launch.[46]

As the foregoing analysis suggests, the *Mu* solid rocket has been recognized by observers both in Japan and abroad as an example of a civilian program that is potentially a missile program. Early in the 2000s moves were made to reconfigure and refine Japan's solid-rocket technology to make smaller, more flexible launch systems. The moves came from both industry and engineers in ISAS, as evidenced in the following: in March 2002, IHI Aerospace Co. Ltd. (IA), with technical support from ISAS, moved more formally to develop a low-cost, small satellite launcher based on components of the *M-V*.[47] The *M-V Lite*, as it was known, would have cost about $35 million to develop and about $13 million per launch. About half the height of the *M-V*, the *Lite*, powered by the second-, third-, and fourth-stage motors of the *M-V*, was to have used a recoverable avionics package and to propel satellites into a 300- by 600-kilometer elliptical orbit. The origins of the *M-V Lite* program can be found in the desire of ISAS engineers to not only continue design and engineering knowledge of solid-rocket booster technology accumulated through more than half a century; it also reflects a defensive measure to compete against the *GX*, which was proposed by some, particularly IHI, as a replacement for the *M-V*. Given the then possibility that the *M-V* program might be phased out at the end of the decade, the *M-V Lite* program was seen by the interested parties as a way of preserving Japan's technical and industrial base for solid missile technology for a new generation of younger engineers. Although the *M-V Lite* did not see the light of day, Japan's solid rocket ambitions did not wither away. As noted above, the *M-V* was retired following the launch of *SOLAR-B/Hinode* in September 2006, but not before it was replaced by the *Advanced Solid Rocket* (*ASR/Epsilon*) program (see Chapter 6).

Being Also Liquid

Until now we have concentrated largely on Japan's solid-propellant rockets, which, as discussed earlier, lend themselves more readily for adaptation to modern ballistic missile technology. However, it is also important to understand the main developments in Japan's liquid-fuel SLV program, which have allowed the country to master high-performance liquid technology and rocket system integration. This program has been carried out in resolute steps, irrespective of whether or not the rockets were able to compete commercially. The government entities involved with early liquid rocket development were the Science and Technology Agency (STA), the National Aerospace Laboratory of Japan (NAL), and from the late 1960s onward, primarily the National Space Development Agency of Japan (NASDA). Research into liquid-fuel rockets in Japan actually began as early as 1954, and led to a series of early rockets that involved MHI.[48]

Then the Japanese turned to technology transfer from the United States in an effort to improve their proposed *Q* and *N* series SLVs at that time. The *Q* project was designed to develop a launch vehicle to put a 150-kilogram payload to a 1,000-kilometer orbit by 1972; the *N* project was to launch a 100-kilogram satellite into a geostationary orbit by 1974. Japan had certainly accumulated enough technology and experience to continue on its own path. But the switch to reliance on U.S. technology can perhaps best be explained by the fact that pre-NASDA entities had made very slow progress with liquid-fuel research. In the process of adopting U.S. technology, the Japanese also got their first clear indication of the ways in which certain technologies would not be transferred and how industrially advanced foreign powers could constrain their own SLV development.

In 1969 the United States and Japan signed an agreement for cooperation in space activities.[49] The United States agreed to transfer unclassified technology and equipment for the development of Japanese *Q* and *N* launch vehicles, as well as communication and other satellites, solely for the purpose of peaceful application. The design, production, or spin-on applications of any launch vehicle or satellite, in whole or in part, was barred by the agreement unless by mutual agreement. Most telling of all, when transferring "unclassified technology to the level of the Thor-Delta vehicles," the United States did so "exclusive of reentry and related technology," which is a critical component of ballistic missiles in the final phase.

If one of the intended effects of the agreement was to stall or divert Japan's overall SLV development away from indigenous solid-fuel technology development it failed. Japan certainly continued to develop its own sophisticated solid-fuel missile technologies rapidly in the years following 1970, resulting in the *M-V*, as seen. Through the technology transfer, limited though it was, Japan also got the best of both worlds—maintenance of its own solid-rocket development program and acquisition of advanced U.S. liquid technology that kick-started new generations of highly sophisticated technologies, leading, as we show, to the *H-IIA*'s advanced cryogenic engines.

Despite, then, the limitations of the transfer, Japan improved both its own capability and the reliability of its liquid-fuel rockets. In October 1969, when NASDA was inaugurated, the first proposed launch vehicle on which the new agency was to work was the *Q* that had been in development since 1965. The *Q* was a four-stage rocket using largely solid-fuel stages (except the third, a liquid stage) that was composed almost entirely of indigenous technologies. The *Q* rocket development lasted a decade. The *Q*, in fact, was superseded (and in effect replaced) by the *N* (*Nippon*) rocket.[50] The *N* series is where NASDA officially begins its own history.

N Building on the *Q* series, as well as U.S. technical guidance, the Japanese (and MHI specifically) turned to formal development of the *N-I* launch vehicle in October 1970. The *N-I* was based on Thor-Delta rocket technology, and it was equipped with the MB-3 engine licensed from the United States for the first-stage motor. Japan also developed and manufactured the LE-3 with American assistance, which was used for the *N-I*'s second-stage motor. Subsequently, MHI used that base to develop the LE-5 (Japan's first liquid oxygen/hydrogen engine), LE-5A, and LE-7, which went on to form the basis for further improvements in rocket propulsion for the *N* and *H* launch vehicles. The design, development, and assembly experience from these early engines also helped improve the propulsion and reliability of Japanese engines, culminating in those manufactured and used today—namely, the LE-5B and LE-7A.

The three-stage *N-I* was about 32.6 meters in overall length and 2.4 meters in diameter at the largest portion (first stage) and weighed a total of about 90 tons not including the payload. It first flew on 9 September 1975, and was launched seven times until September 1982. The objective of the *N-I* rocket was to launch a satellite weighing more than 100 kilograms into geostationary orbit. The first flight successfully launched the *Engineering Test Satellite-I* (*ETS-I/Kiku-I*), with which Japan began to gain satellite tracking and control technology. The subsequent launches were also highly important learning steps, allowing the country to rapidly acquire the ability to launch a geosynchronous (GEO), and more specifically geostationary orbit (GSO), i.e., the ability to place a satellite in a stationary orbit about 36,000 kilometers above a point on the Earth), which was achieved by *N-I-3* with the 130-kilogram *ETS-II/Kiku-2* in 1977. The *N-I* also flew two *Ionosphere Sounding Satellites* (*ISS/Ume*, *ISS-b/Ume-2*). Additionally, despite problems, Japan did begin to test *Experimental Communication Satellites* (*ECS*)—the *N-I-5* with *ECS/Ayame*, during which the third stage of the rocket collided with the satellite and so prevented the achievement of GEO orbit, and the *N-I-6* with *ECS-b/Ayame-2*, which also failed to achieve GEO orbit due to ignition problems in the satellite apogee engine.

The *N-I* was born of old technology. Even as it was making experimental strides, Japan began developing its successor, the 3-stage *N-II*, as part of a publicly stated need to be able to loft heavier payloads and with more advanced guidance systems. The *N-II*, also based on the Thor-Delta rocket, was successfully launched eight times between 1981 and 1987. Around the same size as (length, 35.4 meters; diameter, 2.4 meters) but with greater weight than the *N-I* (135 tons, not including the payload), the *N-II* distinguished itself with an inertial guidance system as opposed to the *N-I*'s radio guidance system that dated back to the 1960s. The third stage incorporated in the *N-II*, again licensed from the United States, was a solid-fuel upper stage with full control and guidance capability that was meant for general space applications. As it was also designed for use with the Thor boosters, the technology for which had been transferred to Japan under the 1969 agreement in part as noted earlier, Japan gained experience in learning how to adapt it for use on a range of launch vehicles to put small and medium payloads more precisely into orbit.

In this way, Japanese technology pushed forward with motor, guidance, and reaction control systems that provided greater attitude stability, more precise control of flight rate, and burnout velocity for orbital payload delivery and Earth-escape missions. The *N-II* also more than doubled Japan's launch vehicle capability to placing 350 kilogram-class satellites in GSO. During its tenure, the *N-II* had a 100-percent success rate in all of its eight planned missions, launching Japan's earliest Geostationary Meteorological Satellites (*GMS-2/Himwari-2, GMS-3/Himawari-3*), *Communication Satellites* (*CS-2/Sakura-2, CS-2b/Sakura-2b*), *Broadcasting Satellites* (*BS-2a/Yuri-2a, BS-2b/Yuri-2b*), and a *Marine Observation Satellite* (*MOS-1/Momo-1*).

H Among the best known of Japan's rockets are its *H* series, which over successive updates have been as much praised as excoriated. We believe they have proved highly valuable and successful ventures in Japan's space progress. From an evolutionary perspective on eliminating Japanese SLV technology dependence on the United States, the *H* series turned out to be on the liquid side of the story what the *Mu* series was to the solid one.

H-I Even as Japan continued to make progress with the *N-II* series, with its attendant constraints under the U.S.–Japan 1969 agreement, it also moved forward with developing its wholly domestic-engineered rocket, making a tremendous technology jump with the *H-I*.[51] NASDA's unstated goal was to reduce dependence on U.S.-licensed rockets, motors, and other technology and to make progress, especially on incorporating a hydrogen-powered cryogenic upper stage (hydrogen-powered means having extra boost for geosynchronous-bound payloads; cryogenic simply means having the ability to handle the hydrogen at extremely low temperatures which, combined with its explosiveness, presents tremendous engineering and technological challenges). In 1981, the same year in which the *N-II* started its launches, Japan began development of the *H-I*, which was to form the basis for subsequently improved and updated rockets in the series. The following year it also began building a facility at Tanegashima for launching the *H-I* rockets.

The three-stage *H-I*'s first stage and Solid Rocket Boosters (SRBs) were the same as the *N-II*. But more than 80 percent of the remainder of the rocket—including the LE-5 engine (Japan's very first liquid oxygen/hydrogen engine) in the second stage and propulsion, the third-stage solid rocket motor, inertial guidance, airframe, and fairing—was indigenous. The *H-I* had also been proposed, as others before it, with the target of doubling the capacity of the rocket. On its maiden launch in 1986, the *H-I* successfully took three payloads into orbit (an *Experimental Geodetic Satellite* [*EGS/Ajisai*]; a *Japan Amateur Satellite* [*JAS-1/Fuji-1*]; and a magnetic bearing flywheel experimentation system) thereby establishing its ability to handle heavier missions. Between 1987 and 1992, the *H-I* took an additional eight satellites successfully into orbit—among them the *Engineering Test Satellite* (*ETS/Kiku-5*), *Communication Satellites* (*CS-3a/Sakura-3a* and *CS-3b/Sakura-3b*), a *Geostationary Meteorological Satellite* (*GMS-4/Himwari-4*), a *Marine Observation Satellite* (*MOS-1b/Momo-1b*), and *Broadcasting Satellites* (*BS-3a/Yuri-3a*, *BS-3bYuri-3b*), as well as the first *Japan Earth Resources Satellite* (*JERS-1/Fuyō-1*). By the time the largely indigenous *H-I* was retired, it had had nine faultless successive launches between 1986 and 1992, and it was capable of putting up to 1,500 kilograms into GTO orbit. The *H-I* represented a leap in indigenous technological advancement in, among other things, using cryogenic propellants and making inertial guidance systems.[52] Like the *N-II*, its faultless launch record spurred Japan's ambition, which was already looking beyond the *H-I* in the early 1980s since international launch service contracts remained of interest.

H-II In 1985 construction started on a rocket launch facility for the *H-II*; formal development on the rocket began in August 1986, which was around the time of the *H-I*'s maiden launch.[53] The two-stage *H-II* was constructed with the twin goals of technological autonomy and, ostensibly, global commercial service entry.[54] It succeeded in the first goal and thus was not subject to the restrictions under the licensing arrangements with the United States. On the pad, the *H-II* was deemed by authoritative observers to be the *most* advanced expendable launch vehicle around, specifically in terms of its integration of modern materials, electronics, computers, and propulsion.[55] Significant advances came in the engine design and improvements for the first and second stages, which also became the showpieces of Japan's technical space prowess. The *H-II* was developed and built at an estimated cost of $2.3 billion.

Of this, about $800 million was for the LE-7—an improved liquid oxygen/hydrogen engine installed in the first stage that was designed to improve propulsion. The second stage continued to use an improved version of the engine designed for the *H-I*, namely, the re-ignitable LE-5A. Additionally, a strapped-down inertial guidance system was also improved to provide better control. The new rocket was designed to be flexible, capable of handling primary missions for intermediate geostationary transfer orbit- (GTO) and low Earth orbit- (LEO) bound payloads, about 4 tons to GTO and 10 tons to a 300 km LEO.

Although the homemade *H-II* was a commercial failure for Japan, it was a technological feat. Commercially, its prospects were doomed for two reasons: unexpectedly high development costs related especially to the much-showcased LE-7 (closed-cycle two-stage combustion) engine that had caused delay and several (and, in one instance, fatal) mishaps; and also the rise of the yen's value following the Plaza Accord, meaning that *H-II* launch costs were roughly double those of the market-leading *Ariane-4*, which had been engineered to be simple and unglamorous as well as reliable. Technologically, the *H-II* also drew attention, for its ostensible suitability for conversion into an ICBM.[56] Certainly there were the pacifist-country rebuttals on this score. But the technical grounds for denial were correct: the *H-II* was entirely unsuitable for an ICBM because it could not be launched speedily. In our analysis, the value of the *H-II* lay in its critical role in Japan's technology acquisition program.

The *H-II* flew on seven missions between February 1994 and November 1999, unfairly earning for itself, and for Japanese technology, a rather poor reputation in doing so—so much so that the last launch of the rocket was actually cancelled.[57] A closer look reveals that these problems started in the late 1990s. In February 1998, due to a problem in the LE-5A second-stage engine, an *H-II* failed to place the communication and broadcasting engineering test satellite *COMETS/Kakehashi* into GTO orbit. Until that time the LE-5 had had a perfect record across the nine *H-I* and five earlier *H-II* missions. In November 1999, the launch of the next *H-II* (at that point already a hybrid of *H-II* and *H-IIA* using a new second-stage LE-5B engine on an otherwise standard *H-II*) also failed by veering off the planned flight plan. This was due to a problem in the first-stage LE-7 engine, and both the rocket and the payload, a *Multi-functional Transport Satellite* (*MTSAT*), had to be destroyed.

But things need to be put in perspective. Before 1998, the *H-II* flew five successful missions, transporting *Geostationary Meteorological Satellites* (*GMS-5/Himawari-5*), *Advanced Earth Observing Satellite* (*ADEOS/Midori*), and an *Engineering Test Satellite* (*ETS-VII/Kiku-7*) as planned. The problem of a highly elliptical orbit with one satellite, the *ETS-VI/Kiku-6*—to date the largest satellite built and designed to conduct Japan's first-ever laser test among several others for inter-satellite and mobile communications—stemmed from the failure of a unit in the satellite's own propulsion system (apogee-engine) rather than anything to do with the rocket.

Perhaps the most important mission for our purposes was the *H-II*'s very first outing, which demonstrated Japan's capabilities for making potentially dual-use warhead reentry technologies with the *Orbital Reentry Experiment* (*OREX*), also known as *Ryūsei*. As we discuss in Chapter 6, by the time the *H-II* was prematurely retired, the critical reentry and guidance control infrastructure was already falling into place.

H-IIA Whatever its technical problems—and, in any event, these should be expected in any ambitious space program over time—we believe the technologies and experience from the *H-II* were indispensable in allowing Japan to develop SLVs to the highest international standards. Instead of going to waste, the *H-II*'s technology was morphed into the *H-IIA*—an "augmented" version that is Japan's main commercial launch vehicle in service today.[58] The two failures coming relatively late in the *H-II* launches overall were actually boons to the subsequent program, exposing deeply buried and hidden problems that have served to strengthen the *H-IIA*'s liquid engine technology to a truly advanced reliable level. Put simply, Japan achieved a high level of technical know-how at the cost of then only two failed launches—this, compared to the numerous engines blown and dozens of launch failures accumulated, for example, by the United States and former Soviet Union in the course of their respective development programs.[59]

We believe that while success in the commercial market was important for the *H-IIA* program, it was not critical.[60] The *H-IIA* was and is primarily needed for independent access to space in line with the Council on Science and Technology Policy (CSTP) goals and as a cutting-edge technology acquisition program for engine design and development as well as guidance and control systems. The *H-IIA* development program began formally in 1996, with the express purpose of entering the highly competitive global commercial launch service industry. As manufacturing and launch operation costs had killed the commercial viability of the *H-II*'s predecessor, the primary public purpose of the *H-IIA* was to become a cheaper *H-II*, halving per-launch costs to about $85 million.[61]

Considerable efforts were made to secure a higher launch rate to help lower costs and foster even more manufacturing efficiencies by encouraging larger multiple-SLV production lots. To legitimize the *H-IIA*'s initial international acceptability, early launch reservation contracts from then Hughes Space and Communications, as well as Space Systems/Loral (SS/L), were secured.[62] Together they would lead to the purchase of twenty launches for a combined potential value of around $2 billion if cost reduction targets were met. Negotiations with Kagoshima-based fishermen's unions to secure longer launch windows from 90 to 190 days, and efforts at reducing the launch process from fifty to twenty days also helped. Finally, in a departure from indigenization concerns to commercial concerns, the *H-IIA* turned to incorporating foreign technologies—Solid Rocket Boosters (SRB-A) composite cases as well as Solid Strap-on Boosters (SSBs) from ATK Thiokol of the United States, core-stage tank domes from Boeing of the United States and Man Technologies of Germany. There were qualms in the United States about exporting solid-rocket motor technology to Japan for the SRB-A, with potential application to missiles, though they eventually quieted.

Unlike the *H-II*, the *H-IIA* family consists of several variants made possible by combinations of SRB-As and SSBs that provide extra thrust. The standard 53-meter *H-IIA202* has a 4-meter diameter payload fairing and two strap-on SRB-As. As Table 4.1 reveals, the *H-IIA* family is designed to be highly flexible. Similar flexibility is apparent in payload fairings, which are designated in both different diameters and payload compartments (single or dual). Increased thrust is provided by the improved LE-5B liquid oxygen/hydrogen re-ignitable engine in the second stage, and technology allows multiple satellites to be put into different orbits. In contrast to the LE-7, the first stage consists of one simplified liquid-fuel engine LE-7A that burns liquid oxygen/hydrogen engines and provides 112 tons of thrust in vacuum. Programmatically speaking, the *H-IIA* forms the technological base for the *H-IIB*.[63] In September 2009 JAXA launched the *H-IIB*, on which Japan was staking its future capability on two fronts—cargo missions using the *H-II Transfer Vehicle* (*HTV*) to support the ISS and reducing launch costs through heavier capacity that allows dual-manifest missions.[64] Apart from enhancements in size, weight, and

Table 4.1. Variants of the *H-IIA*

Designation	*Length (m)*	*Mass (Tonnes)*	*Add-on Modules*	*Launch Capability (kg to GTO)*
H2A202	53	289	2 SRB-A	3,700
H2A2022	53	321	2 SRB-A, 2 SSB	4,200
H2A2024	53	351	2 SRB-A, 4 SSB	4,600
H2A204	53	445	4 SRB-A	5,700
H2B	56	551	4 SRB-A	16,500 (to LEO for *HTV*)

SOURCE: Information is from Japan Aerospace Exploration Agency (JAXA) on space launch vehicles (*H-IIA*) at www.jaxa.go.jp (accessed 1 July 2009).

length, as well as special fairing said to be necessary for the launch of the *HTV*, the *H-IIB* also differs in having two liquid LE-7A engines in the first stage as opposed to only one in the existing *H-IIA* series.

For the *H-IIA*, the biggest design changes came in the LE-7A by MHI and the SRB-As by Nissan. MHI drastically simplified the LE-7A's plumbing, lowering production costs.[65] The LE-5B engine's performance improved after a simplification of the design of its cooling system. The SRB-A redesign was perhaps the most visible change for the *H-IIA* compared to the *H-II* and, in the event, turned out to be problematic. The vehicle's first-stage Nissan SRBs used a radically shorter, squatter monolithic motor case that could be delivered ready for launch—a drastic change from the original SRBs, which required shipping in four pieces and stacking at Tanegashima. The use of the Carbon-Fiber Reinforced Plastic (CFRP) in the SRB-As was also considered to be a very significant technological advance.[66] NASDA stressed that the use of carbon-wound filament casings would lower the weight and cost of the boosters, increasing thrust from 360 tons to 460 tons and allowing them to burn longer.

What about the purported goal of commercialization? As with the *H-II*, the course of development for the new launcher suffered glitches, as is inevitable with something new. The prototype LE-7A engine failed a test when a faulty valve leaked but then went on to perform well subsequently. There were also issues with the next-generation LE-5B engine, which suffered several mishaps including an explosion during a firing test. However, in all, the simplified engine and tank design of the first stage alone were estimated to decrease the costs by about 50 percent, supplemented by other technical measures aimed at reducing prices. For MHI, which stands to gain or lose commercially since it assumed effective control of the launch and marketing of the *H-IIA* in April 2007, cutting costs continues to be an issue for commercial purposes.[67] The mantra of cost competitiveness thus deserves close scrutiny. Estimated launch costs stood at $70 million for the *H-IIA202*, $75 million for the *H-IIA2022*, $80 million for the *H-IIA2024*, and $83 million for the *H-IIA204*. But a price tag of around $70 to $80 million in 2002 still made the *H-IIA* about 15 to 20 percent more costly than the foreign Ariane-5, Delta-4, and Atlas-5 peers.

It was evident in the mid-1990s that the *H-II*A faced stiff competition in a crowded marketplace and especially from the Ariane-5 with its larger capacity. The early commercialization rationale for the development of the *H-II*A suffered a critical blow in 2000 when Hughes canceled its contracts with Rocket System Corporation (RSC) followed by SS/L.[68] Even the Mitsubishi-keiretsu owned Space Communications Corporation (now SKY Perfect JSAT Holdings, Inc.), the obvious customer for the *H-IIA*, also decided to fly the Melco-built *Superbird-7* on the European Ariane-5 rather than the *H-IIA* in 2008.[69] It was only at the start of 2009 that signs emerged that the *H-IIA* was starting to move beyond the historical baggage of the back-to-back *H-II* failures. In January 2009, almost a decade after the cancellation of the Hughes and Loral contracts, the *H-IIA* at last secured one commercial contract, from Korea Aerospace Research Institute (KARI), to launch the *Korea Multipurpose Satellite-3* (*KOMPSAT-3*).[70] While such developments do much to confirm the fact that the *H-IIA* is now widely seen as reliable, it should be noted that *KOMPSAT-3* is hitching a lift aboard a JAXA mission to launch the *Global Change Observation Mission—Water* (*GCOM-W*), and thus most of the launch cost is still borne by the government.

The prospects for the *H-IIA* as a first-choice launcher worldwide are still weak, but it is evident that MHI is still striving in that direction.[71] In April 2007, it took over production and management of the *H-IIA*. It also began advertising its *H-IIA* Launch Services directly as well as through JAXA.[72] Proclaiming that the *Ariane-5* and the *H-IIA* were the most reliable launchers on the commercial satellite launch market, it announced that the two SLVs had been chosen as "backup" launchers for each other. In 2008, MHI announced its aim of slashing launch costs about 30 percent to about $60-$70 million by 2009 to bring it on a par with Arianespace and Boeing-led *Sea Launch*. However, testifying to the chronic oversupply of launchers that has plagued the commercial launch sector since the late 1990s, *Sea Launch* itself filed for bankruptcy in June 2009.

Meanwhile, given the volatility in the global marketplace, it seems reasonable to predict that the *H-IIA's* primary customer will remain the Japanese government for the foreseeable future.[73] Between its inaugural test launch in 2001 to the end of 2009, the standard version *H-IIA202*, as well as variants *H-IIA2022*, *H-IIA2024*, and *H-IIA204*, have flown on sixteen missions, demonstrating increasingly reliable launch capability, except for the ignominious failure of the *H-IIA2024-F6*, which had to be destroyed after launch in November 2003. Contrary to popular perceptions, this did not in the least mean that the technological progress on the launcher was poor or unreliable. The standard *H-IIA202* has flown on five missions, launching a range of payloads such as the experimental *Laser Ranging Equipment* (*LRE*) for flight evaluation, another *ADEOS-II/Midori-II* for Earth observation, and, most famously, the *Information Gathering Satellites* (*IGS*). The *H-IIA2022* has flown three missions, launching a replacement for the *MTSAT-1R/Himawari-6*, the *Advanced Land Observing Satellite* (*ALOS/Daichi*), the *Selenological and Engineering Explorer* (*SELENE/Kaguya*) and the *IGS 5A* high resolution (60 cm resolution) reconnaissance satellite, launched in November 2009. The heavy-lift *H-IIA204* launched one satellite, *ETS-VIII/KIKU-8*, in 2006.

By the end of 2009, the *H-IIA2024* had flown seven missions, and has emerged as the workhorse of the series, at least for Japanese government missions. Between 2002 and 2008, its missions have included the *Data Relay Test Satellite* (*DRTS/Kodama*), the *MTSAT-2/Himwari-7*, the *Wideband InterNetworking Engineering Test and Demonstration Satellite* (*WINDS/Kizuna*), and the *Greenhouse Gas Observing Satellite* (*GOSAT/Ibuki*) along with a number of microsatellites. Three of the four *IGS* missions to date have used the *H-IIA2024*, except for the *H2A202* launch of a single optical satellite in 2006 and another in 2009. The *H-IIA* did, however, suffer a major mishap in November 2003 with the nation's second set of *IGS* satellites when one of the SRB-As failed to separate because of a nozzle meltdown, dragging the rocket off course.[74]

The *H-IIA2024* has also launched other payloads that are no less important for their military aspects, irrespective of the eventual outcome.[75] Its second launch in February 2002 was responsible for the *Demonstrator of Atmospheric Reentry System with Hyper Velocity* (*DASH*), a reentry capsule weighing around 86 kilograms, and armed with its own propulsion system. The name spoke for the goal—to demonstrate Japanese capabilities in reentry systems with speeds exceeding 10 km/sec. Three days after launch, because of improper software writing, it failed to separate from its kick motor and to reenter the atmosphere over the Sahara Desert in Mauritania. This fault was attributed to the maker NEC-Toshiba, which subsequently had to pay a fine of close to a $3 million to ISAS. We believe *DASH* is important for its significance as yet another attempt at a reentry technology test.

This was followed in September 2002 with the *Unmanned Space Experiment Recovery System* (*USERS*), developed by the Institute for Unmanned Space Experiment Free Flyer (USEF). The two-part *USERS* spacecraft was equipped with a service module and an ejectable reentry module. In a unique venture, the service module was designed to continue conducting long-duration technology experiments in orbit while the reentry module would actually return experiment results to Earth. After about eight and a half months of in-orbit experiments and operations, the capsule part of the reentry module (called the recovery vehicle) successfully separated from the service module, de-orbited, and was recovered in May 2003 off the Bonin Islands southeast of Japan. Apart from these less well-known reentry tests, there is another side to the *H-IIA* that has a direct and powerful military implication, namely, the SRB-A and its improvement, which is the basis of the *ASR/Epsilon*, and which, in turn, will form a critical plank in Japan's Space-on-Demand/Operationally Responsive Space (SOD/ORS) capabilities discussed in Chapter 6.

The commercialization (or otherwise) of the *H-IIA* also brings us to an important juncture in our analysis of Japanese launch technologies and their implications for the nation's strategic space capabilities. In our view, the history of the *H-IIA* validates our understanding of the capabilities of Japan's launch vehicle technology. While the loss of *IGS* satellites was acutely painful, the more complex liquid technologies had already been validated. As noted for earlier series, the evolved *H-IIA* is not convertible to an ICBM. The *H-IIA*'s benefit lies in its technology acquisition, total systems integration skills and, from a practical standpoint, capability for launching civilian and military spacecraft.

J-I At the risk of sounding flippant, the *J-I* rocket might represent a launch system that could go straight to the military, bypassing the market altogether.[76] The bifurcation between NASDA and ISAS on launch systems technology came together—at least institutionally—with the *J-I*, development of which began in April 1993 and at a cost of about $100 million.[77] The *J-I* was a three-stage solid rocket capable of sending 1-ton payloads into suborbital or low orbital trajectories. It stood at 33.1 meters, had a diameter of 1.8 meters, and used the radio guidance system that makes ground command over flight control inputs and event timing possible. The *J-I* is known best for combining *H* and *M* solid technologies—the *H-II*'s SRBs for the first stage (using Movable Nozzle Thrust Vector Control or MNTVC) and ISAS's Liquid Injection Thrust Vector Control (LITVC), Side Jet (SJ) thrusters for roll control, second-stage motor (M-23), avionics, third-stage motor (M-3B), and fairing from the *M-3SII* for the second and third stages.

The stated goal was to lower both development costs and time so as to cater to small satellite launches. As a ready-made solid design, it was supposed to be faster and cheaper to launch. To reduce costs further, the *J-I* used the old Osaki launch pad at Tanegashima. All these commercially-oriented themes certainly sounded familiar even if, as in the case of other rockets, they had little bearing on reality.[78] But the integration across some of Japan's most advanced indigenous rocket technology in the service of a new solid rocket also had another less well-known payoff. At least one assessment suggests that the *J-I* technology, assembly, and successful launch gives Japan the potential for an ICBM surpassing the performance of a U.S. Minuteman-3 with a range of about 8,000 miles (about 13,000 kilometers).[79]

Financially, there seemed to be no justification for the rocket's development. As the *J-I*'s estimated initial launch price was $35 million (compared to Russian prices as low as $7 to $12 million) the *J-I* was scheduled for two flights only. It flew the first, but its development was suspended before the second flight. Its first launch, in February 1996, actually cost between $43 and $45 million. But the *J-I* is not memorable for its costs. The *J-I* successfully launched a one-ton 4.4-meter-long miniature shuttle prototype *Hypersonic Flight Experiment* (*HYFLEX*) into a suborbital flight.[80] *HYFLEX* was a precursor for the *HOPE-X*, an engineering demonstrator for an unmanned reusable shuttle. Using *HYFLEX*, the stated goal was to collect actual in-flight and reentry information on a technology flying at hypersonic speed and reentering the Earth's atmosphere, an experiment that could not be conducted on the ground. In the event, *HYFLEX* also allowed for the test of insulated materials necessary for reentry purposes. While *HYFLEX* sank before it could be recovered due to a failure of its flotation devices, its operational, trajectory, heating, and reentry data were already in hand. A second flight of the improved *J-I* was to have launched the *Optical Inter-orbit Communications Test Satellite* (*OICETS/Kirari*), but this was halted after a series of negative reports, notably a very harsh one in 1998 by the Management and Coordination Agency (MCA), the government's internal auditing agency.[81] The halting of the *J-I* program did not, however, halt the continued development of small launcher technology—whether needed or not, whether cost effective or not—as we shall see with the even more troubled development of the *J-I*'s successor, the *GX* rocket. From our perspective, however, the most important thing was that the *J-I* proved highly useful as a technology development program in testing and demonstrating missile-related technologies.

CONCLUDING ASSESSMENT

Taking our cue from the interchangeable natures of SLVs and ballistic missiles as outlined at the beginning of the chapter, we approached SLV development in Japan with a specific goal in mind: to see whether the country has been able to acquire not any one ballistic missile in particular but rather whether it has been able to acquire the components and infrastructure—meaning the propulsion systems, structures, staging ability or large boosters, sophisticated guidance and control systems, reentry vehicles, flight operation skills, and the technical and institutional setup necessary for integration—that would allow it to put together and launch a ballistic missile if needed.

Over the historic postwar period, no aspect of Japan's SLV development was ever guaranteed to be on an onward and upward trend, as this chapter has made clear. There were few outright resources, and initially there was only a knowledge base that descended from the wartime military aircraft and rocket programs. The human element mattered. There were mistakes, disappointments, and some outright disasters, along with the penchant of the media to magnify them. External realities mattered. The United States tried to shackle Japan's burgeoning solid rocket program with less easily convertible liquid-fuel technologies. The goal of commercialization has become an increasingly vociferous mantra even as global markets have continued to remain difficult for Japanese launch services. But we believe Japan's rocket program has succeeded. With comparatively little pecuniary assistance, and sheer determination to indigenize rocket technology, the makers of Japan's SLV program on both the solid and liquid fronts have secured quite a resounding success for Japan from a purely technology acquisition point of view, as Table 4.2 indicates. Research, development, and refinement are ongoing. First Japanese individuals and then intra-competitive Japanese corporations, working with and under diverse government agencies, have been critical to advancing the state of Japan's

Table 4.2. Launch Achievements of Main Liquid and Solid SLVs, 1966–2009[1]

Solid-Fuel				*Liquid-Fuel*			
SLV	*Operational (Years)*	*Capability (to LEO)*	*Record (Successful Launches/ Total Launches)*	*SLV*	*Operational (Years)*	*Capability (to GEO)*	*Record (Successful Launches/ Total Launches)*
L-4S	1966–1970	26 kg	1/5	*N-I*	1975–1982	130 kg	6/7
M-4S	1970–1972	180 kg	3/4	*N-II*	1981–1987	350 kg	8/8
M-3C	1974–1979	195 kg	3/4	*H-I*	1986–1992	550 kg	9/9
M-3H	1977–1978	300 kg	3/3	*H-II*	1994–1999	2 ton	5/7
M-3S	1980–1984	300 kg	4/4	*H-IIA*	2001–Present	2–3 ton	13/14
M-3SII	1985–1995	770[2] kg	7/8	*H-IIB*	2009–Present	4 ton	1/1
M-V	1997–2006	1,800 kg	6/7	*GX*[3]	—	2 ton (est.)	Cancelled
ASR/ Epsilon	—	1,200 kg (est.)	(R&D stage)				

SOURCE: Japan Aerospace Exploration Agency (JAXA), *Jinkō Eisei, Roketto no Genjyō, Kadai, Tenbō* [The Present Condition, Challenges, and Prospects of Satellites and Rockets] (Tokyo,:JAXA, 4 November 2008), pp. 10–11, reference attachment, p. 8.

1. JAXA focuses only on the trajectory of some SLVs that are linked to present programs, but for a summary of all SLVs see earlier in this chapter and in Appendix I.
2. Some sources cite 800 kilograms for the *M-3SII*, such as *Encyclopedia Astronautica* at www.astronautix.com (accessed 21 August 2009).
3. The two-stage *GX* rocket aimed to use a LNG (Liquid Natural Gas) propulsion system, but its development was cancelled. Additional information on the rocket is from "GX Launch Vehicle" at IHI Aerospace, available online at www.ihi.co.jp (accessed 21 August 2009).

rocket technology piece by painstaking piece over the years. From the paltry beginnings of the *Pencil* to the spectacular strides culminating in the *M-3SII*, *M-V*, *J-I*, and *H-IIA* programs, we conclude that Japan has assembled, tested, and successfully acquired a complete "toolkit" of technologies and infrastructure needed to construct intercontinental ballistic missiles.

To our mind, however, these assertions about technical capabilities need to be situated in their proper historical and political realities as highlighted in Chapter 1. We believe a shifting set of such realities is taking Japan ever further away from its postwar pacifist orientations—the long distance the country and especially its youth has come from memories of the war, the subtle but inexorable shift away from the Yoshida doctrine even as military reliance on the United States (with which Japan may not always have the same interests, especially in its Asian neighborhood) continues, the ongoing North Korean missile and nuclear tests, the growing economic and political power of a nuclear-armed China, the chance of a conflict over Taiwan, the competition for undersea gas and other natural resources, as well as a diverse number of unresolved territorial issues that are largely of interest only to Japan.

What, then, is the point of having acquired such capability? In assessing the ballistic missile threat confronting the United States, the representation by the CIA to the Senate Foreign Relations Committee becomes pertinent for a relatively weaker country—like Japan—that confronts potentially far more military powerful neighbors in the region and the world:

> acquiring long-range ballistic missiles armed with a weapon of mass destruction probably will enable weaker countries to do three things that they otherwise might not be able to do: deter, constrain, and harm [their rivals]. To achieve these objectives, the missiles need not be deployed in large numbers; with even a few such weapons, these countries would judge that they had the capability to threaten at least politically significant damage to [their rivals]. They need not be highly accurate; the ability to target a large urban area is sufficient. They need not be highly reliable, because their strategic value is derived primarily from the implicit or explicit threat of their use, not the near certain outcomes of such use. . . . *In many ways, such weapons are not envisioned at the outset as operational weapons of war, but primarily as strategic weapons of deterrence and coercive diplomacy.*[82]

Seen from this perspective, the acquired technology, which amounts to the ability to build a wide variety of missiles all the way up to ICBMs, forms part of a recessed deterrent for Japan. With an eye no doubt increasingly turned to its neighbors' ballistic missile arsenal—particularly China and North Korea—Japan has pushed its rocket program forward in successful bits and pieces that we believe sum up to a ballistic missile capability of its own. All indicators suggest and virtually all knowledgeable observers agree that the Japanese government has the technical pieces in place to use the same technology that puts a heavy geosynchronous satellite into orbit to launch a warhead around the planet.[83]

Of course, there are many things that would have to happen for Japan to convert its SLV technology into a strategic force.[84] The limiting factors are well known, as are the risks Japan faces, including the alarming speed of related technological dissemination and proliferation across Asian borders. In some ways, the changing configuration of cultural, political, industrial, and, now, formal legal aspects have already taken Japan some ways in the direction of that capability. Our point is more that all the pieces in the launch vehicle technology that Japan needed to gain this capability have been independently designed, developed, tested, improved, and proven in plain sight of the public. The saga of Japan's satellite development, to which we turn next, is also highly instructive for understanding any such prospects. A civilian technology carefully and incrementally gathered over the postwar period—to observe, monitor, and evaluate the Earth's topography, resources, and environment, and to do the same for space phenomenon, including cosmic rays, radiation belts, asteroids, lunar, solar, and planetary surfaces, and so on—was converted without much fanfare or blowback into a military instrument to observe, monitor, and evaluate strategic rivals and scenarios. Japan's SLVs may well undergo the same visible market-to-military conversion when Japan so chooses, possibly also in the face of a triggering event.

5 SATELLITES AND SPACECRAFT

OHSUMI, JAPAN'S FIRST SATELLITE, reentered the Earth's atmosphere in August 2003.[1] *Ohsumi* had been designed to test satellite launch technologies, and, despite its poor performance (its signal could no longer even be heard only after its seventh revolution around the Earth), it allowed for data-based orbital estimations. Most important, its indigenous launch on the country's then steadily advancing *Lambda* rocket series gave a tremendous morale boost to Japan's spacecraft effort. From the time that *Ohsumi* flew in February 1970 to its demise more than three decades later, the nation's satellite technologies, like the country's rocket technologies, had undergone a fundamental revolution.

Just months before *Ohsumi* returned to Earth, Japan had already launched its first set of spy satellites into orbit. These so-called *Information Gathering Satellites* (*IGS*), the first two of which flew in March 2000, are perhaps the only well-known and much-publicized part of Japan's now open progress toward space militarization. The program's cost is by far the single biggest for a satellite program in Japan's entire space budget.[2] Since the program's inception in the late 1990s, Japan's *IGS* satellites have been immune to budget pressures, consistently getting budgets needed to stay on schedule. In 2005, for example, unlike other space programs, the classified reconnaissance satellite programs remained unaffected by space budget cuts, got the $1.1 billion request allocated, and also secured initial funding for a next generation of reconnaissance satellites.[3] Even before the Basic Space Law of 2008 and Basic Space Plan of 2009 formally enshrined defensive military space policy as a fundamental rationale for Japan's space programs, the *IGS* became famous for what they implied about Japan's new and proactive security directions. It was as if the Japanese government and defense contractors had suddenly woken up in the wake of the *Taepodong* shock in 1998 (when on August 31 a North Korean *Taepodong* missile flew over Japan, causing an uproar) and moved toward acquiring spy satellites in the interest of national security. Given the trajectory toward a national security component in Japan's space policy as discussed in Chapter 2, however, we believe that as with Japan's SLV development, the military space orientation of Japan's spacecraft development was deeper and more profound than was and is widely believed.

This chapter focuses on the trajectory of Japan's satellites and spacecraft technologies. As with Japan's SLVs, we are not simply concerned with any one single element such as reconnaissance satellite capability; rather, the focus in this chapter is on the range of satellite and spacecraft technologies that have been developed and that will henceforth enable Japan to deploy its military space infrastructure. It is, of course, relatively easy for us to make our case for the market-to-military trend in light of the already deployed military satellite program by Japan. But Japan's *IGS* satellites did not just appear out of thin air. Rather, they are closely linked to the painstakingly acquired progression and sophistication of the country's satellite and spacecraft-related developments. These include the key technologies for an advanced communication and information infrastructure necessary for military operations and even Anti-Satellite (ASAT) systems necessary for Japan's emerging Space on Demand (SOD) and Defensive Counterspace (DCS) capabilities.

The remainder of this chapter is in four parts. The first part briefly sets out the rationale for a focus on satellite and spacecraft technology under the rubric of the so-called Revolution in Military Affairs (RMA). This allows us to frame the role of the related technology in military operations as a force multiplier, as well as the importance of counterspace capabilities for protecting it in the interest of national political and economic security. It also provides a brief overview of the existing satellite infrastructure. The second part then turns more specifically to the *IGS* saga, with a focus on elements (such as already accumulated technologies) and actors (particularly corporations) that shaped their arrival and continue to shape their future. The third part turns to look at the dual nature of other satellite programs that will help form a crucial military infrastructure and that are also bringing newer, less well-established space players into the game. The fourth part turns to a range of spacecraft technologies that have already taken Japan forward in terms of potential counterspace capability.

IN THE SHADOW OF A REVOLUTION

As is now widely known, the nature of battlefield operations underwent a revolution during the first Gulf war.[4] This Revolution in Military Affairs (RMA) has attracted much attention across the world, and not surprisingly within Japan as well. From the perspective of this book, the simple point is that the acquisition of satellite-based intelligence and reconnaissance, as well as the protection of space assets is critical to the RMA. Some concrete demonstration came, for example, in the detection of missile launches by the U.S. Defense Support Program (DSP) satellite system in the Gulf War in January 1991.[5] Along with information from electro-optical and radar imagery satellites, as well as that from aerial imagery and signals intelligence platforms, the DSP data were used to pinpoint and destroy Iraqi Scud missiles and their infrastructure from the first day of war. By that point in time, however, extensive use had already been made of the DSP's infrared sensors to track tactical ballistic missile launches by Iran and Iraq during the roughly eight-year Iran-Iraq war starting in 1980, as well as by the Soviet Union and Afghanistan following the Soviet invasion of Afghanistan in 1979. The DSP's infrared eyes have also tracked the proliferation of ballistic missile capability as Israel, India, and South Africa made test launches in the 1980s and North Korea continues to do so.

The RMA involves much more, however, than simply the relaying of communications via space-borne assets across continents to conduct military operations. Above all it is about the integrated and speedy conduct of warfare—the detection, identification or designation, and attack and destruction of enemy targets, or the spotting of missile launches. From a broad and strategic standpoint, the RMA is also about the way that leading-edge technologies and information warfare have allowed such operations to become possible. One contemporary U.S. vision of military revolution focuses on information-technology RMA (IT-RMA) components (such as satellite-backed information and communication, internet-based tasking orders, and so on), which can be exploited to great advantage in the battlefield. The most ambitious IT-RMA goals are to create armed forces that are wholly "network-centric," in which all military assets become nodes in a network of several grids (i.e., sensor, information-processing, support, weapon, or strike grids). The application of IT-RMA started in earnest during the Gulf War, and the benefits of revolutionary improvements in situational awareness are widely thought to have carried the day. Some also believe that continued progress in this direction took place in 2003 in the early combat phases of Operation Iraqi Freedom, where U.S. forces moved beyond jointedness to actual networked operating models.[6]

Japan too has reoriented its capabilities in line with the IT-RMA, seeking to improve its technological and qualitative (rather than quantitative) lead over rivals and to acquire a force structure capable of greater interoperability with U.S. forces.[7] With the underlying emphasis on information-as-advantage, Japan's main efforts have thus far concentrated on upgrading its Battle Management Command, Control, Computers, and Intelligence (BMC4I) system. The Japan Air Self-Defense Force (JASDF), for example, has announced the replacement of the Base Air Defense Ground Environment (BADGE)-integrated network of radar posts and air defense installations with the almost $935 million project called the Japan Aerospace Defense Ground Environment (JADGE).[8] JADGE is designed to integrate Japan's ballistic missile sensors and interception systems and to protect Japan from ballistic missile attacks by improving early warning systems. JADGE forms the crux of Japan's BMD infrastructure through centralizing and automating search, detection, tracking, and interception of ballistic missiles and is linked to U.S. communications satellites, which allows the United States and Japan to share data. JASDF's supreme command authority, the Air Defense Command, which is the headquarters for BMD and the new JADGE, has also relocated to the U.S. base in Yokota to improve coordination between the two forces despite the lack of any formal agreement. As discussed briefly in Chapter 3 and more extensively in Chapter 6, Japan is already contemplating its own version of a space-based network-centric defense system; and in a less visible way, we believe, it is developing its own Operationally Responsive Space (ORS) capability, called SOD.

All this helps bring Japan's space policy, particularly related to satellites and spacecraft, in sharper focus. Setting aside any attempt at evaluating the efficacy of RMA-based warfare models or the development of ORS/SOD in Japan or elsewhere, we concentrate primarily on their underlying discrete and especially space-related components. Needless to say, satellite and spacecraft for broadcasting, reconnaissance, warning, surveillance, and communication are critical for network-centric warfare, and the revolution is fuelled fundamentally by their technological advances and integration. For this reason even though IT-RMA is still a work in progress, it has several dimensions that make it especially relevant for our purposes in examining the changing parameters of Japan's space policy. These dimensions include reconnaissance and surveillance, communication systems, electronic warfare, precision striking, digitization and modularization of ground forces, and provisioning forces in combat. The very presence of such dimensions supporting an IT-RMA suggests several things—the importance of space assets to provide information, surveillance, and communications, the emergence of cyberspace as a new domain of conflict, and also, of course, the critical importance of protecting any such assets.

In the remainder of this chapter, we show how Japan has moved in these directions with a focus on its satellite and spacecraft technology, concentrating initially on the space-based intelligence and communication infrastructure that can affect its potential operations in a military or conflict scenario. As implied, for Japan to develop military space capabilities it needs an infrastructure: the necessary observation satellites to give it the ability to spy on its neighbors both for strategic purposes and for tactical battlefield situations; the ability and bandwidth to relay that information to whomever needs it, that is, to fixed and portable receivers; an on-orbit network to cope with shunting the information around; and the ability to protect the underlying infrastructure, both in space and on the ground. Of course, there are other areas for development, and the contours of some have just emerged concretely. As we discuss in the next chapter, if Japan had its own GPS system, which it is attempting to attain, for example, it could use that information not only to move troops, but also to target forces and munitions. One stage beyond this, Japan is now also committed to developing the ability to independently detect and intercept incoming missiles and to possess its own counterforce missile capability for response purposes.

EXISTING INFRASTRUCTURE

At present, as Table 5.1 shows, Japan already has a basic reconnaissance and communication infrastructure in place. It has a known range of sophisticated Earth Observation (EO), communication, positioning, and military satellites operational or in development. In fact, even a decade ago, the country's sophisticated land, marine, and geostationary meteorological satellites' capabilities for EO and communications were drawing attention in the United States.[9] Echoing the importance of applied space technologies for collecting and analyzing information for defense, disaster management, conflict prevention, environment monitoring, and agricultural production, the Kawamura initiative discussed earlier also fundamentally emphasized Japan's ability to "see" and "hear" preemptively.[10]

Table 5.1. Japanese Satellites and Spacecraft, 2009

	Status	
Type	*Currently/Recently Operational*	*Under Development*
Earth Observation Satellites	*Greenhouse Gas Observing Satellite* (*GOSAT*), *Ibuki* *Advanced Land Observing Satellite* (*ALOS*), *Daichi* *Earth Observation Satellite, Aqua* (USA–Japan sensor AMSR-E) *Tropical Rainfall Measuring Mission* (*TRMM*) (Joint U.S.–Japan precipitation radar [PR]) *Innovative-technology Demonstration Experiment* (*INDEX*), *Reimei* *Aurora Observation Satellite* (*EXOS-D*), *Akebono* *Magnetospheric Observation Satellite* (*GEOTAIL*) (Joint Russia, Europe, Japan)	*Global Precipitation Measurement* (*GPM*) (Joint USA–Japan dual-frequency precipitation radar [DPR]) *Global Change Observation Mission* (*GCOM-W* [water], *GCOM-C* [climate]) *Earth Cloud, Aerosol and Radiation Explorer* (*EarthCARE*) (Joint Europe, Japan Cloud Profiling Radar [CPR])
Communication, Positioning, and Engineering Test Satellites	*Wideband Internetworking Engineering Test and Demonstration Satellite* (*WINDS*), *Kizuna* *Engineering Test Satellite VIII* (*ETS-VIII*), *Kiku-8 Optical Inter-orbit Communications Engineering Test Satellite* (*OICETS*), *Kirari* (Joint ESA tests) *Data Relay Test Satellite* (*DRTS*), *Kodama* (Joint ESA tests) *Experimental Geodetic Satellite* (*EGS*), *Ajisai* *Small Demonstration Satellite-1* (*SDS-1*)	*Quasi-Zenith Satellite System* (*QZSS*, *Michibiki*)
Astronomical Observation Satellites	*Solar Physics Satellite* (*SOLAR-B*), *Hinode* (Joint UK–U.S.–Japan) *Infrared Imaging Satellite* (*ASTRO-F*), *Akari* *X-ray Astronomy Satellite* (*ASTRO-EII*), *Suzaku*	*Radio-Astronomical Satellite* (*ASTRO-G*) *X-ray Astronomy Satellite* (*ASTRO-H*)
Lunar and Planetary Exploration Satellites	*Asteroid Sample-return Space* (*MUSES-C*), *Hayabusa* *Selenological and Engineering Explorer* (*SELENE*), *Kaguya* (operation completed)	*Venus Climate Orbiter* (*PLANET-C*, *Akatsuki*) *Mercury Exploration Mission* (*BepiColombo*) (Joint ESA–Japan)
Military[1]	*IGS-1* *IGS-2* *IGS-3* *IGS-4*	*ASNARO*, *Sasuke*[2]

SOURCE: With the exception of the military satellites discussed in the text, information on satellites and spacecraft shown here is as identified by JAXA at www.jaxa.jp (accessed 17 August 2009).

1. See also future development projections in Table 5.2 and Figure 5.1.
2. *ASNARO* (*Advanced Satellite with New System Architecture for Observation*)/*Sasuke* (*Small Advanced Satellite for Knowledge of Earth*), a provisional name, may potentially be a series for military uses as well.

Japan's efforts do not end here. Like the United States, Japan is confronted with strong possibilities that other powers' counterspace capabilities could render its space-based intelligence and communication infrastructure useless and cripple its political and economic security. For this reason, we go on to examine how Japan is taking potential steps to protect precisely that infrastructure by acquiring technologies that may well equip it with significant response capabilities of its own in the near future. As we show, Japan is moving beyond preemptive seeing, hearing, and unified network integration to developing ORS and potential counterpace capabilities—and perhaps to acquiring the ability to soon shoot down satellites. These moves are, to some extent, a logical extension of the IT-RMA based vision. We next turn to the details and saga of Japan's satellite- and spacecraft-related technologies, beginning with the *IGS*.

FROM PRECURSORS TO *IGS* AND BEYOND

The story is now often told, but it is worth recounting briefly.[11] On 31 August 1998, North Korea launched a *Taepodong* rocket that flew over Japan and then fell into the Pacific Ocean near the Sanriku coast in northeast Japan. Less visibly, this incident focused attention on the necessity of Japan having indigenous space assets and systems, as well as the role of corporations and their various competitive interests in actually making those assets and systems.[12] More visibly, and better known, the incident brought Japan's defense needs to the forefront of public debate, providing a groundswell of support for boosting Japan's military capabilities, which was reflected in the country's move to acquire reconnaissance or spy satellites. Below we provide a brief overview of Japan's moves toward their acquisition and then turn to the corporate and technology foundations that actually made that move possible in line with our market-to-military themes.

Japan Spies

In retrospect, the "*Taepodong* shock" was perhaps more a *trigger* that legitimated Japan's long-standing wish to upgrade its information and intelligence-gathering infrastructure. This had certainly been on the minds of government officials for general crisis and disaster management. But it was no doubt also on the minds of defense planners for military operations, given the uncertainties over North Korea's ballistic missile capability since at least 1993 and the launch of *Nodong-1*. One of Japan's first institutional responses was the establishment of the Defense Intelligence Headquarters (DIH) in early 1997 to improve Japan's autonomous gathering and analysis of intelligence.[13]

In the aftermath of the launch, the speed with which Japan made a political decision to institutionalize a spy satellite program made headline news around the world.[14] In fact, the rapiditiy of the decision can at least partially be accounted for by the fact that the *Taepodong* trigger actually followed several significant calls for the development of such spy satellites, with political support expressed at the highest levels.[15] For example, in 1996 the Japan Defense Agency (JDA), backed by the prime minister, had seriously considered building its own military reconnaissance network. Other agencies were also involved. According to the budget requests by the Ministry of Foreign Affairs (MOFA), Japan had been studying spy satellites since at least 1997.

But the context in 1998 made all the difference in legitimating the lurch toward the military angle to the public. Spurred on by the mood of the crisis, and by the much-ballyhooed dependence on U.S. intelligence, the government's positive decision on military reconnaissance satellites made possible what had hitherto been a politically untouchable subject dating back to 1969.[16] There was little doubt expressed all around about being able to stretch the Peaceful Purposes Resolution (PPR).[17] With the country's own satellites, the Self-Defense Force (SDF) could then take part in and benefit from the program. Even though the public appeared to support the program, what made it even more palatable perhaps was the "dual-use" nature of the *IGS*—certainly for military purposes but also for peaceful purposes such as monitoring natural disasters and weather patterns, all of which was stressed from the start of their development.

This emphasis was thought useful for several reasons. It meant the *IGS* could be operated in accordance with a 1985 government statement that prevented the SDF from using any satellite technology that was not also readily available on the commercial or civilian side. It also helped counter negative reactions by North Korea and China.[18] North Korea warned that Japan was entering a dangerous phase of militarization that hearkened back to its past when it invaded Asia. China, which viewed itself as the target, openly charged that the larger purpose of the spy satellites was to actually allow intelligence-gathering in the service of a possible U.S.–Japan Theater Missile Defense (TMD) system—and subsequently, these charges would prove to be correct.

In March 2003, the first two *IGS* satellites—one for optimal imagery at nominally 1-meter resolution and another for radar imagery at resolutions ranging from 1 to 3 meters—were launched.[19] These were successfully followed by a second optical satellite in September 2006 and a second radar satellite in February 2007. In the meantime, the Japanese government pressed ahead with plans for the replacement of the original pair of *IGS* satellites, which were nearing the end of their design life.[20] As is to be expected, however, the specific details of the *IGS* program remain shrouded in secrecy—so much so that despite their much-proclaimed multipurpose nature, they are not registered with the United Nations.[21] The United States moved to classify the orbital parameters of Japan's first two *IGS* (although they are in an orbit that is low enough that they are visible to ground observers).[22] Even in 2003, a mere four satellites was not going to be the end of the story, with some then calling for at least 16–20 satellites to provide around-the-clock surveillance of regional hotspots like North Korea.[23] These echoed earlier calls in 1996 by the JDA, which had called for a minimum of ten satellites for such operations. As indicated in Table 5.2, the Japanese government has also moved forward with ambitious plans for increasing the quality and quantity of observation satellites all around, including national security/military ones, and has identified Pan-Asian observation specifically as a top priority for EO purposes. If all goes as planned until the year 2020, the Strategic Headquarters for Space Policy (SHSP) estimates the total cost of public and private development for these satellites to be around $25 billion.[24]

Meanwhile, there are both budget constraints and technical realities with which to contend. Whether or not budgetary pressures cause Japan to scale back these overall developmental ambitions in the next decade remains to be seen. But if the institutional and programmatic emphasis on the *IGS* is any guide, the government will probably continue to heavily favor EO, communications, positioning, and military satellites. As with the technicalities in Japan's other space technologies, there have also been delays, setbacks, and failures that have marred the *IGS* program: the *H-IIA* launch failure in which a second pair of optical and radar *IGS* were lost in November 2003, the inability of the first optical satellite to achieve its 1-meter resolution by July 2003, the electrical problems of the first reconnaissance satellite in March 2007 leading to its abandonment, and so on. Such technical problems will continue to crop up. But behind the pecuniary and technological problems lie two points in line with our market-to-military thesis. First, long before Japan's official Basic Space Plan in 2009 stressed the need for extra satellites, the *IGS* had become a vital institutional part of Japan's military space paradigm. Second, although

Table 5.2. Estimated Number of Satellites/Spacecraft by Function, 2003–2020[1]

Function (satellite type)	Military/Dual-use		Civilian	
	Series	*Number*	*Series*	*Number*
Pan-Asia Observation				
Land and water observation[2]	*ALOS/Daichi 1, 2, 3*	3		
	ALOS/Daichi type projected	2–4		
	ASNARO/Sasuke Optical Test	1		
	ASNARO/Sasuke Radar Test	1		
	ASNARO/Sasuke type projected	2–4		
Data relay	*DRTS/Kodama*	1		
	Other *DRTS* projected	1–2		
Earth Observation				
Planetary environment observation			*Aqua*	1
			TRMM	1
			GOSAT/Ibuki	1
			GOSAT/Ibuki type projected	1
			GCOM-W	1
			GCOM-W type projected	1
			GPM	1
			EarthCARE	1
			GCOM-C	1
			GCOM-C type projected	1
Weather observation	*MTSAT/Himawari 6, 7, 8, 9* (including standby)[3]	4		
Advanced Communication (including Experimental)	*OICETS/Kirari*	1		
	ETS-VIII/Kiku-8	1		
	WINDS/Kizuna	1		
	Next-generation projected	1		

(continued)

Table 5.2. *(continued)*

	Military/Dual-use		*Civilian*	
Function (satellite type)	*Series*	*Number*	*Series*	*Number*
Positioning (including Experimental)	*QZSS/Michibiki*[4]	at least 3		
	QZSS/Michibiki operational	up to 6		
National Security/Military	*IGS Optical 1, 2, 3, 4, 5*	5		
	Optical series projected	at least 1		
	IGS Radar 1, 2, 3, 4	4		
	Radar series projected	at least 1		
	Experimental/test	2		
	Early warning	at least 1		
	SIGINT	at least 1		
Space Science				
Exploration missions			*MUSES-C/Hayabusa*[5]	1
			MUSES-C/Hayabusa, SCOPE (magnetosphere) small type projected (for solar system exploration)	at least 2
			SELENE/Kaguya (Moon exploration)	1
			Planet-C/Akatsuki (Venus exploration)	1
			BepiColombo (Mercury exploration)	1
Astronomical observation			*SOLAR-B/Hinode*	1
			ASTRO-EII/Suzaku (X-ray)	1
			ASTRO-F/Akari (Infrared rays)	1
			Astro-G (Electric wave)	1
			Astro-H (X-ray)	1
			SPICA (Infrared rays)[6]	1
			Other *Astro-H, SPICA* type projected	at least 2
Small-scale missions			*INDEX/Reimei*[7]	1
			Ikaros[8]	1
			Small-scale satellite type projected[9]	at least 6

Small-Scale Satellites (including Experimental)	*SERVIS-1, 2*[10]	2	
	SDS-1, 2	2	
	Small-scale satellite type projected[11]	at least 9	
	Private, Academic[12]	18	
	Private, Academic type projected	10–20+	
Estimated Totals		90+	30+

SOURCE: Strategic Headquarters for Space Policy (SHSP), *Uchū Kihon Keikaku: Nihon No Eichi ga Uchū o Ugokasu* [Basic Space Plan: Wisdom of Japan Moves Space] (Tokyo: SHSP, 2 June 2009), Appendix 1, 2. The division between military/dual-use and civilian use largely follows from the discussion in the text about the IT-RMA. Existing and projected spacecraft/missions under the manned space program, the solar photovoltaic/power programs, and foreign commercial/government satellites are not included.

1. Estimates also take into account Japan's joint satellite development/use (either operational or in progress) as identified earlier in Table 5.1.

2. Estimations do not include the U.S. Terra in this category, on which Japan flew the optical ASTER (Advanced Spaceborne Thermal Emission & Reflection Radiometer) sensor until 2004.

3. JAXA lists the *Geostationary Meteorological Satellite* (*GMS*) series, also known as the *Himawari* series, as a past project. After *Himawari-6*, the series was replaced by the *Multi-functional Transport Satellite* (*MTSAT*) series to broaden its scope for climate observation. Its operations were also taken over by the Japan Meteorological Agency (JMA), which bills it as the next generation of satellites covering East Asia and the Western Pacific, coverage of which is also backed by the U.S. National Oceanic and Atmospheric Administration (NOAA) through an agreement. The *MTSAT/Himawari* series is multifunctional in that it also has an air-traffic control mission for the Civil Aviation Bureau of the Ministry of Land, Infrastructure, and Transport (MLIT). The prime contractor for the series is Melco, which claims that the satellite series is also a next-generation global-scale air traffic safety system made up of communications, navigation, tracking, and air traffic control. See JAXA, "Satellites and Spacecraft: Past Projects," at www.jaxa.jp (accessed 8 August 2009); Japan Meteorological Agency (JMA), "Satellite Home Page: Meteorological Satellite—MTSAT Series," and "JMA, NOAA Sign Satellite Back-up Arrangement," press release, 24 February 2005, both at www.jma.go.jp (accessed 8 August 2009); and for Melco's continued economic stakes in the series, see "Japan's Mitsubishi Electric to Build Two New Himawari Satellites," *BBC Monitoring Asia Pacific*, 18 July 2009.

4. One *QZSS/Michibiki* system relies on at least three satellites.

5. *Hayabusa* is designed to show Japanese competence in, among other things, autonomous navigation systems, asteroid-landing and sample-collecting systems, and also spacecraft/reentry capsule systems—some of this technology could be transposed toward advancing that on autonomous proximity operations, as discussed in Chapter 6. See a general assessment in A. J. S. Rayl, "Hayabusa: Itokawa Beckons as Japan's Spacecraft Searches for Places to Touch Down," The Planetary Society, 16 September 2005, available online at www.planetary.org (accessed 8 August 2009). According to ISAS, the SCOPE (cross-Scale COupling in the Plasma universe) mission, in conjunction with U.S. NASA's MagCon mission, which will scatter tens of small satellites in the magnetosphere, is concerned with understanding the causes of the magnetospheric plasma (plasma particles and waves, electromagnetic fields) through the satellites. See Masaki Fujimoto, "New Aspects of Magnetosphere Dynamics," *Reports & Columns: The Forefront of Space Sciences*, 5 January 2004, available online at www.isas.ac.jp (accessed 17 August 2009).

6. The Space Infrared Telescope for Cosmology & Astrophysics (SPICA) mission aims to launch a Japanese infrared astronomical satellite with a large cooled telescope that can make high-resolution infrared observations at medium to long distances.

(*continued*)

Table 5.2. *(continued)*

7. The *INDEX/Reimei* (*Innovative-technology Demonstration Experiment*) is a cutting-edge small-scale (approximately 60-kilogram) satellite designed and made by JAXA to test a number of technologies that are expected to pave the way toward ever smaller scientific satellites. See relevant information through JAXA at www.jaxa.jp (accessed 17 August 2009); and Shinichiro Sakai, "Better Development of Attitude Control Systems for Satellites: Development Diary of REIMEI Satellite," *Reports & Columns: The Forefront of Space Sciences*, 28 May 2007, available online at www.isas.ac.jp (accessed 17 August 2009).

8. The *Interplanetary Kite-craft Accelerated by Radiation of the Sun* (*IKAROS*) mission aims to test the world's first interplanetary solar-powered spacecraft using photon propulsion and solar power generation.

9. Projections are for about three small satellites every five years beginning around 2012.

10. The *Space Environment Reliability Verification Integrated System* (*SERVIS*) project is managed by the Institute for Unmanned Space Experiment Free-Flyer (USEF) and the New Energy and Industrial Technology Development Organization (NEDO), both operating with METI's cooperation and/or support, to develop commercial-off-the-shelf (COTS) technologies for increasing technical competence and cost competitiveness for LEO satellite buses and equipment.

11. Projections are for about one small-scale governmental satellite per year beginning around 2012.

12. Although small satellites developed by private or academic units can have commercial or scientific uses, they can also be used to test or develop potential military/dual-use purposes. Between 2003 and 2010, approximately eighteen private/university small satellite projects/missions were completed. Based on past trends, projections are for numerous (say, at least two) small/ultra-small satellites per year beginning around 2012.

Japanese spy satellites are criticized as being inferior to U.S. and even commercial satellites in terms of their ground resolution, at this stage we believe the gamble on an indigenous four-satellite reconnaissance program has by and large paid off—at least from the perspectives of the defense contractors who pushed it, help maintain it, and, not to be forgotten, can also improve it. And it is to their involvement that we now turn.

Behind the Spying

From a cost and development perspective, it may well have made sense for Japan to purchase satellite systems from U.S. contractors as Washington wanted. But, as we saw above, Japan chose to develop independent spy capability, with an emphasis on indigenous production that incorporated some U.S. components and imagery-analysis help. The focus on foreign defense contractors brings us to a consideration of the private forces behind Japan's moves to spying openly. The fact is that in the high drama of the *IGS* there was little opposition to the push by Keidanren and Japanese defense contractors for indigenously building the spy satellites. Satellite makers such as Melco had not been sitting around. They were ready, having slowly put the basic observation technology in place over time. As the following discussion shows, the saga of the roughly $2.1 billion development of the *IGS* satellite nicely encapsulates our market-to-military themes: the outright shift from commercial to military uses, especially of indigenous satellite imagery; the pivotal role of Melco, which was able to make the *IGS* satellites and rose to become the prime contractor; and the painstakingly acquired civilian satellite technology base that already existed and that allowed Melco and Japan's spysat program to actually move forward. We turn to each of these below in succession.

Satellite Imagery As discussed above, Japanese uneasiness in relying on commercial or U.S. defense satellite imagery has played a role in the push for domestic capabilities such as independent reconnaissance and, more recently, even early warning (see Chapter 6). We believe that the move toward the *IGS* in particular grew out of Japan's long-standing interest in satellite imagery, which dates back to the 1980s. In its quest for data and information, Tokyo started buying imagery from the U.S. LANDSAT satellites with 30-meter resolution in 1984 and also French Satellite pour l'Observation de la Terre (SPOT) satellites with 10-meter resolution in 1987.[25] As foreign commercial technology improved, private enterprise played a key role as conduit and reseller irrespective of qualms about military and security concerns.[26] Throughout the 1990s, Mitsubishi Corporation and Hitachi saw opportunities in the growing high-resolution commercial imagery market and were involved in launching a series of private imaging services to exploit near-spy quality imagery.[27] In October 1996, Hitachi set up a consortium with about fifty companies to do just this, expecting the business to generate ¥30 billion. The following year, Hitachi signed a deal to purchase and resell image data from U.S.-based EarthWatch's (which morphed into DigitalGlobe in 2002) EarlyBird satellite, and Mitsubishi Corporation formed an agreement with the Space Imaging Eosat consortium, which included Lockheed Martin, Raytheon, and Eastman Kodak.

The open secret was that the targets of all this imagery were various actors in the Japanese government—MOFA, the Maritime Safety Agency, and, most controversial, the JDA. Opening the doors on the defense side was, as previously discussed, the 1985 government decision allowing the purchase of foreign commercial satellite images for military intelligence, which was determined not to breach Japan's ban on the military use of space.[28] By 1996, egged on by the Europeans to loosen their dependence on the United States, the Japanese were openly strengthening their cooperation and relations with them regarding high-resolution satellite imagery.[29]

Not surprisingly, the JDA sought the most accurate satellite imaging available. One source was Mitsubishi Corporation's Japan Space Imaging Coroporation (JSI), the regional affiliate for Space Imaging that famously launched its Ikonos satellite with its 1-meter resolution in September 1999, revolutionizing the commercial side of the market.[30] When Space Imaging was bought out by Orbimage in 2006 and both reemerged as GeoEye, Mitsubishi Corporation also signed JSI as the regional affiliate in April 2008.[31] GeoEye-1, launched in September 2008, provides a ground resolution of 0.41 meters, although this resolution is only available to the U.S. government and any foreign government the U.S. government designates. Commercial customers, such as Google, have access to resolutions of 0.5 meters. The GeoEye-2 satellite, currently in development for a debut in 2011–2012, is projected to have a resolution of 0.25 meters. Again, restrictive licensing will mean that only the U.S. government and some of its allies, possibly Japan, will have access to imagery at the full design resolution. When it goes up, Geoeye-2 will join GeoEye-1 and Ikonos; and assuming that all of these satellites continue to operate, this will give GeoEye three sub-meter-resolution satellites in orbit and the ability to offer its customers daily revisits. Another commercial source for Japan's defense establishment is through HitachiSoft (Hitachi Software Engineering), which also provides spatial imagery as a master distributor for DigitalGlobe, GeoEye's main competitor.[32]

Through HitachiSoft and JSI, Japan can certainly continue to acquire ever more sophisticated foreign-based satellite imagery, potentially now even down to an accuracy of 25 centimeters. But even this is seen as an inferior option to gaining direct control of image acquisition. In 2004 the JDA was estimated to be spending about ¥1 billion a year to buy images based on European and U.S. satellites. More important was the irritant noted earlier that the Japanese defense establishment continued to be dependent on U.S. intelligence, which could be delayed or even unforthcoming—an element which has been and remains a concern.

Pivotal Corporations In 1993 a defense advisory panel recommended that Japan develop its own satellite-based reconnaissance system.[33] In 1994 the JDA, Space Activities Commission (SAC), and the nongovernmental Defense Research Center (DRC) all issued findings that photographic reconnaissance capability was a logical extension of Japan's space activities. And in 1997 feasibility studies were actually funded for assessing that possibility at MOFA. The fact that the *IGS* satellites did not suddenly materialize out of thin air was another major factor in influencing the speedy shift toward autonomous intelligence gathering. Where exactly did all this technology come from in Japan's peaceful-purposes-only space program?

This points to the pivotal role of corporations and to the evolution of the underlying technology that they helped build. Most of the technologies needed for the first generation of satellites were already in the portfolios of major players like the Communications Research Laboratory (CRL, now the National Institute of Information and Communications Technology, NICT), Melco, Nippon Electric Corporation (NEC), and to a lesser extent, Toshiba.[34] They had already been involved in constructing a series of Earth Observation (EO) satellites, beginning with the *Geostationary Meteorological Satellite* series, *GMS/Himawari* (1977, 1981, 1984, 1989, 1995); the *Marine Observation Satellite* series, *MOS/Momo* (1987, 1990); and the *Japanese Earth Resources Satellite, JERS-1/Fuyō-1* (1992) for land observation. The *JERS-1/Fuyō-1* is noteworthy because it required the manufacturing of an optical sensor and Synthetic Aparture Radar (SAR), for example.[35]

As it turned out, by the 1990s both NEC and Melco had mapped out realistic proposals for spy satellite systems architecture, and the processes and contents of their actions reveal the concerted effort by defense contractors to move space assets toward more military uses.[36] As discussed in Chapter 3, Melco won out over its rivals and moved forward almost single-handedly with what became the *IGS* system. Although the exact details of the *IGS* constellation are classified, a window of opportunity in the months immediately after the *Taepodong* incident, when the Science and Technology Agency (STA) was openly discussing candidate designs, is instructive about the importance of corporate competence. In October 1998, Melco briefed the Cabinet on the company's proposals, which were, we believe, largely taken up by the government.

It is important to understand Melco's ideas about the *IGS* and how closely they matched the actual satellites, as it reveals the great degree to which the Japanese government's security policies are dependent on what the corporations can actually do.[37] According to the Melco proposal, the *IGS* were floated not only as a spy satellite constellation but also for general-purpose uses such as disaster alerts, coastal security monitoring, resource information, marine and agriculture information, weather monitoring, and precise geodesic mapmaking systems. However, Melco's highest stated priority was the preservation of national security through surveillance and observation, requiring extremely high levels of integration between ground and space assets, with coverage not only of the Korean Peninsula, but also of major Chinese and Russian missile and military installations surrounding Japan. Melco even raised the possibility of the *IGS* linking with the U.S. DSP early warning system tied in with Japan's TMD system to which, incidentally, Japan had yet to committ officially. Now, a decade later, the long-term early warning system has become an official military space policy priority, as revealed in the new Basic Space Plan.

Melco's configurations are also of interest, because many of them closely resemble the actual system on orbit: *IGS* would provide twice-daily wide-area coverage to detect and monitor, for example, missile deployment, troop movements, coastal surveillance, and three-dimensional mapping capability. As for technical capability, Melco cited its design heritage back to *JERS-1/Fuyō*; it also pointed to its history of cooperative experience in building stable, on-orbit platforms for sensor pointing control for the *Advanced Land Observing Satellite* (*ALOS/Daichi*), infrared sensor technologies for the optical sensors in the *Advanced Earth Observing Satellite* (*ADEOS/Midori*), phased array radars for the JDA for use in radar satellites, MISTY 128-bit encryption algorithm for secure transmission, and high-bandwidth (200 Megabits/second or Mbps) communications experience. The *Data Relay Test Satellite* (*DRTS/Kodama*), which the company was also building, is capable of 240 Mbps.

Melco's proposed system was for four satellites (two radar, two optical) in approximately 500-kilometer sun-synchronous orbits. The 1.5-ton radar satellites would provide optimum resolution in 10-kilometer swaths and lower-area observation of between 5- to 10-meter resolution in 60-kilometer swaths. The optical satellites would weigh 1 ton and provide 1-meter resolution for 14-kilometer swaths. The radar satellites would revisit the same spot at least twice a day (lunchtime and night) while the optical satellites would revisit once per day each. The proposal also outlined the building of ground stations in Okinawa and Hokkaido, as well as three signal links for imagery analysis to a government ministry, a government agency (which could mean the JMA or the JDA), and an additional facility. Ideally, this reconnaissance system could make an observation of a designated target in as little as thirty minutes, process and transmit the data in thirty to sixty minutes, involve image analysis in one to two hours, and produce three-dimensional maps within two to four hours. The first satellites were to be launched in fiscal year 2002 (by 31 March 2003) with the program costing ¥197 billion over the five-year operational life of the constellation. The actual satellite construction would cost ¥157 billion and operational costs would be ¥6 billion annually.

As Melco became a prime contractor for the constellation, it is believable that government planners accepted most of its technical proposals. For one thing, Melco's plans went on to closely resemble those later developed by then STA, MITI, and MPT, which were officially charged with developing technologies for the constellation, as well as the Cabinet Office (CO), which was charged with general usage and directions.[38] For another, both the satellite weights and orbits ended up closely matching Melco's proposal.[39]

Earth Observation (EO) Technology As Melco's proposal had hinted, by the the time the spy satellites became a reality, the technological foundations for that move had been laid down in Japan's long trajectory of EO technology.[40] Corporate competence in this area was a major enabler in making that shift, and to some extent the *IGS* program represents nothing more than the evolution of peaceful EO technology into national security usage. Japan has had an EO system in place since the late 1970s.[41]

As with SLVs, the process began with basic satellite technology acquisition and steadily evolved to competency based on indigenous technology and integration skills. In July 1977, there was the launch of the *GMS-1/Himawari-1* for cloud imagery and sea- and cloud-level temperatures. In October 1978, the Earth Observation Center (EOC) was opened to encourage satellite remote-sensing technology, and today the EOC is administered by the Earth Observation Research Center (EORC), which was established in April 1995 as JAXA's principal center for EO data acquisition, processing, and research. Launches following *GMS-1/Himawari-1* were as follows: in August 1986, the *Experimental Geodetic Satellite* (*EGS/Ajisai*) for precision mapping all of Japan and, especially, its islands; in February 1987, the *MOS-1/Momo-1* for ocean-related phenomenon, which was followed in 1990 by its successor satellite *MOS-1b/Momo-1b*; and in February 1992, the *JERS-1/Fuyō-1* for surveying resources, environmental phenomenon and disasters, geographical and coastal topography, and to allow further testing and development of the SAR. Subsequently the *GMS-1/Himawari* series acquired a broader mission and new name as the Melco-built *Multi-functional Transport Satellite* (*MTSAT-1, MTSAT-2*) series when it was transferred to the Japan Meteorological Agency (JMA) in February 2005.

By the late 1990s, Japan had moved forward with establishing advanced satellite and EO technologies: in 1996, the *ADEOS/Midori* for observing all kinds of environmental changes and allowing development of future technologies, such as inter-orbit communications for data relay; in December 2002, the highly evolved *ADEOS-II/Midori-II* for environmental monitoring, as well as observation of meteorology and fishery; in November 2007, the *Tropical Rainfall Measuring Mission* (*TRMM*), built jointly with the United States, for weather forecasts and climate change research; the *Earth Observation Satellite* (*Aqua*), a joint project with the United States and Brazil, for improving land and ocean surface observation at night or in cloudy weather; and in 2009, the *GOSAT/Ibuki*, which is bearing the brunt of on-orbit measurement of global carbon dioxide emissions. This is especially important after its U.S. equivalent, NASA's Orbiting Carbon Observatory (OCO) failed to reach orbit due to a launch failure in February 2009.[42] All in all, the fact is that the contribution and importance of Japanese EO technology is recognized globally.[43]

It is precisely this incremental mastery of EO technologies that enabled Japan to decide to wield those technologies for military purposes, and this militarization trend is now picking up pace.[44] All the technological developments and improvements—the testing of satellite technologies such as remote sensing, SAR, ground resolution improvements, inter-orbit communications, and so on—in the service of harmless-sounding EO ventures such as meteorology and weather forecasts, environmental and climate changes, disaster surveillance, resource surveys, and precision mapping of land and coastal topography could also of course be transposed to serve military purposes.

Technological maturity to at least begin military applications began with the *ALOS/Daichi*, which was designed a decade before its launch in January 2006. Building on *JERS-1/Fuyō-1* and *ADEOS/Midori*, the large 4-ton satellite *ALOS/Daichi* has improved remote-sensing instruments—the Panchromatic Remote-Sensing Instrument for Stereo Mapping (PRISM) using three optical systems to get three-dimensional land-surface data with a 2.5-meter spatial resolution; the Advanced Visible and Near Infrared Radiometer Type 2 (AVNIR-2) for obtaining ground surface data with a 10-meter spatial resolution; and the Phased Array type L-band Synthetic Aperture Radar (PALSAR) with 10- to 100-meter resolution to detect, regardless of daytime or weather, signals resulting from changes in Earth topography and geology.[45] *ALOS/Daichi* also features major improvements in the speed and volume of data transmission, as well as in the positioning and attitude of the spacecraft itself. To date, *ALOS/Daichi* has proved highly useful in a broad swath of applications. In 2003 the CO moved to develop, for example, an emergency response infrastructure that would combine data from *ALOS/Daichi*, *IGS*, Ikonos, Quickbird, and the Israeli Eros-A1 satellites to provide quickly updatable situation maps of disaster hit areas. But *ALOS/Daichi* is most notable for being engaged in an ever-improving quest for precision cartography—part of an ambitious project to map all land areas around the globe, with a stated special focus on those in the Asia-Pacific region. As of December 2007, it had captured 1.8 million images, and the bus, sensor, and attitude control systems were reported to be working fine.

In terms of dual-use application, PRISM in particular proved useful. Within a week of the *Taepodong* launch, *ALOS/Daichi* technology emerged as a candidate reference design at the STA, with a senior NASDA engineer claiming that redeveloping *ALOS/Daichi* for sharper resolution would not pose great difficulties.[46] Much of the optical and data storage technology, high-precision attitude control system, radar, and three-dimensional imaging technology was also redeveloped from *ALOS/Daichi*. In an effort to improve spatial resolution from previous spacecraft, the Melco proposal was to offer the Cabinet improvements on the satellite's PRISM sensor to 1-meter resolution from a 500-kilometer orbit.

IGS technology thus had a clear heritage. The first generation *IGS* were built from *ALOS/Daichi* technologies: redevelopments in the PRISM sensor by NEC; upgrades in the AVNIR sensor by Melco (that came out of *ADEOS/Midori*) and improvements in the PALSAR sensor (evolved from *JERS-1/Fuyō-1*). In fact, so vital was *ALOS/Daichi* to the *IGS* program that as early as October 1998, STA officials began to lobby the CO to disburse extra money to fast-track the satellite's development, which was then at the basic design phase.[47] Those actually in charge of the *ALOS/Daichi* project also admitted that NASDA sought to similarly fast-track the satellite's development in order to help advance the *IGS* design along several dimensions—sensor resolution, data recording and transmission (gigabit speeds), and more precise attitude-control technologies.[48] The basic point of all this: it was only when a range of civilian satellite technologies—increasingly improved by corporations and extensively tested by the government—were in place that Japan could seriously think about reducing its dependence on U.S. intelligence and begin going down its very own path to their military uses.[49] In addition, as discussed in the next chapter, this narrative is now being repeated with METI's SOD program.

COMMUNICATIONS SATELLITES

Continuing with the emphasis on information-as-advantage, the IT-RMA discussed earlier made clear the importance of communication technologies to military operations.[50] The crucial technologies needed for militarily useful satellite assets include the ability to maintain a constellation of satellites that can reliably relay high-bandwidth data around orbit so that communications can be looped to ground stations at all times, and the ability to transmit high-bandwidth communications and data to mobile terminals in the field. Military units need a means of secure communication to receive and supply information in places and during operations where terrestrial modes of communication are not accessible, impossible, or unreliable. Any aspiring satellite-based communication system in the service of such military ends obviously then needs to protect against jamming. Additionally it needs to have both the speed and flexibility to extend services around the world and to also have the ability to reallocate system capacity when needed.

Apart from the public call to build a network-centric military space communications systems we alluded to earlier, there is little information about Japan's efforts to directly acquire secure military communication satellites.[51] But we connect some dots below that may prove instructive with respect to any such moves in the near future. Once again, as with the *IGS* saga, this takes us into the civil advancements made in Japan's satellite communication technologies.[52] It is thus helpful to briefly step back and consider that here too Japan has come a long way from when it launched its first set of satellites in the aid of improving communication and broadcasting—the first experimental *Communication Satellite/Sakura* (*CS/Sakura*) in 1977, the first practical communication *Broadcasting Satellite/Yuri* (*BS/Yuri*) in 1978, and the first geostationary *Experimental Communication Satellite/Ayame* (*ECS/Ayame*) in 1979. As the country's communication technologies began to advance, it was a foregone conclusion that satellite-based military communications would be as much of interest to Japan's defense establishment in the 1980s as they had been to the U.S. military with the growth of the U.S. space program in the 1960s.

As is well known, the JDA (and now the Ministry of Defense, or MOD) already made use of satellites for military communications via transponders, such as on Space Communication Corporation's (SCC) *Superbird* satellites.[53] Indeed the move to allow the JDA to use satellites for communications purposes was one of a significant series of steps taken toward the military use of space before the full-fledged move to build the *IGS* constellation. As with the *IGS*, a justification was found for this move in 1985; here, the government maintained that the use of domestic satellites for communication purposes did not violate the PPR, which of course also incentivized domestic satellite producers. Over time, they have been key players in advancing the state of related technologies.

In terms of advancing and securing satellite communications, Japan rates among the world's top developers and is even a pioneer. Below we examine criss-crossing sets of Japan's satellite communications technologies. These give an overall snapshot of the advanced present state of Japan's satellite communication technologies, which have, over time, also enabled Japan to make the decision to deploy an indigenously built military space infrastructure.

The *ETS-VI/Kiku-6* and the *OICETS/Kirari*

As a background to Japan's efforts, there was the ambitious T-SAT (Transformational Satellite Communication System) program begun in 2004 in the United States at an estimated cost of about US$15.5 billion.[54] It was designed to transform the military communication system to achieve a networked force of soldiers and systems operating together seamlessly. Using a constellation of five satellites with satellite-to-satellite laser links, it was to have been the backbone of the U.S. military's soaring bandwith needs over the next decade. Until it was recently restructured due to cost and technical issues, the T-SAT program sought to advance the volume, speed, and connectivity of military communications as never before by integrating internet-like and laser capabilities.

But optical communications, or the use of lasers in particular for optical Inter-Satellite Links (ISL), still remain of keen interest to both the U.S. military and NASA, because they offer distinct advantages.[55] Unlike radio frequency-based communcation, laser communications allow higher data volume over greater link distances at lower size, weight, and power. Their other major advantage is that they cannot be jammed, and as such they are free of interference problems, which ensure secure transmission between Low Earth Orbit (LEO) to LEO, LEO to geosynchronous orbit (GEO), GEO to GEO, and GEO to ground links. Additionally, unlike the radio frequency spectrum, the optical spectrum is largely unregulated.

In March 2008, the importance of laser communications to military space communications was being underscored by a series of "flawless" in-orbit experiments between the TerraSAR-X satellite sponsored by the German government and the Near Field Infrared Experiment (NFIRE) satellite sponsored by the U.S. Missile Defense Agency (MDA).[56] The laser communication terminals (LCT) for the experiment were built by a German firm, Tesat-Spacecom, and provided to MDA under a cooperative U.S.–Germany agreement. The experiment successfully demonstrated bi-directional transmission at a data rate of more than 5.5 Gigabits per second (Gbps) between the two LEO satellites operating at a range of about 5,000 kilometers. The astonishingly high data rate demonstrated by the ISL shows how important laser communications could be to military space planners in both Europe and the United States, a point that is surely not lost on other militaries around the world. As of August 2009, in the United States, for example, the Missile Systems Center of the U.S. Air Force Space Command has reportedly moved to solicit information on whether the industrial base in the United States could provide a laser communications payload and host spacecraft by 2011 and demonstrate it in orbit by 2015.

What is less well known is that similar and successful experiments had actually already been carried out by Japan with technologies studied since 1985 and designed and built in the early 1990s.[57] During the 1990s, the Communications Research Laboratory (CRL) and MPT were awash with plans for communications constellations, demonstrating high-bandwidth optical intersatellite communications that were LEO to LEO, GEO to GEO, GEO-LEO -GEO and, in addition, GEO and LEO to ground. At present, the fact is that Japan has already demonstrated cutting-edge laser communications technologies.

The key technologies were developed through a series of satellites, starting with the Toshiba-built *ETS-VI/Kiku-6*, upon which work was begun in 1991. It was to test advanced attitude control, high-power electrical systems, advanced bus structure, an advanced ion engine system to replace conventional thrusters, both S- and K-band intersatellite technologies, and satellite-to-mobile communications—all of which made it highly advanced for its time.[58] As noted in an earlier chapter, a stuck valve on its IHI-built apogee engine condemned the satellite to a highly elliptical orbit, but both CRL and NASDA nevertheless managed to test its systems largely to great success. With *ETS-VI/Kiku-6* Japan first deployed and tested GEO-to-ground optical communications equipment in 1994. In 1995, Japan allowed NASA researchers to transmit signals to the satellite, thus successfully conducting two-way links from the Table Mountain Observatory in California to the satellite—the first known time lasers were used to provide two-way communications with space.

JAXA then decided to build the *OICETS/Kirari* to test laser ISLs even further.[59] The 570-kilogram *OICETS/Kirari* was originally scheduled for launch aboard a *J-I* as early as 1997 to test optical ISL links with the European Space Agency's (ESA) Advanced Relay and Technology Mission Satellite (ARTEMIS), precise laser ranging with NICT, as well as advanced star tracking sensors, among other things. The *OICETS/Kirari*'s principal communication equipment was the part-electrical/part-optical Laser Utilizing Communication Equipment (LUCE), designed by CRL/NICT and manufactured by then NEC Toshiba space systems. Taken together, these technologies related to increased visibility, as well as pointing, communication, tracking, acquisition, and so on that had evolved over a decade. The *OICETS/Kirari* was also supposed to link with the *Communications and Broadcasting Engineering Test Satellite* (*COMETS/Kakehashi*), ARTEMIS, and a series of ground stations. After many twists and turns, the *OICETS/Kirari* was eventually launched by a Dnepr rocket (by the International Space Company Kosmotras (ISC Kosmotras), a joint project, between Russia, Ukraine and Kazakhstan) from Baikonur into a 610-kilometer LEO orbit.

In December 2005 the *OICETS/Kirari* achieved the world's first known bidirectional optical ISL with ARTEMIS over a distance of about 40,000 kilometers. More than 100 laser experiments were carried between the two satellites, and most of the inter-orbit links were carried out without failures. The 50 Megabits per second (Mbps) ground-satellite and 2 Mbps *Kirari*-ARTEMIS performance satellites were described as excellent. In March 2006 the *OICETS/Kirari* also managed another first—laser optical communication between a LEO satellite and the ground, establishing an optical communication link with NICT, which involved tipping the satellite 180 degrees to face Tokyo. This was followed up in June with a successful hookup with the German Aerospace Center, some 600 kilometers below the speeding satellite. *OICETS/Kirari* successfully wound up its mission in Septmber 2009.

The *COMETS/Kakehashi*, the *DRTS/Kodama*, the *ETS-VIII/Kiku-8*, and the *WINDS/Kizuna*

Through the 1990s, Japan was also busy taking steps to develop other advanced conventional communications satellites, which are outlined below. These developments have kept Japanese technology up to date with U.S. technologies and ensured that Japan can now build its own military space communications infrastructure.

Launched in 1998, the NEC-built *COMETS/Kakehashi* was a geostationary satellite developed from the *ETS-VI/Kiku-6* bus.[60] At the time, it was billed as the most advanced communications broadcast research satellite ever built in Japan. It was designed to demonstrate, for example, acquisition and tracking, data transfer between LEO and GEO satellites, orbit determination by satellites by tracking data from the links, a Ka-band sytem test for high-definition television broadcast capabilities, Ka- and S-band data relay for a docking test (for the *ETS-VII/Kiku-7* discussed below), and Ka-band transmission to mobile terminals. Despite being dumped into the wrong orbit by the misfiring of the *H-II*'s second stage, after some corrections, the *COMETS/Kakehashi* was able to successfully perform many of its planned tests.

Importantly, the *COMETS/Kakehashi* became a technology demonstrator for the Melco-built *DRTS/Kodama*, which was Japan's first data relay satellite.[61] Originally there were two *DRTS* (East and West) to provide Japan's first data relay network, which were also supposed to be interoperable with those of NASA and ESA. But they were restructured into a single satellite following the *H-II* launch failures. The *DRTS/Kodama* was launched by *H-II*A in 2002, and after 6.5 years in operation in a geostationary orbit over the Indian Ocean completed its mission in October 2009. *DRTS/Kodama* has verified a high-performance tracking and acquisition system and communication links with the target spacecraft. One of the satellite's critical infrastructural roles is inter-satellite communications. Equipped with an antenna that has a steering mechanism to orient itself toward spacecraft in LEO or MEO orbits, it can receive data from them and relay it to ground stations. It has successfully performed such tests with the *ADEOS-II/Midori-II* (S-band and Ka-band) and with the *OICETS/Kirari* (S-band and laser beams). It has reached a data transmission data rate capability of more than 240 Mbps, achieved a global-leading speed of 278 Mbps with *ALOS/Daichi*, and linked with the *Japanese Experiment Module* (*JEM/Kibō*) on the International Space Station (ISS). Finally, in joint tests with the ESA, the satellite has also successfully demonstrated interoperability with European spacecraft, namely the data relay satellite ARTEMIS and EO satellite Envisat (Ka-band), as well as JAXA's *Small Demonstration Satellite-1* (*SDS-1*).

Given its performance, the *DRTS/Kodama* is a pivotal technology for any national operational space communications infrastructure that will be capable of meeting military needs for high volume and uninterrupted communications. The original *DRTS/Kodama* program was a test case for a more advanced communication system, such as the *Advanced Data Relay Test Satellite* (*ADRTS*) and *DRTS-X*, proposed in the mid-2000s by NICT.[62] According to some players, Japan had planned a much more capable and advanced *DRTS*, provisionally called the *ADRTS*, which would demonstrate optical intersatellite communications beyond both its predecessor and the *OICETS/Kirari*. The satellite was to be based on the estimated 1.5-ton *IGS* bus developed by Melco, and, with its advanced features, it was likely a prototype for a secure and exclusive military communication satellite. The Basic Space Plan of 2009 does not refute this possibility, and Japan may well launch just such a satellite in the 2010s.

Finally, both the *Engineering Test Satellite* (*ETS-VIII/Kiku-8*) and the *Wideband Internetworking Engineering Test and Demonstration Satellite*

(*WINDS/Kizuna*) are important in showcasing Japan's continuing interests in its regional focus on Asia. The development of *ETS-VIII/Kiku-8* took place over six years, beginning with the conceptual design in 1992.[63] It was seen as a demonstrator for bus technologies for Japan's first large geostationary satellite (3-ton class), large deployable antenna reflectors (LDRs), mobile satellite communications for handheld terminals, and fundamental technologies for satellite-positioning systems. Although it has had some anomalies, such as those related to the communication equipment and the ion engine, it was used experimentally in Feburary 2007 to test mobile phone-based communications as well as a satellite-based multimedia broadcasting system for mobile devices. It is also designed to enable direct communications with a single geostationary satellite to cover all of Japan. The huge antennae are supposed to provide the technology for Mbps-speed communications to mobile terminals across a wide swath of Asia. In addition, the satellite carries a highly advanced atomic clock for extremely accurate time signals. This will help conduct positioning experiments to obtain basic satellite positioning system technology, crucial for the *QZSS/Michibiki* system discussed in the next chapter.

With the theme of offering ultra-high-speed communications to enable an internet society within and around Japan, the *WINDS/Kizuna* was successfully launched in February 2008.[64] It evolved from a prior CRL project called the *GIGABIT Satellite*, which was intended to push up Ka-band speeds to the gigabit class. Following the *H-II* accident with the *COMETS/Kakehashi*, the *WINDS/Kizuna* also took on some of *Kakehashi*'s responsibilities. The reconfigured mission is designed to provide up to 155 Mbps (receiving) speeds to 45-centimeter household antennas and 1.2 Gbps to 5-meter office dishes; more instructively, the Ka-band active phased-array antenna can transmit and receive beams nearly everywhere in the Asia-Pacific region. In May 2008 the satellite carried out the world's fastest known satellite data communication speed test to date—a 1.2-Gbps connection between a 2.4-meter-diameter vehicle-mounted antenna in Kushiro, Hokkaido, and a 5-meter-diameter antenna in the NICT Kashima Space Research Center in Ibaraki prefecture. This adds up to the ability to communicate very quickly, all around Asia, via the satellite and without large-scale ground stations. The *WINDS/Kizuna* was designed to develop data relay between existing secure terrestrial networks and satellites, airplanes, helicopters, and vehicles, which would then allow real-time video transmission from a site to a government or command center. Its national security role has been seen as a test platform for the construction of a new disaster-resistant network. In addition, it is seen as being part of a phased approach to developing highly advanced satellite information and communications networks around the region, which could also be useful in military operations.

COUNTERSPACE AND POTENTIAL ANTI-SATELLITE (ASAT) SYSTEMS

To understand an area as complex as counterspace and ASAT systems requires an examination of both the larger political environment concerning their potential use as well as some of the newer changes in potential space weaponry design. There has been a long-standing debate about the weaponization of space, with many experts, ourselves included, questioning its utility in the long run.[65] The general understanding of space weapons, including ASAT weapons, is quite broad, in that they can very well be "any system whose use destroys or damages objects in or from space."[66] Such capabilities run the gamut from mere denial and deception (for example, camouflage), electronic warfare (for example, satellite signal jamming or interference), satellite ground station attacks, satellite sensor blinding or dazzling, and LEO satellite attacks with pellet clouds to direct satellite attacks with sophisticated microsatellites and hit-to-kill weapons (for example, missiles fired from the ground, planes, or orbit) and even high-altitude nuclear explosions.[67]

Importantly, the use of such weaponry is not just limited to military and intelligence satellites but can also be used against commercial satellites and civil assets, thereby crippling both political and economic security. With the high degree of dependence on satellites, especially in the advanced industrial countries, this prospect certainly raises legitimate concerns. With this as a departure point, we examine the potential for Japanese capabilities in this area by placing the country's space technologies in the context of developments worldwide. The evidence suggests that Japan's long history of developing dual-use space technologies can now be used for defensive counterspace, ASAT, and possibly even offensive counterspace systems. We focus on two sets of potential Japanese capabilities in particular—destroying satellites with missile interceptors and damaging them with other satellites.

Direct-Ascent Missiles

In Chapter 4 we highlighted the fact that the long trajectory of Japan's rocket technologies over the postwar period also means that Japan today certainly has the capability to develop Intermediate-Range Ballistic Missiles (IRBMs) and Intercontinental Ballistic Missile (ICBM) technologies. The flip strategic side to this is also the ability to shoot down such missiles. In March 2009, for example, faced with yet another *Taepodong* ballistic missile launch (or, as the North Koreans claimed, an SLV launch for a satellite test), Japan's leaders authorized the shooting down of the missile or other object if it was deemed to threaten the country.[68] As a precautionary measure to detect, track, and intercept the North Korean projectile when it was actually launched in early April 2009, Japan had activated its Ballistic Missile Defense (BMD) shield for the first time. Japan deployed Aegis destroyers equipped with the Standard Missile-3 (SM-3) interceptors, units for intelligence and surveillance operations, and also air defense units equipped with PAC-3 Patriot missiles. When asked point-blank about whether Japan was capable of actually shooting down the missile or any such object, Defense Minister Yasukazu Hamada responded that Japan had obviously been preparing to do just that, and he had little doubt that Japan could do it. In light of several missteps by MOD, which included publicly embarrassing false alarms, Hamada may have been a little optimistic. But, as explained briefly below, such optimism does have a basis in technological reality.

The connection between Japan's SM-3 missiles and an ASAT system requires a brief detour. The use of ASATs has certainly drawn more attention on the world stage in recent years, especially in the aftermath of China's much-publicized use of them to destroy one of its own aging weather satellites in January 2007.[69] Nor was this China's first use of such weapons, with reports suggesting that the latest ASAT test actually came after three failed ones that had taken place between September 2004 and February 2006. In September 2006, more ASAT tests by China against U.S. space assets were also reported; this time China had fired ground-based lasers in an effort to demonstrate the ability to blind U.S. spy satellites flying over its territory.

The United States, increasingly sensitive to its own space security, responded most concretely a year later in 2008 by shooting down one of its own falling National Reconnaissance Office (NRO) satellites in order, the official explanation went, to prevent its fuel tank full of hydrazine from harming human life back on Earth.[70] This was a high-profile move after roughly a twenty-year hiatus in which the United States had officially banned ASAT testing. The United States had last shot down a satellite in September 1985 when an F-15-launched experimental ASAT missile also intercepted and destroyed an obsolete spacecraft. Whether right or wrong, the U.S. actions in 2008 were widely seen as an ASAT test because many experts doubted that the falling satellite actually represented any real danger and because U.S. actions came close on the heels of the full-scale Chinese test.

Japan too was watching this belligerent exchange, and no doubt defense planners in the then only days old MOD after China's ASAT test must have also come to the same conclusion about the vulnerability of the country's space assets. We will come back to BMD in Chapter 6, but here we highlight a potential way in which Japan already has access to the basic technology for an ASAT weapon with the U.S. demonstration. The United States brought down its errant spy satellite with the very same navy missile that forms part of the U.S. Navy's Aegis BMD system, namely the SM-3. Raytheon, the defense company responsible for designing and building that missile, had temporarily modified three SM-3s (and three ships) to shoot down the satellite at an estimated cost of $30 to $40 million. Raytheon's adaptation showed not only specifically that the SM-3 technology was adequate enough to intercept a very high-velocity target and at a higher altitude than that for which the missile was originally designed and built.[71] It also showed, how a defensive BMD system could be relatively easily reconfigured into an offensive ASAT weapon for destroying a satellite.[72]

This missile is also of course part of the Japanese defensive BMD program and one on which Japan has been cooperating with the United States.[73] In 2005 the Japanese government made a decision on joint development, and in June 2006 the United States and Japan started the SM-3 Block IIA Cooperative Development (SCD) project for which Mitsubishi Heavy Industries (MHI) is the prime contractor on the Japan side. As of 2009, Japan has had one successful firing test with the SM-3. Both Japan (MHI and MOD) and the United States (Raytheon and MDA) have finalized the joint system design review for the next-generation SM-3 missile and can proceed with co-development. The new SM-3 is expected to provide greater engagement and warfighting capabilities against ballistic missile threats. If the past is a guide, with modifications the new SM-3 could also double as an ASAT weapon against satellites and other projectiles for both the United States and Japan.

Microsatellites

Having shot down their national satellites in a global drama, the United States and China stood warned; each understood that actually destroying another nation's satellite or spacecraft is the same as sinking a ship or downing an airplane and, as such, tantamount to an act of war. If such a warmongering act did occur, there stood to be not just military but also huge economic losses on both sides. And yet each side, and all other spacepowers, also understood that they could not afford to blithely assume their space assets would never be targeted in the future. If anything, the open Chinese actions and U.S. reactions only focused attention on alternative ASAT weapons.

As outlined above, direct-ascent ASAT weapons such as those used openly by China, and now the United States, represent only one possible facet of counterspace efforts. There continues to be a range of destructive (as opposed to merely jamming) ASAT systems, which includes high-altitude nuclear explosions as the ultimate option.[74] Although impressive, the Chinese and U.S. direct-ascent ASAT tests, using missiles equipped with kinetic-energy vechicles, are technologically among the more doable ways to destroy a satellite. There are also ground- or space-based laser ASAT weapons that could take out satellites in LEO. Despite decades of research, however, their battle-readiness continues to be marred with problems of attenuation, dispersal, and concentration of any such beams as they pass through atmospheric distortions; adaptic optics, such as that incorporated in the U.S. Starfire system, which seek to narrow and thus increase the density of a beam, remain a work in progress.

At present, the high ground of ASAT weaponry rests on using precision maneuverability and advanced guidance technology to make satellites, particularly hard-to-detect microsatellites.[75] In the early 2000s, the United States had expressed outright concerns about the advances in satellite miniaturization, particularly dual-use micro- or nanosatellites ranging from 10 to 100 kilograms with rapid development timeframes of six to thirty-six months. These would be low-cost, and thus attractive to university researchers, smaller spacepowers, and governments. The micro- or nanosatellites would produce lightweight, capable systems to conduct what are now called "autonomous proximity operations" around other satellites in space.[76] On the commercial side, this meant they could rendezvous and dock, inspect, maintain, repair, refuel, resupply, salvage, rescue, reposition, de-orbit, service, and remove debris concerning other satellites. However, on the military side, they could also potentially intercept, image, inspect, jam, dirupt, disable, damage, ram, smash, destroy, or bump other satellites from orbit. The offensive uses of this technology may well allow a country to move from mere space control to space superiority. Although the most cutting-edge programs involve smaller microsatellites, what goes for them also goes for larger satellites and component technologies. As a space militarization expert put it quite clearly, it is worth understanding in full what satellite-based ASAT technologies might entail, whether big or small:

> With regard to . . . microsatellite programs, any microsat or other sized sat that can perform autonomous proximity operations around another satellite or that could dock with another satellite, for any reason, could be kitted out to serve as an ASAT. It could use kinetic energy simply to run into the target, or it could be equipped with RF or laser jammers/blinders (providing they can miniaturize the jamming technology to fit the sat). It could even be kitted out with explosives, but . . . that would get heavy and might require a larger sat. The technology for these sorts of sophisticated maneuvering sats is decidedly dual-use. This is why the USAF is so terrified of the Chinese work with Surrey Sat. Inc. to develop microsat technologies. This fact presents a bit of a conundrum for space policy-makers and for military space strategists. Because things like inspection or repair of a satellite could be very, very useful, and in some cases even work to dampen a crisis (if you can know for sure that your satellite wasn't attacked but suffered a malfunction)—but at the same time, they raise suspicions and fears . . . how can anyone else tell that your inspection satellite isn't secretly a microsat ASAT?[77]

There is sure to be a commercial side, though it has yet to prove its actual utility.[78] The idea of an Orbital Maintenance System (OMS) involving, for example, de-orbiting, refueling, and repairing, has certainly been seen as a business opportunity by market players. In Europe, there was the London-based Orbital Recovery, which developed the ConeXpress Orbital Life Extension Vehicles (CX-OLVEs). CX-OLVEs were to have attached themselves to existing motor fixtures of geosynchronous telecom satellites and boost them back into their orbits.[79] This enterprise morphed into the Orbital Satellite Services, with the renamed Smart Orbital Life Extension Vehicle (SMART-OLEV), which offered the promise of extending the life of large telecommunications satellites by helping them maintain their proper orbital slots. Having established its corporate headquarters in Sweden in 2008, Orbital Satellite Services then awaited orders in order to finalize the design of the vehicle.

In the United States, similar sounding efforts have also taken place but have always had military overtones.[80] There was the Orbital Express program, run by the Defense Advanced Research Projects Agency (DARPA), which sought to deploy a robotic spacecraft to service and refuel U.S. spy satellites, and thus in theory, allow a constellation of twelve radar surveillance satellites to do a job that otherwise would require twenty-four spacecraft. It was not always clear that DARPA would prevail. After all, the Demonstration of Autonomus Rendezvous Technology (DART) experiment was only partially successful, given the fact that the guided DART vehicle accidentally bumped into the retired spy satellite it was supposed to approach (although it did demonstrate satellite orbits could be disrupted as well).

The general goal of the Orbital Express project was to demonstrate the technical feasibility of autonomous on-orbit refueling and reconfiguration of satellites for national security and commercial purposes. In April 2007, tests began involving the servicing satellite named the Autonomous Space Transport Robotics Operations (ASTRO) and the serviceable satellite (NextSat). The autonomous proximity operations were all pronounced a success, referring to refueling, battery replacement, unmating-flyapart-remating, circumnavigation, and grapple-and-capture by robotic arm. In July, according to DARPA, having completely met their mission success criteria, the two satellites were decommissioned. The pronounced success and lessons from the Orbital Express program may also cross over into other similar satellite-based programs in the United States, if for no other reason than as a technical boost. The other well-known programs deserve attention, once again less for their commercial value and more, at least at this stage, for their military usefulness.

These include, for example, programs for which funding continues to be requested: the Near Field Infrared Experiment (NFIRE) program, run by MDA, involving a maneuvering satellite that was stripped of its kill vehicle but continues experiments with fly-bys near targets; the Experimental Satellite System (XSS) program, run by the U.S. Air Force (USAF), for which two microsatellites were already launched in 2003 and 2005 to conduct autonomous proximity operations and which may involve space-based kinetic and/or directed energy for offensive counterspace missions in LEO; the Autonomous Nanosatellite Guardian for Evaluating Local Space (ANGELS) program, also run by the USAF, which focuses on using nanosatellites to improve space situational awareness for host satellites and which may incorporate improved threat warning sensors to protect GEO and LEO satellites; and the Front-end Robotics Near-term Demonstration (FREND) program, run by DARPA, which is another advanced robotic manipulator technology satellite combined with detailed stereo photogrammetric imaging and which is designed to interact with, service, reposition, or decommission GEO-based military and commercial spacecraft.

Put in the light of the brief background above, some components of Japan's civil satellite technologies may also be thought of in a very different way than the official version. The first glimmers of this appeared with the introduction of the OMS concept by Japan dating back to at least the mid-1990s, which, although tenuous at that stage, nevertheless still had a powerful allure, both for the commercial and military spheres.[81] The commercial prospects for Japan's public OMS movement were based on the loss of the *ADEOS/Midori* satellite in 1997 and reinforced also by the loss of *ADEOS-II/Midori-II* in 2003.[82] In the aftermath of these losses, it is not hard to imagine that supporters of OMS were easily able to conjure up a market. By the early 2000s, CRL was solidly investigating a government-funded OMS system; as elsewhere, the laboratory too emphasized its commercial aspects with a focus on rescue and removal as well as satellite-servicing, primarily for large telecommunication satellites.[83] In one of its most ambitious incarnations, OMS evolved into an innovative concept called OMS for *Next Generation Leo System* (*NeLS*), which went through several iterations. In one version, OMS for *NeLS* was proposed as having a small satellite-based maintenance system for a 120-satellite *NeLS* constellation, involved in such services as de-orbiting dead or malfunctioning satellites. The OMS concept later helped fuel the *Micro Labsat 1*, which was successfully launched in 2002, and the *SmartSat* program, which was shut down in 2008. As it is, the OMS concept remains a highly challenging innovative technology program but, to our knowledge, researchers have yet to release any, let alone convincing and thoroughly argued, business models for the commercial utility of the program.

In light of current understanding about the programs discussed above, Japan's OMS emphasis could easily be construed from a military point of view as a dual-use potential ASAT technology development program. It is important to know that there are literally dozens of projects in Japan today seeking to develop micro-, nano-, and picosatellite technologies, all of which might be thought of as a grassroots, bottom-up movement to speed development. We will examine the phenomenon concerning such satellites in Chapter 6. For the present, the following section is limited to a number of the most high-profile, historic, and government-funded initiatives to develop inherently dual-use on-orbit satellite technologies. To be sure, Japan has not demonstrated nor even remotely hinted at any sort of full-blown ASAT technology tests. But as discussed below, component by component in its civil satellite program, Japan already has considerable expertise in the new high ground of ASAT systems—involving autonomous co-orbital and proximity operations—whose commercial value is less clear as compared to military prospects at least at this stage.

The *ETS-VII/Kiku-7* Japan's strategic plans to develop autonmous rendezvous and docking capabilities, as well as manipulator control competence for future robot satellites go back to the early 1980s- representing an astonishingly ambitious target.[84] At that point, the last time such docking had taken place in space had been by the United States in 1966 and by the then Soviets in 1968. Remote manipulator control competence was also novel. The Russians had certainly pioneered automatic, remote-controlled rendezvous and docking systems for the Mir space station, but even they had allowed for the possibility of human involvement from space.

The resultant roughly $300 million *ETS-VII/Kiku-7*, launched in November 1997, spent eighteen months testing safe approach and docking technologies necessary to enable the *H-IIA Transfer Vehicle* (*HTV*) to dock with the ISS later.[85] The *ETS-VII/Kiku-7* was an extraordinary experiment in that it consisted of a satellite that was designed to divide into two twin parts, fly apart, and recombine in order to test remote controlled and automatic docking technologies, thus de facto testing elements of complex defensive and offensive counterpace technologies. In JAXA's own description, one part was a box-shaped main "chaser" bus, and the other was a much smaller square panel-shaped "target" satellite. For publicity value, the chaser-target satellites were named *Orihime* (target) and *Hikoboshi* (chaser), famous in the Tanabata festival as lovers who were separated by an angry God and could only meet once a year on the seventh night of the seventh lunar month.[86]

Seven maneuvers were planned, which required close visual and data monitoring to prevent possible collisions. Replete with its own internal navigation systems and senors, the *Orihime-Hikoboshi* system was able to compute the relative positions and speeds of the separated satellite units from 9 kilometers out to docking. This system also included close inspection and circling, miss and hit-abort and an "r-bar" approach, where the chaser would maneuver into position from below and behind the target. Even more astonishingly, the *ETS-VII/Kiku-7* conducted and advanced Japan's future capabilities for unmanned in-orbit construction and repair capabilities. It had a stated focus on large antenna assembly technology, carrying a 2-meter, 140-kilogram Toshiba-built main robot arm to be used for refueling and orbital repair simulations, as well as a series of smaller experiments by national research institutes and government agencies. The satellite mounted a then MITI-sponsored Fujitsu-built advanced robotics hand and an NEC antenna assembly mechanism for the CRL.

The program in general and the satellites in particular provided high drama for all concerned.[87] The mission fought through the loss of the *COMETS/Kakehashi* satellite that was placed in a faulty orbit and that was to be used for data relay purposes between the ground station in Tsukuba and the satellite. Eventually NASDA successfully substituted the U.S. Tracking and Data Relay System Satellite (TDRS), despite the fact that the *ETS-VII/Kiku-7* was not designed for it. There were also problems caused by the satellite's extremely complex system of computers.[88] The satellite used seven computers with advanced 32-bit microprocessors and also used an order of memory and computing power that put it in a different league compared to prior NASDA satellites. But the satellite, to the credit of its designers, was designed to be extensively programmable.

Most importantly, despite some mission hitches that were corrected, the *ETS-VII/Kiku-7* worked.[89] On 7 July 1998 (in line with the Tanabata myth) *Orihime* and *Hikoboshi* conducted their first automatic separation and docking experiment, which NASDA claimed was the first in the world followed by medium (500 meters) and long range (2,000 meters) follow-ups. The *ETS-VII/Kiku-7* teleoperations robotics experiments were also successful, including change-outs with orbital replacement units. Athough the mission was pitched as a romantic galactic saga thanks to the Tanabata festival reference, it is a fact that Japan publicly tested technologies that could double for defensive and offensive counterspace and ASAT technologies in full view of the world and, to our knowledge, nearly a decade before U.S. versions.

The *Micro LabSat* (*Micro-OLIVe Experiment*) The *ETS-VII/Kiku-7* program represented a complex mission that yielded target, approach, maneuver, and docking from relatively short distances in LEO orbits. Following this, the *Micro LabSat* program, which consisted of a microsatellite that flew on a piggyback mission, tucked under the *ADEOS-II/Midori-II*, was from our point of view a critical paradigm-shifting technology-platform enabler.[90] For more than thirty years Japan had spent considerable effort learning how to build bigger and more complex satellites; now, in line with technology and policy trends, it needed to learn how to also build small ones. What is clear at this stage is that with technology now opening up whole new vistas, the basic frame and mission descriptions below show that the idea of ever-smaller satellites has become increasingly institutionalized in Japan's government agencies.

Micro LabSat 1 made that paradigm shift possible in the early 2000s.[91] The stated purposes behind its development were to nurture younger engineers, acquire small satellite bus technology, and obtain high performance at low cost and short time frames. Seen by its makers as a first step toward spreading small satellite technology that was inexpensive and mass produced with commercial-off-the-shelf (COTS) components, it also helped spawn research and development (R&D) cooperation between NASDA, CRL, NAL, Toshiba, and the University of Tokyo. *Micro LabSat 1* was merely a 70-by-50-centimeter octagonal-shaped 50-kilogram microsatellite that flew in an 800-kilometer orbit. More remarkable than its dimensions was the fact that it helped demonstrate that Japan could equip its satellites with technologies for autonomous capabilities, something which is considered the most important and fundamental issue in space robotics.

One of the missions on board *Micro LabSat 1* was named *Micro OMS Light Inspection Vehicle* (*Micro-OLIVe*).[92] The CRL, which was studying and pushing an on-orbit satellite maintenance system, devised a technology program that required three steps: an OMS Light Inspector satellite with a high level of autonomy to find, recognize, rendezvous, inspect, and image a target; an OMS Light Re-orbiter designed to do all this and capture the target; and an OMS Light Repairer that was supposed to have the ability to repair its target. *Micro-OLIVe* thus acted as a pre-demonstration of the software and hardware technologies for autonomous image-processing technologies, including a multichip module for controlling the OMS, camera units, and software. As an actual on-orbit experiment, *Micro-OLIVe* tested an image-processing computer as well as an inspection-monitoring camera, and the mission was judged successful. At this stage at least, it appears that small satellite technology is being promoted specifically in the context of servicing or repairing satellites. This emphasis is in line with the OMS concept discussed above. But as the microsatellite-based counterspace paradigm made possible by *Micro LabSat 1* shows, it also signifies military uses in which themes of space situational awareness (SSA) loom large.

The *SmartSat-I* and *II* Until it was restructured, a move that contributed to the resignation of the project's leader, we believe the *SmartSat* program would have represented Japan's first direct demonstration of critical ASAT technologies using small satellites.[93] The restructured mission, which is of one satellite, consists of a communications experiment and an observation experiment looking at the sun's corona. As of 2008, the official reason for the restructuring (or perhaps stalling of the ASAT technology demonstration function) is lack of funds. However, whether it continues or morphs into a similar program, the elements of this program deserve attention given the country's increasing emphasis on OMS.

Led by NICT and MHI, *SmartSat* was to combine *ETS-VI/Kiku-7* and *Micro LabSat* capabilities in stages:[94] *SmartSat-1* was a 200-kilogram-class mini satellite consisting of a 150-kilogram *SmartSat-1a* (chaser service satellite), and one 50-kilogram-class microsatellite called *SmartSat-1b* (target satellite) that could, given their size, be launched piggyback style into a geosynchronous transfer orbit (GTO)—a highly elliptic orbit useful for counterspace applications. The basic public objective of the program was to develop autonomous rendezvous technologies using microsatellites. *SmartSat-1* was to have been equipped with devices for image acquisition for target recognition and visual guidance. The program was designed to work in a series of steps, including target acquisition, rendezvous, fly around, and capture and removal of the satellite from the orbit. From a military perspective, this would mean redeveloping the *ETS-VII/Kiku-7* into a modern counterspace technology paradigm. Put simply, the experiment was to have featured autonomous maneuver planning in which the chaser satellite autonomously developed a plan of action to reach the target satellite based on the orbit position and the detected relative target location.

As with other small satellite programs under way under the rubric of OMS, the *SmartSat* program can be seen as a high-utility repair system or an aggressive ASAT technology demonstrator.[95] We believe the restructuring of the *SmartSat* program should not be taken as an indication of a lack of official interest in microsatellite development. In fact, Japan's next-generation ORS program is going to be built on small satellites. In this context, the building and launching of microsatellites has been officially encouraged and funded in the new Basic Space Plan 2009.[96] It is also instructive to know that MOD's newly minted Committee for the Promotion of Outer Space Development and Use (CPSDU) used its concerns with China's ASAT test in January 2007 to mark satellite protection as one of its military space policy objectives, and urged the importance of investigating launching small satellites across government agencies with its own needs in mind.[97]

The *Mu Space Engineering Spacecraft* (*MUSES-C/Hayabusa*) There are other discrete components of Japan's civil program that have not attracted much attention but that continue to advance Japanese potential for ever more sophisticated and autonomous robotic operation capabilities. On point is the *Mu Space Engineering Spacecraft* (*MUSES-C/Hayabusa*), equipped with autonomous navigation and a newly minted ion engine.[98] It began its voyage to near-Earth asteroid 25143 Itokawa in 2003. Although the mission has had significant problems, the *MUSES-C/Hayabusa* did arrive at the asteroid in 2005 using autonomous navigation; it touched down and is now en route back to Earth for a projected arrival date of 2010. It is a significant technological achievement, no matter what the delays or problems. To facilitiate its scientific observations, it was equipped with such things as a telescope wide-view camera for taking images and a laser pulse for detecting and conducting range measurements. It was also designed to pick up asteroid samples during its thirty-minute touchdown on the asteroid, but this is to be verified on return.

CONCLUDING ASSESSMENT

Throughout this chapter, we have attempted to situate the flow of Japan's civilian and spacecraft technologies in the context of evolutionary conceptual shifts that affect their progress and their acquisition worldwide. Japan is no exception in today's global space trends. On some dimensions, in fact, it is an innovator. Taking our cue from the RMA framework as outlined at the beginning of the chapter, we began first with a look at space-based assets as critical national security components for information, surveillance, and communications information, surveillance, and communications in peacetime and as force multipliers in combat. From there, we also traced the follow-on necessity of protecting the infrastructure of those very same assets that is pushing the world, including Japan, into as yet undeclared contests in space. The importance of such an infrastructure to an IT-RMA framework has been aptly stressed as the "eyes, ears, and voice" of American power but applies with equal measure to all other aspiring space powers, including Japan:

> What are in space are the sensory organs, which find and fix targets for [military] forces, and the nervous system, which connects the combatant elements

> and permits them to operate cohesively. These assets permit [military] forces to detect and identify different kinds of targets; exchange vast and diverse militarily relevant information and data streams; and contribute to the success of combat operations by providing everything from meterological assessment, through navigation and guidance, to different platforms, weapon systems, and early warning and situational awareness.[99]

As discussed earlier, the Kawamura initiative that set Japan more openly down the path toward space security consciousness as never before also placed a strong emphasis on Japanese capabilities for seeing and hearing preemptively. Today, Japan certainly has those capabilities, and we believe these are only likely to get stronger amid the new geopolitical concerns. Some sense of the scope of such capabilities can be gleaned from Figure 5.1, which, if nothing else, leaves a strong impression of interest in acquiring progressive satellite series.

From *Ohsumi* onward, Japanese agencies and corporations have been busy with acquiring an impressive toolkit of cutting-edge technologies that are now parlayed in the country's civilian EO and communications satellite programs. On both fronts Japan has forged ahead individually; in some cases it has also partnered up with Europe and the United States in development and testing. Only these painstaking steps have allowed and prepared Japan to think about deploying a national security space infrastructure at the present stage on the scale of that represented in Figure 5.1. In the absence of clear commercial markets, discrete elements of Japan's spacecraft development have been incrementally militarized in the interest of national defense, equipping the country not just for reconnaissance, but also communication-, navigation-, and meterological-based support for military operations, and from there even ASAT technologies.

The saga of the spy satellites, particularly, illustrates our market-to-military themes quite clearly. Japan's *IGS* satellites had a clear heritage in Japan's EO technological base and, more specifically, the *ALOS/Daichi* satellite. Although not representing the cutting edge of military capabilities at this stage, they demonstrate the duality of space technologies. As it turned out, some technologies that were designed to monitor and survey the Earth's land, ocean, and air phenomena could also, as the actual construction of *IGS* technologies indicated, be transposed for military purposes. The birth of the *IGS* was an illustration of the ways in which civilian-use technology built by corporations, when developed appropriately, could act as direct precursors to military versions.

Figure 5.1. Japan's Military/National Security-Related Satellite Development Plan, 2010–2020

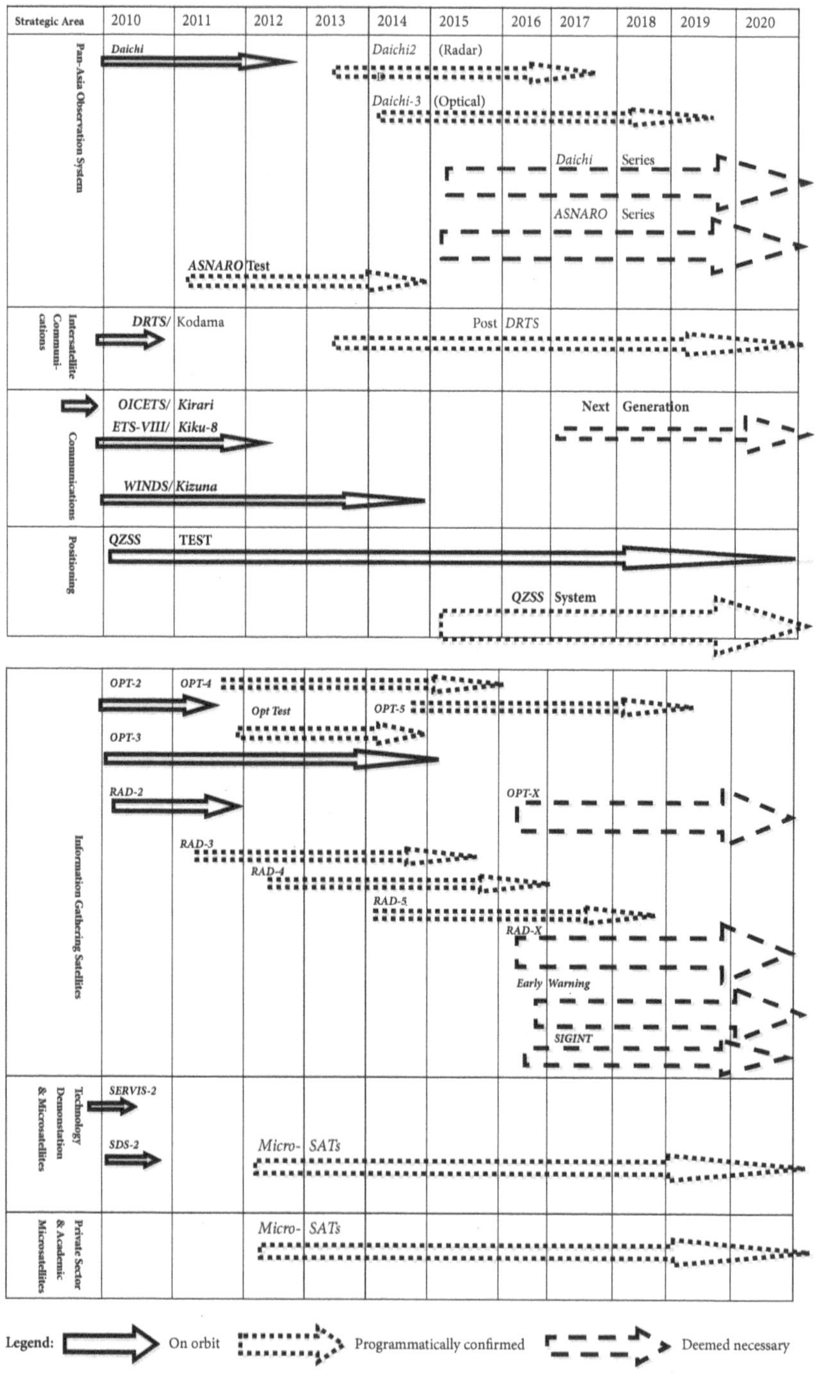

SOURCE: Taken from SHSP, *Uchū Kihon Keikaku; Nihon No Eichi ga Uchū o Ugokasu [Basic Space Plan: Wisdom of Japan Moves Space]*, Tokyo: SHSP, 2 June 2009, appendix 2.

Japan's communications satellites have also commanded attention in the new high ground of military-related communications, namely extremely high bandwith and unjammable optical inter-satellite links, which are of keen interest to militaries around the world. From the *ETS-VI/Kiku-6*'s demonstration of the world's first successful laser-based two-way space communication experiment to the *OICETS/Kirari*, which marked the world's first successful bi-directional optical inter-satellite link experiment, Japanese satellites have remained abreast and sometimes ahead of the most cutting-edge research and development. Japan's new space communication infrastructure also shows a regional focus on the Asia-Pacific region. The *ETS-VIII/Kiku-8*, demonstrating bus technologies for Japan's first large geostationary satellite, is designed to enable direct communication covering all of Japan and wide areas of Asia with a single geostationary satellite. Additionally, it is designed to obtain technologies for a basic satellite positioning system. The *WINDS/Kizuna* satellite recently carried out the world's fastest satellite data communication speed test to date and also demonstrated Japanese capabilities to do so without large-scale ground stations around Asia.

The recent and open test of destructive direct-ascent ASAT technologies by both China and the United States also necessitates a focus on Japan's capabilities in this area. It is a fact that in building its defensive BMD program with the United States, Japanese corporations have been cooperating on the development of the SM-3—the very same missile that the United States used, with modifications, in its ASAT test. Japan and the United States are also now cooperating on building a more advanced version of this missile. More impressive is the fact that the high ground of ASAT technologies today also finds resonance in the long trajectory of Japan's civilian efforts toward autonomous and rendezvous capabilities. Astonishingly, in the *ETS-VII/Kiku-VII* Japan tested counterspace and ASAT technologies dressed as a romance myth in full public view. From there, Japan has already taken solid steps toward miniaturization of these technologies with the *Micro LabSat 1/Micro-OLIVe* programs. Had it not been stopped, the *SmartSat* program would have constituted a highly aggressive dual-use ASAT technology demonstration program. The point is, the wherewithal to put this together is on the benches of Japan's leading space communications laboratory; all it needs is assembling. To some extent, these technologies were also on display in the *MUSES-C/Hayabusa*, the asteroid probe that is now en route back to Earth. Over time, these capabilities were acquired under the rubric of OMS, with the stated commercial purpose of servicing and repairing on-orbit satellites with operations such as visual inspections, imaging, equipment change-outs, de-orbiting, refueling, and so on. But, in keeping with current understandings, OMS can also be seen as a thin veneer for a state-of-the-art ASAT program. As explained in the next chapter, Japan's highly dispersed but very active research on micro-, nano-, and picosatellites is being ramped up while ORS and space control technology development have become official.

With MOD now directly mentioning satellites, it is likely that the technologies in these space assets are going to play an important role in Japan's military space infrastructure in the near future. From a national standpoint, they are a legitimate means of securing Japan's security concerns given the multilayered threats continuing to stem from North Korean behavior.[100] Just as the firing of the *Taepodong* legitimized the indigenous development of the *IGS*, so the continued missile firings by North Korea at present have also sanctioned open moves toward other elements of strategic defense—not just missile defense, or space surveillance and ASAT systems, but also early warning systems. In the next chapter we set out the directions of these emerging technologies and others to show that the development of Japan's SLVs and spacecraft technologies has not reached a plateau.

6 EMERGING TECHNOLOGIES

IN THIS CHAPTER we turn to some of the emerging technologies that have evolved directly from Japan's long experience with rockets and satellites. The chapter is divided into two main parts, one focusing on rockets and space launch vehicles (SLVs) and the other on satellites and spacecraft. We are mindful that ongoing programs may not see fruition or may evolve in new and different directions; moreover, not every space-related technology development program of interest can be forced into our thematic focus. We therefore focus on ways that the emerging technologies discussed below speak to the market-to-military trend that characterizes Japan's space policy more openly than ever before.

ON THE ROCKET FRONT

We concentrate here on developments related to next-generation SLVs, such as the *H-IIB*, *Galaxy Express* (*GX*), the *Advanced Solid Rocket* (*ASR/Epsilon*), and those resulting from Ballistic Missile Defense (BMD). We also highlight some additional aerospace activities, such as the work on reusable launch vehicles (RLVs) and scramjet engines—all of which is keeping Japan on the forefront of research and activities around the world.

Beyond the *H-IIA*

The departure point for understanding Japan's SLV directions in the future has to begin with the reality of the present launcher, the standard *H-IIA*. As discussed in Chapter 4, Mitsubishi Heavy Industries (MHI) completely took over the production, management, and commercialization of the *H-IIA* from the Japanese Aerospace Exploration Agency (JAXA) formally at the start of April 2007. Despite the *H-IIA*'s high level of technological accomplishment, its growing reputation for reliability, and ongoing cost cutting, the realities of the launch market mean this superb launch vehicle may have a difficult time attracting many commercial customers around the world.

Looking forward, MHI has upgraded the *H-IIA* to the *H-IIB*, Japan's first two-engine clustered booster, which uses two LE-7A engines on a thicker first stage to yield a much heavier launch capacity of 8 tons to geostationary transfer orbit (GTO) for maximum launch capacity.[1] The *H-IIB* was successful in its September 2009 maiden voyage, successfully lofting the first demonstration flight of the *H-II Transfer Vehicle* (*HTV*), Japan's unmanned cargo transfer spacecraft for the International Space Station (ISS).[2] The *H-IIB*'s larger launch capacity will allow it to launch more than one satellite simultaneously, which is a useful sales point for the heavy launch market. In the tradition of lowering development costs and breaking into the international launch market, the *H-IIB* builds on the *H-IIA*'s second-stage, Solid Rocket Booster-As (SRB-As), and guidance control systems. Given the new design, the rocket was subjected to extensive testing.[3] However, in the same vein as the *H-IIA*, despite the likelihood that it will prove to be of excellent design, it is unlikely that the *H-IIB* will have a significant competitive advantage in the heavy launch market anytime soon.

It may well, however, spur on cargo missions, not just to the ISS but also the moon. MHI would like to continue the *H* series, including a possible *H-III*, replacing the second-stage LE-5B with the joint Boeing-MHI MB-XX engine that produces twice the thrust, and possibly reengineering the LE-7A.[4] Beyond that, the *H-X* design would be a redeveloped version of the *H-III* capable of launch on demand. Either of these new vehicles would have a minimum of 10 tons to GTO and 20 tons to low Earth orbit (LEO) capacity. More capacity would be added by clustering rocket cores together.

Perhaps the most interesting aspect of these plans are comments made by MHI officials that speak to Japan's lunar ambitions or manned space policy.[5] Specifically, they suggest how easy it would be for the company to upgrade the *H*-series technology to carry astronauts. In their view, this is indicative of the astonishing array of technologies Japan has already accumulated to make manned spaceflight possible—a prospect that could be made more concrete if only an appropriate budget was made available by the government. Certainly the *Japanese Experiment Module* (*JEM/Kibō*) aboard the ISS, and the *Seleno-*

logical and Engineering Explorer's (*SELENE/Kaguya*) lunar study (billed as the world's largest lunar missions since the U.S. Apollo program) deserve attention in this respect.[6] MHI officials believe that it would take as few as five years working at a cautious, bureaucratically controlled pace to upgrade the *H* series for that purpose. The pressure for Japan to build a manned space program has grown in recent years for a number of reasons, the most prominent of which is China's manned space activities.[7]

From *J-I* to *J-IU/J-II* and Then *GX*

The tortuous development and eventual cancellation of the *Galaxy Express* (*GX*) rocket in December 2009 is a long and complicated story that demonstrates how industry pressures and emerging priorities can transform official government policy. Starting in 1995, the National Space Development Agency of Japan (NASDA) made moves to replace the *J-I*—which combined technologies from both the *H-II* and *M-3SII*—with a much more cost-effective vehicle for smaller payloads.[8] In 1998 the *J-I* program was delivered a fatal blow by the Management and Coordination Agency (MCA), the Japanese government's main auditing agency at the time, which stated outright that unless costs could be cut on both the *M-V* and especially the *J-I* programs, they should be suspended. Given the concerns about the *J-I*'s launch price, officials set specific criteria for reducing development and per-launch costs. The *GX* thus originated in the eventual cancellation of the *J-I*.

Mindful of its duty to fund technology programs, the Science and Technology Agency (STA) had in fact already conducted its own reviews in 1997 and, by the time of the MCA investigation, was looking for a *J-I* replacement. In May 1997 the STA set an informal target of developing a vehicle that would be able to launch 1 to 2 tons to LEO at a cost of $10 to $20 million a launch (compared to the *J-I's* nominal cost of $35 million per launch for a 900-kilogram-to-LEO payload, according to estimates using U.S. dollars at the time).[9] Following the MCA report, the STA saw Japan's options for a new launcher as, among others, developing a completely new vehicle that used proven foreign components, upgrading the *M-V*, launching smaller payloads as piggyback on the *H-II*, or simply relying on proven foreign launchers.[10] This opened a three-way competition for the *J-IU* (*U* for "upgraded") vehicle. Nissan offered an SRB-A first stage with several options for the second stage, among them one based on a methane or solid propellant; Ishikawajima-Harima Industries (IHI) offered a liquid booster design, powered by a Russian liquid-kerosene engine marketed by a U.S. company; and MHI countered with the adapted application of its own LE-5B engine, then being improved for the second stage of the *H-IIA*, for the first stage of the *J-IU*. It also incorporated the use of an existing second stage from Boeing. Eventually NASDA chose an IHI-led consortium for the *J-IU* (also called the *J-II*) that was expected to fly in 2005 and to enter service in 2006. For the first stage, the approved design at that time included a Russian-designed NK-33 engine supplied by Aerojet, which is highly rated for its cost, convenience, power, and throttling ability; and hardware supplied by Lockheed Martin, including lightweight fuel tanks used for its Atlas family of launchers. Meanwhile, NASDA said it needed the rocket to launch a small experimental NEC-made satellite called the *Mission Demonstration Satellite* (*MDS-2*).

As discussed in Chapter 3, the *J-I* replacement was seen by IHI as a chance to join MHI as a complete rocket systems integrator.[11] To exploit the opportunity afforded by the *J-IU*, IHI made two major moves.[12] First it formed IHI Aerospace in 2000, buying up Nissan's space business altogether in that year. In 2001 it established Galaxy Express Corporation (GALEX) to coordinate the development and manufacture of the *J-IU*, which then became the *GX*. More importantly, GALEX focused on marketing the *GX* launch services, a move advocated as international cooperation with the participation of Lockheed Martin as a shareholder.[13] As a private venture, the construction of the *GX* rocket was eventually led by IHI in partnership with these companies. The *GX* rocket itself was an expendable, medium-sized SLV standing about 48 meters long and measuring 3.3 meters in diameter. It was originally estimated to have about one-third the liftoff capacity of the *H-IIA* rocket, capable of launching about 4 tons into LEO orbit (200 kilometers), 2 tons to Sun Synchronous Orbit, or SSO (800 kilometers), and 1.8 tons to GTO. In terms of the actual launch market, the focus was on mobile communications, Earth observation (meteorological, land, sea, and information gathering), navigation, space science, and planetary missions were forecast.

The development of the *GX* after 2001 was slowed by several factors. One was budget, and the second involved continuing technical difficulties. IHI's original understanding was that the Japanese government would underwrite one-half to two-thirds of the total development cost, a ratio that was disputed by the government.[14] Meanwhile, beginning in 2002 the Space Activities Commission (SAC) began a long review of the *GX* program that eventually became a radically different design, one that basically broke all the original parameters for the *J-IU* laid out by the STA. In fact, when the SAC finally authorized the program in March 2003, it was for a design that was completely different from both the original request for a quick, cheap launcher and even the original design proffered by IHI.[15] The 2003 green light was dependent on a compromise between then NASDA (then JAXA) and GALEX to pay for the Liquid Oxygen/Liquid Natural Gas (LOX/LNG) second-stage propulsion development, meaning that a certain portion of the projected $420 to $450 million development costs would have to be borne by private industry.

The *GX* program is of interest to us because it was primarily a technology development program for which the commercial case was never convincingly made.[16] The second stage of the *GX* was to have used an extremely challenging new technology using LOX/LNG, which is on the frontier of research by the USAF.[17] The so-called LNG Propulsion System, for which JAXA had prime responsibility, was touted as a critical next-generation space transportation technology that would have allowed the *GX* to increase performance with a far more compact size than possible otherwise. However, cost cutting is rarely possible when developing new rocket technologies, given high uncertainties in both time and pecuniary investments. The SAC's earlier reluctance on funding is thus better understood from this point of view, in that it was much too hard to justify money spent on Research and Development (R&D) for the LNG engine which could have turned out to be problematic in actual tests over time. In fact, by the time the program was cancelled in December 2009, developing the LNG engine had proved difficult: developmental problems with the second-stage LNG engine tripled the costs and had pushed the SLV's debut to 2011, thirteen years after the original request for a cheap, fast replacement for the *J-I*.

Instead of the NK-33, the first stage of the *GX* came to be based on Lockheed Martin's Atlas-III technology, which represented the culmination of about forty years of steadily refined launch technologies.[18] At its debut the Atlas-III featured the Russian-made RD-180 engine, which is widely considered an outstanding design. The Atlas-III was retired in 2005 with a 100-percent success rate, but the RD-180 continues to be incorporated into present upgrades such as in the Atlas-V 400 and 500 series. Importing this engine technology would have allowed Japan to update its rockets to another front in existing U.S. rocket technology. Indeed, there were concerns in the United States about missile-technology proliferation, especially on the part of NPO Energomash Khimki, the Russian company responsible for the RD-180 used in Atlas-III and also Atlas-V (which, along with Delta-IV, is responsible for launching the vast majority of U.S. national security payloads). But by the end of 2002, the U.S. State Department approved Lockheed Martin's application to export the Atlas-III first stage to Japan. However, to be clear, the *GX* was primarily a hybrid of a commercially-based first stage and an experimental second-stage design and wholly unsuitable to be used militarily. A long and serious effort would have been needed to get to the point where the SLV's systems could be redeveloped as a ballistic missile even after or if the rocket had proved reliable.

Finally, as an aside, we would like to point out that MHI saw the *GX* program as diverting budget and resources better focused on the *H-IIB*; the company, therefore, insisted as a condition of accepting the privatization of the *H-IIA* that it be designated Japan's primary and first-choice launcher.[19] Our conclusion is that the *GX* was primarily a technology development program. Indeed, over time, to justify its existence the tenor of the *GX* development saga shifted from a commercial basis to one that could also make a contribution to Japan's safety and national security by giving the country another option for independent access to space.[20]

Advanced Solid Rocket (*ASR/Epsilon*) and Air Launch

In 2006, JAXA formally decided to abandon the *M-V* rocket in favor of a cheaper design more suited to smaller payloads. As discussed in Chapter 4, the *M-V* is still considered one of, if not, the best and most sophisticated multistage solid-fuel rockets around—so much so that it also caught the attention of the United States for its comparability to the MX Peacekeeper ICBM. The knowledge of its design and construction remained intact of course, as did the Institute of Space and Astronautical Science's (ISAS) continued ambition to secure the next frontier in solid-fuel rocket technologies.

This takes us to the *ASR/Epsilon*, the very name of which speaks for its designation. With no commercial narrative for its development, the *ASR/Epsilon* has been designed primarily to refine extant solid technologies to develop a cheap, reliable, accurate and responsive (quick and easy to launch) vehicle. From the perspective of the Japanese space-related agencies, the *ASR/Epsilon* is designed to be the summation but also the continuance of Japan's solid rocket technology.[21]

Approved by the SAC in 2007, the *ASR/Epsilon* is a work in progress.[22] Built by JAXA in conjunction with the former ISAS/Nissan cadre of engineers who worked with ISAS and are now at IHI Aerospace, the three-stage *ASR/Epsilon* combines the tried and tested *H-IIA*'s first-stage SRB-A, as well as improved second and kick-motor stages taken from the second and third stages of the *M-V.* Weighing 90 tons, the 2.5-meter diameter, 24-meter tall *ASR/Epsilon* will have about two-thirds of the payload capacity of the *M-V,* specifically 1.2 ton to LEO and 0.6 ton to SSO. With guidance and control on all stages, and with an optional fourth stage with guidance and control for enhanced orbital insertion accuracy, the estimated development cost for the new rocket was expected to be about a third of that of the *M-V.* However, costs have gone well beyond the initial figures suggested in 2006, and final estimates are presently set at about $200 million for development and about $25 to $30 million per launch.[23] As with any new technology direction, whether it combines existing or past technologies (as the designers of the *J-I* tried to do), cost estimates should be treated with caution.

Across the *Mu* series, the *M-V*'s performance is recorded as the highest with respect to the payload ratio. The *ASR/Epsilon* is billed as even better, but the chief focus is on fast launching: The *ASR/Epsilon* will have autonomous pre-launch self-examination—mobile computer-based control systems and shared onboard computer equipment across rocket families. Developers want to cut the *ASR/Epsilon*'s launch pad preparation time to six days as opposed to forty-seven days for the *M-V* (and, as examples, twenty-five days and ten days, respectively, for the U.S. Operationally Responsive Space (ORS)-designated Minotaur-1 and Falcon-1 SLVs). While an *M-V* launch needed sixty to seventy ground control personnel, the designers of the *ASR/Epsilon* are ambitiously seeking the idea of mobile launching—the ability to check and control rockets anywhere in the world with the use of a single laptop.

Given that the *H-IIA* is the only viable Japanese SLV for satellite launching, the *ASR/Epsilon* is expected to make a debut sometime soon to provide balance on the solid-rocket side, possibly 2011, though no date has been set officially. If all goes well, the *ASR/Epsilon* may turn out to be the top choice for launching small-satellite missions, such as the fourteen requests that were under consideration at JAXA as of 2008. The *ASR/Epsilon* also stands to play a role in Japan's Space on Demand (SOD) initiative that echoes the growing ORS activity that has suffused national space policy in the United States more formally since 2005.[24] The key point about the ORS initiative is that it seeks to dramatically improve the reliability, responsiveness, and cost of space transportation in order to defend national security along multiple dimensions, including economic ones. From the perspective of the U.S. Air Force, for example, responsive launch capability is the keystone of ORS; without it, the improvement and defense of space assets and infrastructure would be impaired.

Given the *ASR/Epsilon*'s design, we cannot overlook its potential as a dual-use system, but, as noted, the *ASR/Epsilon*'s primary utility will be its low cost and flexibility. In Japan's case, the *ASR/Epsilon* looks as if it will be focused on launching small, micro-, or picosatellites as well as future small science satellites.[25] In addition, it has been slated by the Ministry of Economy, Trade, and Industry (METI) as the chosen launch vehicle for Japan's new constellation of high-resolution optical satellites that are being developed under the Institute for Unmanned Space Experiment Free Flyer (USEF) as the *Advanced Satellite with New System Architecture for Observation* (*ASNARO/Sasuke*) project, discussed later.

Ballistic Missile Defense (BMD)

In early July 2009, unconfirmed reports emerged that the Ministry of Defense (MOD) was considering introducing the U.S. Terminal High Altitude Area Defense (THAAD) system as a third layer, in addition to SM-3 and PAC-3.[26] These reports emerged just after North Korea launched a volley of ballistic missiles into the Sea of Japan. Whether or not Japan does adopt a fully functional THAAD system, which would significantly improve the geographical coverage of BMD, the reports at face value seem to confirm the common narrative that the story of Japan's BMD is widely thought of as one driven primarily by external security concerns.[27] This narrative, as highlighted at many junctures in this book, most notably begins with North Korea's *Nodong* launch in 1993, which alerted even the Japanese public more visibly than ever before that it was potentially within range of a tangible ballistic missile threat.

Table 6.1 outlines the events that led to the development of Japan's BMD system. As it shows, the Japanese government has certainly been aware of BMD and, more specifically, Theater Missile Defense (TMD), thanks to U.S. interest in protecting itself and its own forward-deployed bases and personnel since the inception of the Strategic Defense Initiative (SDI) in 1983, better known as "star wars." What then President Ronald Reagan suggested as a long-term research initiative—to find the ability to intercept and destroy strategic ballistic missiles before they struck the United States or its allies—remains the guiding technical essence of BMD systems today.[28] Even at its inception,

Table 6.1. Trajectory of Events Related to the Development of Japan's Ballistic Missile Defense (BMD) System, 1983–2015

Date		*Event*
1983	Mar	Inception of Strategic Defense Initiative (SDI) in the United States
1984	Jan	U.S. Presidential Directive 119 establishes SDI to explore possibility of developing missile defenses as alternative means of deterring nuclear war
1985	Mar	United States invites allies, including Japan, to participate in SDI
		Japan requests clarification on principles for participation (for example, framework for private industry participation and withdrawal, government participation, and so on)
1986	May	Japan presents further conditions for participation in SDI (such as not pursuing superiority over USSR, strengthening Western deterrence, continuing nuclear disarmament, not undermining the ABM Treaty, and discussing SDI development and deployment with allied countries)
	Sep	Japan expresses intention to participate in SDI studies
1987	Jul	Japanese government signs Memorandum of Understanding with United States on SDI participation (United Kingdom, West Germany, Israel, and Italy sign separate agreements)
1988	Aug	Corporate tenders for SDI research program for Western Pacific (WESTPAC) Missile Defense Architecture study submitted to USDOD
1989	Jan	WESTPAC Missile Defense Architecture Study begins under a group of leading Japanese and American defense contractors
1991	Jan	First time ever anti-missile missile (Patriot air defense missile) intercepts and destroys ballistic missile (Iraqi Scud missile) under combat conditions.
1992	Jun	United States proposes Theater Missile Defense (TMD) deployment in Japan (and South Korea)
1993	Apr	WESTPAC Missile Defense Architecture Study ends
	Mar	North Korea announces withdrawal from Non-Proliferation Treaty (NPT)
	May	North Korea conducts ballistic missile test over the central Sea of Japan (*Nodong*)
	Dec	Japan and United States agree to establish a Japan–U.S. Working Group on TMD
1994	Aug	Special advisory panel (Higuchi Panel) drafts security policy vision for the twenty-first century, recommending both that Japan cooperate with United States on BMD and develop military reconnaissance satellites
	Sep	Japan–U.S. Joint Study on BMD established

(continued)

Table 6.1. *(continued)*

Date		*Event*
1995	Jan	Japan–U.S. Joint Study on BMD initiated
	Apr	JDA establishes Office of Ballistic Missile Defense Research (BMDR) to work with U.S. Ballistic Missile Defense Organization (BMDO) and U.S. Pacific Command to analyze threat perceptions and TMD development options; Japan Defense Agency (JDA) commences three-year comprehensive study of Japan's air defense system and continues study of BMD jointly with United States
	Jul	China conducts missile tests off Taiwan
		China conducts underground nuclear test, then declares moratorium on nuclear testing
	Aug	China conducts another series of missile tests and naval exercises near Taiwan
		JDA issues report on BMD, arguing in its favor
1996	Mar	China carries out further missile tests and military exercises in waters close to Taiwan
1997	Jan	Establishment of Defense Intelligence Headquarters
	Sep	Japan and United States establish new form of defense cooperation with Security Consultative Committee (SCC)
1998	Aug	North Korea launches ballistic missile over Japanese airspace (*Taepodong*)
	Dec	Government of Japan (GOJ) approves Japan–U.S. Cooperative Research Project Research on Ballistic Missile Defense Technologies (aimed for part of sea-based upper-tier system)
1999	Aug	Exchange of notes take place between Japan and United States to bring into effect Japan–U.S. Memorandum of Understanding on Cooperative Research on Ballistic Missile Defense (focused on Navy Theater-Wide Defense [NTWD] system, and four major components of advanced interceptor missiles)
2002	Jan	Initial intercept missile test from Aegis warship successful
	Jun	United States withdraws from the Anti-Ballistic Missile (ABM) Treaty (so as to allow defense of the homeland against ballistic missile attack)
	Dec	JDA supports future development and deployment at Japan–U.S. Defense Summit meeting
		United States announces deployment of missile defense
		North Korea announces resumption of nuclear facilities
2003	Jan	North Korea withdraws from NPT
	May	Prime Minister Koizumi and President Bush agree to speed up cooperation on ballistic missile defense at the Japan–U.S. summit meeting
	Aug	JDA requests BMD-related budget for FY 2004 from Ministry of Finance (MOF) for the first time

	Oct	North Korea announces completion of reprocessing of spent nuclear fuel rods
		China becomes third country in the world to launch manned spacecraft into orbit
	Dec	GOJ formally issues decision on the introduction and deployment of the BMD system
2004	Jul	United States begins stationing of intercept missiles for long-range missiles in Alaska
	Nov	Chinese nuclear submarines intrudes in Japanese territorial waters
	Dec	GOJ shows strong commitment to BMD through defense plans
		Japan and United States sign BMD Framework Memorandum of Understanding for integrated cooperation; Japan simultaneously exempts U.S.–Japan missile defense (MD) development and production from the three principles of weapons exports
2005	Jul	Japanese Diet concludes legislation for response to ballistic missiles; JDA centralizes joint staff command structure allowing greater joint operational posture between the United States and Japanese military forces (chairmanship of Joint Staff Council upgraded from ceremonial position to one directly conveying commands from Japan's defense minister)
	Oct	China launches *Shenzhou-6*
	Nov	United States and Japan agree to deploy a mobile X-band radar system to enable advanced targeting discrimination technology to detect both cruise and ballistic missile threats
	Dec	GOJ approves Japan–U.S. Joint Development of Interceptor Missiles Having Improved Capability of Ballistic Missile Defense (SM-3 joint cooperative development)
2006	Mar	Japan and United States successfully conduct joint performance test of SM-3 interceptor missile off Hawaii
		Partial amendment of Defense Agency Establishment law, including on point that related to measures for destroying ballistic missiles
		SDF establishes Joint Staff Office to integrate air, ground, and maritime forces for operational purposes
	Apr	Japan and United States begin research projects on advanced technologies for sea-based radar and combat command system as part of bilateral cooperation in BMD, with goal of building advanced phased-array radar with greater capability to detect ballistic missiles
		Japan Air Self-Defense Force (JASDF) permanently links information gathered from its Base Air Defense Ground Environment (BADGE) system with U.S. 5th Air Force in Yokota Air Base, seeking to share/exchange its air defense information with that on regional ballistic missiles gathered by the United States through its advanced spay satellites and patrol aircraft
		United States and Japan conduct Joint Cooperative Test One, testing successful launch of an SM-3 Block 1A and Japanese-designed nose cone (updated to split in half and thus make it possible to deploy larger warhead)
	Jun	SM-3 Cooperative Development Project (SCD) begins: Japan and United States formalize agreement to cooperate on joint development of next-generation advanced-capability missile interceptor, a variant of SM-3 known as SM-3 Block IIA designed to use a unitary kinetic warhead to intercept longer-range missiles (United States focuses on kinetic warhead development, Japan focuses on nosecone and rocket motor); new missile designed also to be structurally bigger (allowing for more sensitive computers and sensors, and larger kill vehicle with bigger infrared seeker and also bigger propulsion/maneuver system), speedier, more agile, and with greater operational range (considered here to be a game-changer because it greatly expands the area a single ship can defend against long-range threats)

(*continued*)

Table 6.1. *(continued)*

Date		*Event*
		Japan reaffirms that arms exports to the United States related to the joint MD project are exempt from Japan's weapons export ban (On the table: multiple kinetic vehicle for integration into SM-3 in the future called SM-3 Block IIB variant)
	Jul	North Korea launches seven ballistic missiles into the Sea of Japan
		GOJ and United States conclude memorandum to transfer weapons and weapon technologies for the joint development of BMD
	Aug	U.S. Navy deploys Aegis ship (*Shiloh*) with SM-3 missiles to Yokosuka base
	Sep	X-band radar relocated and activated in Camp Shiraiki in Aomori
	Oct	North Korea conducts underground nuclear weapon test
2007	Jan	JDA transforms into the Ministry of Defense (MOD)
		China conducts successful anti-satellite weapon test on its own aging weather satellite
	Mar	Essential Emergency Response Guidelines for destruction of ballistic missiles approved by GOJ, allowing MOD minister to launch interceptors without the prime minister's approval under certain conditions, such as an incoming ballistic missile
		Japan deploys its first interceptor in history, the Patriot PAC-3, at Iruma Air Base, Saitama
	Oct	Joint Tactical Ground Station (JTAGS) deployed to U.S. Air Force Base in Aomori to allow mobile information-processing system to pick up and analyze satellite data on ballistic missile launches
	Nov	Japanese Aegis destroyer *Kongo* successfully takes part in SM-3 Block IA tracking and surveillance exercise
	Dec	Japanese Aegis destroyer *Kongo* successfully test fires and intercepts midrange ballistic missile target in space with an SM-3 Block IA for the first time
2008	Feb	United States uses SM-3 Block IA to intercept and destroy a failing U.S. intelligence satellite, demonstrating its utility as an ASAT weapon for the first time
	Mar	Japan completes BMD system for Tokyo region, with four PAC-3 missile batteries installed around capital
	Jul	MOD conducts first full-fledged MD exercise in Tokyo using PAC-3 interceptors
	Sep	PAC-3 test successfully intercepts target at White Sands Missile Range, New Mexico
	Dec	Japanese Aegis destroyer *Chōkai* unsuccessful in SM-3 firing test
2009	Mar	Japan issues first destruct order for incoming missile or object based on Self-Defense Force (SDF) Law
	Apr	North Korea launches flying object (missile or satellite launch controversy)
		SDF takes steps toward tracking and surveillance but does not execute measures to destroy the flying object
		Japan terminates destruct order for incoming missiles or objects
	Sep	PAC-3 test to be conducted at White Sands Missile Range in New Mexico

	Four new (FPS-XX) and seven upgraded (FPS-3) radar units to improve missile surveillance network expected to be deployed
2010	Four *Kongo*-class Aegis destroyers expected to be upgraded and equipped with SM-3 Thirty PAC-3 launchers are to be deployed at nine other Air Self-Defense Force (ASDF) bases
2011	Integrated SM-3 Block IIA ground tests expected
2012	SM-3 Block IIA aerial tests expected
2013	SM-3 Block IIA end-to-end flight test expected
2014	Initial operational capability of SM-3 Block IIA for United States and Japanese navies expected
2015	Full production and operational capability of SM-3 Block IIA system expected

SOURCES: "MD Kyōdō Kaihatsu ni 30 Oku En" [¥30 Oku Towards Joint Development of MD], *Asahi Shinbun*, 1 September 2005; Ministry of Defense (MOD), *Nihon no Boeishō 2007* (Tokyo: MOD, 2007), pp. 186–192 (table III-1-2-1), pp. 454–465; MOD, "Overview of Japan's Defense Policy," November 2007, available online at www.mod.go.jp (accessed 27 May 2008); MOD, "Dandō Misairu Bōei Shisutemu no Seibi Nado ni Tsuite" [About the BMD System], available online at www.mod.go.jp (accessed 27 July 2008); Missile Defense Agency (MDA), "Missile Defense Timeline 1994–2004," available online at www.mda.mil (accessed 27 June 2008); "Chronology of Tension," *Globe and Mail*, 12 March 1996; Reiji Yoshida, "PAC-3 Patriot Missiles Debut at Iruma Air Base," *Japan Times*, 31 March 2007; "MSDF Ship Passes Interceptor Test," *Japan Times*, 19 December 2007; "Second PAC-3 Missile Defense System Deployed," *Japan Times*, 30 November 2007; "Missile Drill Held in Tokyo," *Japan Times*, 30 July 2008; "SDF Ranks Would Run Defense Ministry Bureaus Under Radical Revamp," *Japan Times*, 13 July 2008; "SDF to Upgrade, Deploy New Radar to Detect Missiles," *Japan Times*, 11 September 2005; "Lower House Passes Missile Defense Bill," *Japan Times*, 15 June 2005; Michael Mecham, "X Marks the Spot: U.S.–Japan Agreement Boosts Cruise and Ballistic Missile Defenses with New Radar," *AWST*, 7 November 2005, p. 39; "U.S. Eyes More Radar Against North," *Japan Times*, 23 August 2006; "U.S. Relocates X-Band Radar in Aomori to Watch North Korea," *Japan Times*, 29 September 2006; "ASDF Now Giving Radar Info to U.S.," *Japan Times*, 13 May 2007; and "Cabinet OKs Terms to Expedite Missile Shield," *Japan Times*, 24 March 2007. All additional details and latest test results on the SM-3, PAC-3, and so on, are from MOD, "Dandō Misairu Bōei ni Tsuite" [About BMD], available online at www.mod.go.jp (accessed 31 August 2009).

when the SDI initiative became embroiled in controversy, two things were clear—that the research and development, as well as the upkeep, of any such system would be extremely costly and that it would eventually put the Anti-Ballistic Missile (ABM) Treaty in jeopardy.[29] Looking back, neither of these factors deterred progress. The development and deployment of a multilayered BMD system was urged on more specifically with a directive from President George Bush to the U.S. Department of Defense (USDOD) in December 2002, the same year that the United States withdrew from the ABM treaty.[30]

It is helpful to have some overall sense of the state of BMD systems in the United States, the world's most advanced developer and deployer of its related technologies, so as to know exactly how and where Japan's proposed system fits in.[31] In general, BMD systems are designed to counter ballistic missiles in one of three phases: the first, and most desirable phase, is when the ballistic missile is still in the boost phase in the originating territory, which usually lasts about 1 to 5 minutes. Setting aside problems of being at the right place at the right time, if sensors are close enough to monitor the actual launch, which can be detected because of the missile's exhaust plume, two types of boost defense elements can be brought into play. While a considerable budget has been expended on boost phase interception with the Airborne Laser (ABL) and Kinetic Energy Interceptor (KEI) programs, this type of interception has been deemphasized, for now, in the United States in favor of more mature technologies.

The second or midcourse phase takes place once the missile booster burns out and its payload separates and coasts in space for as long as twenty minutes toward its target. Because it has the relatively longest intercept window and a more predictable glide path and occurs in space, this phase is considered a major opportunity for the destruction of the ballistic missile or its payload. If all goes well, such actions would take place outside the Earth's atmosphere and falling debris should be destroyed on reentry. However, the long intercept window also means that the attacker has more times to deploy countermeasures against the two principal midcourse defenses. These defenses include the Ground-based Midcourse Defense (GMD) and also what is now called the Aegis Ballistic Missile Defense System. The GMD uses a ground system to connect, coordinate, and control the necessary sensors/early warning radars, and an interceptor called the Exoatmoshpheric Kill Vehicle (EKV) that uses, along with three solid-fuel boosters, only speed and force—"hit-to-kill" technology—to ram into and destroy the warhead in space. Interceptor missiles have already been placed in Fort Greely, Alaska, and Vandenberg Air Force Base, California, with others planned for deployment.

The Aegis Ballistic Missile Defense is a sea-based defense that builds on the capabilities and technologies of the existing Aegis weapon system, Standard Missile-3 (SM-3) capability, and the U.S. Navy's overall resources. Already tactically certified and deployed both in the United States and Japan, it is designed to intercept Short-Range Ballistic Missiles (SRBMs) and Medium-Range Ballistic Missiles (MRBMs) in both the ascent and descent midcourse phase using Long-Range Surveillance and Track (LRST) capability as well as engagement capabilities with the SM-3. As of December 2007, the Aegis BMD defense had a track record of approximately fourteen successful intercepts out of sixteen attempts (including one by Japan), with more tests planned in the near future.

The third or terminal phase is the least desirable for defensive purposes, both because it lasts only thirty seconds to a minute, leaving virtually no margin for error, and because at that point the warhead is over the homeland territory. There are several terminal defense systems as outlined here. Backed by the X-band radar and fire control/communication units, THAAD, in which Japan has now reportedly shown an interest, basically consists of a mobile truck-mounted launcher that has an interceptor with hit-to-kill endo- and exoatmospherphic lethality. While development and tests still continue, THAAD has completed five successful intercept tests, including a successful intercept of a separating target ballistic missile in June 2008. As a system, it is completely transportable and is billed as being able to rapidly deploy to anywhere in the world within hours.

Perhaps the most well known, mature, and battle-tested element of BMD is the land-based Patriot Advanced Capability-3 (PAC-3) system, which builds on the previous Patriot air and missile defense systems and is currently operated by the U.S. Army. Several hit-to-kill PAC-3 interceptors are mounted on wheeled vehicles and, with the use of advanced radars, provide 360-degree coverage for taking out ballistic missiles as well as other objects, such as aircraft, cruise missiles, and Unmanned Aerial Vehicles (UAVs), at short range. Finally, there are other systems, such as the Arrow system developed jointly by the United States and Israel for use against SRBMs and MRBMs, as well as the mobile Medium Extended Air Defense System (MEADS) that builds on the PAC-3 platform and is being developed jointly by the United States, Germany, and Italy.

We now turn more specifically to Japan, showing exactly where Japan is on the spectrum of BMD systems and what potentially that has to do both with its space industry and the market-to-military trend.[32] At the outset, it is helpful to know that Japan's MOD identified two space-related BMD projects prior to the establishment of the Basic Space Law in 2008—namely, missiles from the Aegis BMD and sensors for conducting surveillance of ballistic missiles flying through outer space.[33] Our purpose in examining BMD as part of a space-related military structure is to make three larger points—that Japanese corporations have long been more critical to the shift toward BMD than the attention given primarily to external geostrategic factors; that the very same SLV makers examined in this book have been central to the advent of a BMD system in Japan, where they could parlay their space-based knowledge; and that, as with much of Japan's other space-related developments, the realities of BMD technologies also fit uneasily within pacifist constraints.

Even a cursory examination of these players' interests and lobbying shows, as in Table 6.1, that the politics behind Japan's decision to acquire a BMD system do not stand in a historical vacuum as far as the business of its space-related corporations go. Indeed, in line with our emphasis on the market-to-military trend, it is intertwined with the economic woes and fortunes of exactly that industry. As we discussed in the opening chapter, Japanese industry faced considerable economic pressures in the 1990s, and military-oriented space projects began to loom as concrete options for its very livelihood. BMD was one such project. It was all the more important because a BMD system designed to counter the threat of incoming ballistic missiles promised to be extensive—with its integrated span across detection capabilities (for example, through land-, sea-, and space-based sensors and radars), destruction capabilities (for example, through ABL, KEI, Aegis SM-3, multiple kill vehicles, THAAD, and PAC-3), and, importantly, management and engagement capabilities (through a geographically dispersed control and communication infrastructure with discrete parts). Importantly also, no single military service in any one country alone can detect, destroy, manage and engage all ranges of ballistic missiles—which means the need for an integrated multilayer BMD system spread across countries increases exponentially, irrespective of whether key components within it work or not.[34] At that time, then, the estimated \$10 to \$50 billion BMD system (with \$30 billion for R&D alone), the acquisition of which could go on for decades, made perfect corporate sense.[35] For the Japanese aerospace industry, consideration of BMD was almost an imperative, given that defense contracts accounted for around 80 percent of its work and that it was increasingly dependent on demand for defense materials. The drive for corporate profitability also became very much cloaked in the rhetoric of defense of the homeland, whether in Japan or elsewhere.

In Japan's case in particular, some sense of this can be gauged from corporate—and, we believe, also government—interest in BMD from the start, especially if SDI is seen as its predecessor. It was only in 2003 that the Japanese government moved to formally commit itself to the research, development, and deployment of a BMD system.[36] However, as Table 6.1 reveals, Japan has been involved with the study of SDI, and subsequently BMD virtually from its inception. The option of BMD for Japan, then, did not just appear out of nowhere in the aftermath of the *Nodong* and then *Taepodong* incidents; there had already been a clear lineage of joint studies, consultations, and dialogues between Japan and the United States spanning close to two decades. To be sure, as observers pointed out, the subsequent external security environment, as well as bureaucratic, political, diplomatic, and strategic contingencies, were important in moving Japan toward a BMD system. But as we discussed in Chapter 1, also central to the shift were the space-related industries overall who were able to parlay their considerable rocket-cum-missile (and later satellite) expertise in the interest of national defense and who had been attempting to do so visibly as the SDI initiative got under way.

Here it is necessary to gain some historical perspective on corporate activity. By the end of 1986, it was clear that there was a scramble among defense contractors around the world for defining the SDI architecture for a TMD system for countering tactical ballistic and cruise missiles—one that had been lent even greater credibility with Raython's Patriot air defense system to intercept a surface-to-surface missile.[37] Japanese contractors were similarly so interested. In January 1988, as U.S. proposals for designing a new air-defense system for Japan as part of the SDI came to the fore, six Japanese electrical/electronic companies (MHI, Melco, Nippon Electric Corporation [NEC], Fujitsu, Hitachi, and Japan Radio company) ambitiously set out to contract directly with the U.S. SDI bureau rather than be subcontractors of U.S. firms.[38] Interestingly, the reported main feature of the U.S. plan was a reliance on satellites to shoot down enemy missiles with laser weapons in space, along with calls for an anti-air missile, a more sensitive radar network, high-speed computers, and an electromagnetic weapon called the "rail gun."

The part of the plan more directly applicable to Japan focused on the destruction of medium- and short-range missiles and hypersonic aircraft in the atmosphere. By August 1988, the top Japanese and U.S. defense corporations were moving forward to develop the SDI research program, which was formalized under the Western Pacific (WESTPAC) Missile Defense Architecture Study. Japanese companies were particularly enthusiastic. MHI, for example, bid for the tender for conducting SDI research to USDOD, leading a study group that included fourteen companies (five American, seven Japanese), among them Melco, Mitsubishi Corporation, NEC, Hitachi, Fujitsu, Boeing, Raytheon, McDonnell Douglas, and Lockheed Martin.[39] In November 1988, the MHI team was one of two multinational industry teams awarded a $3 million contract by SDIO to study TMD concepts for the Western Pacific.

Throughout this time, and even after the *Nodong-1* incident, the JDA reportedly remained concerned about the cost, political sensitivity, and political support of a joint TMD system as it continued discussion with the U.S. government.[40] None of this appeared to deter the corporate front, which continued to operate as if BMD was going to come to fruition for Japan.[41] By September 1998, angling for BMD-related contracts in the aftermath of the *Taepodong-1* incident, which elevated fears of North Korean ICBMs reaching the United States, Melco and Lockheed Martin had formed an alliance to exchange information on defense technologies and to propose and develop new products and technologies of interest to the JDA.[42] As high-technology electronic devices were seen to be a lead growth sector in defense equipment, Melco was keen to receive contracts—for infrared sensors and missile guidance systems, for licenses to make Lockheed Martin equipment, for high-tech weaponry, and Aegis destroyer-type system maintenance. Such moves could potentially give it a considerable edge over domestic rivals like NEC and Toshiba.

In the United States federal funding continued to come through. In October 1998, despite the fact that the Ballistic Missile Defense Organization (BMDO, the predecessor of the MDA) was criticized for weak program management and despite warnings by the U.S. Joint Chiefs of Staff that the requisite technology was not mature enough, approximately $1 billion out of an $8 billion-plus military budget was appropriated for research, development, testing, and evaluation of missile defense systems under the BMDO's supervision.[43] As the initial Memorandum of Understanding on joint technology in August 1999 made clear, U.S. and Japanese industry were to advance technologies to improve the capabilities of four major guided-missile components.[44] These included, more specifically, sensor technology for the radar and missile, an advanced kinetic energy warhead, an advanced second-stage propulsion for the Standard Missile SM-3, and a lightweight nose cone—some of the very same basic technologies to which Japan's experienced rocket makers could contribute and also learn from.

In 1997 the U.S. Navy proposed the "Theater Wide" (formerly known as Navy Upper Tier) missile defense, with a focus on protecting a large area and having the ability to kill missiles near their origin. In conjunction with the then BMDO it also advocated that that system could protect Japan from ballistic missile attack from North Korea. As shown in Table 6.1, in 1999, spurred on by the *Taepodong* missile that also made a dent in the public consciousness, the Japanese government finally embarked on research and development (but, at that point, not production) with the United States on BMD technologies. The focus then was on the Navy Theater-Wide Defense (NTWD) system, also known as the Aegis BMD System. Japan's objective, of course, was then and is now the same: a multilayered BMD system that can identify and track ballistic missiles, destroy them, and be integrated across a real-time command, control, and communication infrastructure to actually make that possible from start to finish.

It was industry actors who could make, provide, and upgrade any such related technologies. On the Japan side, MHI was named the prime contractor in the proposed system, with Melco, Toshiba, KHI, Fujitsu, Nissan Motors, and IHI also taking part.[45] According to press reports in June 2001, MHI's enthusiastic pursuit of R&D with the United States on missile defense grew out of its sense of mission that it was essential to Japan's defense and that it would give Japanese companies an advantage in dealing with U.S. licensed technologies.[46] This enthusiasm was all the more remarkable given that the Japanese government, at that point, had yet to officially confront the legality of Japan's participation in collective self-defense and had also yet to officially make a concrete decision to deploy any missile defenses given the economic doldrums then wracking the country. It is simply, then, not true that U.S. pressure led Japan to build a missile shield; Japanese companies were heavily vested in seeing it come to fruition both in their own, and, as they constantly stressed, in Japan's national interest.

For Japan the BMD-related research and development needed to be within pacifist constraints, and at first blush the BMD system going into place in

Japan seemed admirably suited to the country's defense-only orientation. In December 2003 Japan officially announced it would acquire and deploy two elements of the missile defense system—an upgrade of all four of the Maritime Self-Defense Force's (MSDF) Aegis cruisers to accommodate the SM-3 (the ballistic missile interceptor version of the weapon), and preparation for deliveries of the PAC-3, to be produced under license by MHI, to all six of the ASDF's then upgraded Patriot surface-to-air missile units.[47] As noted above, this meant that Japan was seeking to concentrate on the mid-course and terminal phases of missile defense, in theory making it consistent with the country's defensive orientation.

However, the balance between defensive and offensive BMD system technologies is a fine one, and the BMD system raised concerns from the start in Japan. To begin, it would be difficult to determine immediately whether a ballistic missile had been fired at Japan or another location, which still necessitates greater clarification of collective self-defense as opposed to individual self-defense.[48] Clarity on this issue becomes even more critical as moves have already taken place to allow the Japanese joint chiefs to work directly with their U.S. counterpart under a unified and redesigned command structure linked directly to the Japanese defense minister.[49] Although the United States has urged Japan to use its BMD system to counter missiles also headed for the United States, and there is an influential and growing chorus within Japan to defend an ally under attack consistent with international law, the Japanese government has yet to provide a clear and compelling constitutional interpretation for collective self-defense. As in other areas, the reality perhaps is that the law will catch up when the technology is in place. Some indication of this comes in another area. As the BMD system got under way for Japan, there was the thorny issue of weapons export. It was necessary for Japan to ease the extremely strict weapons export regulations in order to allow its corporations to export BMD technology (such as that related to Lockheed Martin's PAC-3 and Raytheon's SM-3 Block II) to the United States over time. Here, legal relaxation did take place in December 2004, much to the satisfaction of industry actors in both the United States and Japan.[50]

Finally, Japan's continued development and testing of several MD-related technologies shows a trend toward, or at the least interest in, acquiring greater offensive capability.[51] Some of this is unavoidable given the as-yet untested nature of a fully functioning BMD system and the more important fact that long-range incoming ballistic missiles can also employ countermeasures such as penetration aids, decoys, and modified flight trajectories (which, of course, also takes Japan well beyond a mere TMD system). The fact is that nobody can say for certain how things will turn out once there is an incoming ballistic missile on its way. It thus makes practical sense that the best way to prevent ballistic or cruise missile attacks on a country, such as Japan, is by having that country acquire the capability to strike enemy facilities directly on foreign soil in the boost phase.

Technologically, at least, Japanese industry continues to make progress even here. Already in 2000, for example, Japan had begun development of an Advanced Infrared Ballistic-Missile Observation Sensor System (AIRBOSS), an infrared missile detection and tracking system that was specifically designed to detect and track ballistic missiles in the boost phase.[52] This system was tested successfully again in December 2007, when the Japanese Aegis destroyer *Kongo* carried out Japan's first—and successful—intercept of a ballistic missile target with the SM-3 variant. At that point, the AIRBOSS was tested on the SM-3 itself—an act that suggests Japan may well go on to acquire the capability of first tracking, and from there possibly destroying, a ballistic missile launch in the boost phase on foreign soil.[53]

As we saw previously in Chapter 5, that very same SM-3 Block 1A was also used by the United States in a successful intercept of a failing U.S. intelligence satellite in early 2008—a move that was widely seen as a response to an anti-satellite test (ASAT) conducted by China in early 2007.[54] Although the United States officially denied it, the SM-3 could also be successfully used as a direct-ascent ASAT weapon—a technology to which Japan has access, alongwith the underlying software and processes it has developed and can duplicate for further ASAT missions in the future. The SM-3 missile itself, which can presently intercept SRBMs and MRBMs (1,500 kilometers or less), is also in the process of being upgraded through a joint U.S.–Japan research and development program.

Given its rocket background (focusing on second- and third-stage boosters, nose cone, as well as second-stage steering and control systems) along with U.S. industry (focusing on improved kinetic warhead and its infrared optics to hone the missile's warhead-detection capabilities), MHI will play a leading role in the development of the proposed SM-3 Block IIA to improve not just the speed but also the range of the interceptor to counter LRBMs, IRBMs and ICBMs—a move that calls into question the purely "defensive" nature of Japan's BMD system in the future.

It is also important to note that Japan has expressed interest in another boost-phase interceptor technology, the Airborne Laser (ABL), which may arm it with the capability to use lasers to disable ballistic missiles even on enemy soil or in enemy airspace.[55] In 2004, Japan's National Institute for Defense Studies was reported to endorse the idea of a preemptive strike on North Korean bases and to urge the necessity of having the capability (such as the ABL) for countering any threat emanating from any such foreign missile base. ABL remains a work in progress, although the Pentagon has reportedly talked to Japan about its industrial opportunities. Meanwhile Japan's MOD has recently moved to request R&D funds for a ground-based laser weapon system after high-level official talks between the two countries.

Summing up, Japanese interest in BMD can be traced back to the original Reagan-era "star wars" initiative. Rather than emerging in a historical vacuum, BMD is, in our view, just another stage in Japan's space militarization trend. In fact, the politics of Japan's BMD acquisition is consistent with the country's long march to acquire launch vehicles or rockets. As we have suggested in earlier chapters, launch vehicle development in Japan, as elsewhere, has been hard to isolate from the development of military technologies—a fact that is nowhere more clear than in Japan's involvement with BMD development. The SM-3, for example, is really a three-stage, ship-launched rocket; and Japan is now heavily vested in parlaying its SLV technology and experience to design and develop elements such as the clamshell nose cone and the second-stage rocket in future variants of this missile. This is especially important to note as there remain serious technical questions about the effectiveness of MD. Some see its interceptor-missiles less as a defense against incoming ballistic missiles and more as an offensive ASAT space weapon aimed at orbiting enemy satellites, which can be technologically easier to pinpoint because they may be stationary, have known orbits, and are better lit against a dark background.[56] Overall, it can hardly be said that constitutional, legal, or normative constraints held sway as BMD development has proceeded. Nor has industry race for lucrative contracts in this area, whether in the U.S. or Japan, abated.[57] What is truly remarkable is that top Japanese officials no longer shy away in global audiences from making their positions clear on national security involving space-related technologies, going so far in some cases as to bluntly threaten the preemptive use of missiles to deter foreign aggressors.

Other Aerospace Activities

There are several other activities that are worth mentioning, especially those focused on Reusable Launch Vehicles (RLVs) and scramjet research. These represent cutting-edge R&D efforts in Japan's broader aerospace sector.[58] They are noteworthy because, in line with our main themes, they exemplify the militarization of Japan's space technologies and programs.

Reusable Launch Vehicles (RLVs) Japan's reusable launch vehicle program has been considerably downscaled and deemphasized since the halcyon days of the late 1980s and early 1990s when the country hoped to build a space plane.[59] The goal, ultimately, was to develop the *HOPE* (*H-II Orbiting Plane*), an unmanned reusable shuttle that was scrapped in 1997.[60] Subsequently, the goal shifted to a smaller prototype version called *HOPE-X* (*H-II Orbiting Plane Experimental*), the development of which was also stymied as funding was withdrawn in 1998 and has not reappeared.

The more important point about the *HOPE* projects is that they have allowed the testing of dual-use technologies.[61] In 1994 the wok-shaped (more like a flattened nose cone) *Orbital Reentry Experiment* (*OREX/Ryūsei*) allowed flawless testing of reentry survivability. In 1995 the *Space Flyer Unit* (*SFU*), the first retrievable (and, in theory, reusable though never so used again) space science platform, launched by *H-II-3* for a series of in-orbit experiments, allowed also for the testing of advanced guidance techniques for payload injection and orbital stability. In early 1996 the suborbital *Hypersonic Flight Experiment* (*HYFLEX*) on the *J-I* allowed testing of hypersonic flight for the first time and, although the vehicle itself was not recovered, also allowed further experience with reentry-vehicle technology, such as flight control technologies and protection from aerodynamic heating on reentry. In late 2002 and mid-2003, the *High Speed Flight Demonstration* (*HSFD*) allowed, among other things, further improvements in design technology for reentry vehicles' transonic guidance, navigation, and controls systems.

While the totality of Japan's tested RLV technologies may not yet be moving forward toward an actual space plane, research and testing still continue on various development streams both in JAXA and MHI.[62] The trends to obtain RLV technology in small steps should certainly not be disregarded. In mid-1996, for example, the *Automatic Landing Flight Experiment* (*ALFLEX*) confirmed basic automatic landing technologies given conditions of steep glide paths for landing and constant threat of spin. There is always the motivating concern, in Japan as much as in Europe, that the United States is still the only dominant player in RLV technology. Japan has been engaged in some initiatives with European partners in this area.[63] From the perspective of this book, however, more important is that Japan's quest for RLV technologies thus far has allowed the country to successfully develop and test reentry and advanced guidance technologies applicable to ballistic missile programs—a trend we examine more closely later.

Scramjet (Supersonic Combustion Ramjet) and Hypersonic Aircraft Japan's interest in RLVs and space planes also resonates in its R&D efforts related to scramjet technology, which is billed as critical to advancing propulsion technologies. In October 2005, JAXA reported that it had successfully flown and landed a scaled experimental supersonic transport (SST) model plane.[64] This initiative, which focuses on next-generation propulsion technologies, seeks to develop an environmentally friendly hypersonic transport system capable of Mach 5-plus speed that could cut travel times to a fraction of those of today's jet planes.

Japan's interest in hypersonic vehicles powered by scramjet engines (leading to air-breathing vehicles that, instead of carrying oxidizers in tanks, "breathe" oxygen from the atmosphere to burn fuel) has its roots in the early 1980s.[65] In 1981, Japan, along with Australia, Germany, and the United States, gave funds to the University of Queensland, Australia, for research and testing of hypersonic scramjets. At a commercial level, like the other interested players, the goal of Japan's research was the development of a piloted single-stage-to-orbit (SSTO) space plane or launch vehicle (much like the National Aero-Space Plane, the NASP, in the United States which actually lost funding). By 1994, Japan had tested its first scramjet engine, reaching an altitude of 20,000 meters and a speed of Mach 4. JAXA started research on supersonic aircraft in 1997 and has plans to develop a thirty- to fifty-passenger aircraft with reduced sonic booms and noise pollution.[66] The goal is to build a viable supersonic aircraft in the 2020s and a Mach 5-capable hypersonic plane by 2025. It is also exploring pre-cooled turbo engines, which are supposed to produce massive thrust by using liquid hydrogen to cool the air entering the engine. In turn, this reduces carbon dioxide emissions and makes the engine environmentally friendly.

The dream of hypersonic flight remains a distant goal at this stage: Japan's flight test in April 2006 for a scramjet engine resulted in failure, but research continues around the world either in national programs or in collaboration with foreign (defense lab) partners.[67] The United States and United Kingdom are clear about the primary military purpose of scramjet research—the propulsion technology is considered the likely choice to power a hypersonic missile, the sheer speed of which turns it into a lethal kinetic weapon.[68] The U.S. Department of Defense (USDOD) underscores the interchangeability of the technology between the market and the military: ramjet/scramjet propulsion and materials technologies to enable hypersonic missiles can also potentially be optimal for an (air-breathing, first-stage in a two-stage-to-orbit) SLV design.

ON THE SATELLITES AND SPACECRAFT SIDE

We turn next to the latest ideas and developments related to satellites and spacecraft programs that may feed further into Japan's security interests. These include the *Quasi-Zenith Satellite System* (*QZSS/Michibiki*), the future *IGS*, the *ASNARO*/Sasuke project, missile launch early warning systems, and, finally, a range of microsatellite programs that can lead to dual-use possibilities. We end with a snapshot of UAVs, which are under consideration for providing more integrated information, surveillance, and reconnaissance (ISR) networks for Japan.

Quasi-Zenith Satellite System (QZSS/Michibiki)

Japan now plans to launch its own positioning system, as the new Basic Space Plan makes clear.[69] This is through the *Quasi Zenith Satellite System* (*QZSS/Michibiki*), which is described as a highly precise positioning service providing constant coverage across all of Japan regardless of physical or urban terrain. The system relies on multiple satellites in orbit, one of which is always above or near Japan, providing total and clear lines of sight at all times. Billed as an international project, the ambition of this system is to go on to provide space-based positioning, navigation, and timing (PNT) services throughout the Asia-Pacific region, all of which can be put to multiple uses in the public realm. Such PNT uses range from civilian ones, such as traffic control and topographical surveys, to crime fighting and personal safety, and on to (but not mentioned) the more precise ways in which the military can target systems.[70] This is all the more important under the envisioned rotating system of satellites in which one (of at least three) satellite picks up tracking an object when others move away in its figure-eight rotation. PNT, in short, is as indispensible for a vast range of civilian and commercial concerns as military ones, the latter of which are likely to loom large given the national security concerns in Japan's space policy.

If the launch set for 2010 on the *H-IIA* goes as planned, the *QZSS/Michibiki* begins to put Japanese technology on par with the United States' Global Positioning System (GPS), the European Galileo global navigation satellite system of Europe, and the Russian Global National Satellite System (GLONASS). Through extensive negotiations through 1998, the political pathway toward the *QZSS/Michibiki* was eased by a joint U.S.–Japan GPS statement, one that established a cooperative mechanism for annual meetings and working groups to allow the system to go ahead.[71] Recognizing that *QZSS/Michibiki* would be going ahead in some form, the United States signed off on the system's interoperability with GPS formally in November 2004. The United States expressed strong support for Japan's plans to develop the *QZSS/Michibiki*, as it would provide significantly improved regional service to PNT users in Japan and surrounding areas, strengthen cooperative relations between the United States and Japan, and help accelerate Japan's leadership in space technology.

The *QZSS/Michibiki* system as it is now being built is designed to be both complementary and augmentative to GPS: in its complementary mode, a single *QZSS/Michibiki* satellite overhead can act as an "extra" GPS satellite, and as an augmentative system it can work to enhance GPS accuracy, which is estimated to range from 1 meter or better.[72] The service will also have the capability to be augmented with geostationary satellites using the *MTSAT*-based Augmentation System (MSAS), making it similar to a geostationary design of the U.S. Federal Aviation Administration's Wide Area Augmentation System (WAAS). Certainly the *QZSS/Michibiki*, when combined with a geostationary hub satellite, can make a useful system for Japan and surrounding countries, as Japanese government agencies and corporations have stressed.[73]

Although the *QZSS/Michibiki* itself is a product of the 2000s, the system as a whole represents the culmination of efforts to develop a regional GPS system dating back to the late 1980s.[74] Like a lot of the other space-based technologies discussed in this book, this one has had a long trajectory. In the aftermath of the first Gulf War, in which GPS-based capabilities were used to great effect, the Japanese government became more explicit about developing its own PNT capability in case it was denied access or remained dependent on U.S. technology.[75] There were specific concerns about the availability of military-grade GPS signals, specifically from U.S. policy denying civilian users the system's most accurate signals. While the United States was considered a friendly ally, the widespread feeling all around was that therefore Japan needed to develop its own GPS system or at least acquire the technology to do so.[76] In the 1990s, the STA, METI, and the Ministry of Posts and Telecommunications (MPT), in particular, looked to develop a positioning system that would cover a large swathe of Asia, from the Kurile Islands to the north, China to the east, and Guam in the south. In March 1997 the then STA asked NASDA to move ahead with research into the highly accurate, satellite-mounted atomic clocks needed for a high-precision GPS. At the time this was billed as a matter of economic security.

As with other space-based assets, corporate interests have been critical in moving the country toward a GPS alternative, such as in the *QZSS/Michibiki*.[77] Fuelled by satellite-based navigation applications in the 1980s estimated at close to $500 million per year, many of the country's key manufacturers, such as Hitachi, Honda, and Melco, began to pursue concerted avenues for development of an improved GPS and commercial services. In 1999, Keidanren had already proposed a *QZSS/Michibiki*-type system focused on communications, following up in 2002 with the establishment of a special Promotion and Investigative Committee for the *QZSS/Michibiki*. The company had also optimistically projected that the *QZSS/Michibiki* system would develop about ¥1.7 trillion in business revenue creation in its first five years of service out of equipment, broadcasting, communication, and car navigation-related sales.[78] In 2000 the *Japanese Regional Advanced Navigation Satellite* (*JRANS*) concept was proposed by Itochu, NEC, and Toshiba, and it too came out of concerns about depending on the U.S. GPS system as an only source for PNT services in Japan—services widely pitched as fueling the country's economic and social system.[79] The *JRANS* project members publicly identified a formative *QZSS/Michibiki* as a building block toward an independent regional system while stressing full compatibility and interoperability with the GPS.

In 2002, more than fifty Japanese companies, including Melco, NEC, Hitachi, and GNSS Technologies, founded the Advanced Space Business Corporation (ASBC) to facilitate public participation and investment in the *QZSS/Michibiki* project. Even when ASBC was shut down after failing to attract sufficient private sector investment because of the lack of a commercial market, it was succeeded almost immediately by the Satellite Positioning Research and Application Center (SPAC) in February 2007.[80] As of 2009, SPAC is now coordinating the satellite-based PNT efforts across the public and private spheres with the aim of advancing a geospatial information society. Throughout all this, Melco, in particular, has been perhaps the major proponent of the *QZSS/Michibiki* system, with an interest in another geostationary bus design. In fact, while making the case for its ubiquitous and seamless communication services to all aspects of public, business, and consumer life, it has gone so far as to bill the *QZSS* as its very own system in promoting its utility to audiences worldwide.[81]

Overall *QZSS/Michibiki* program costs were estimated to be about ¥170 billion, of which ¥90 billion was to be funded by ASBC and the remainder by the government—a major procurement by Japanese standards.[82] It should be noted that these figures put the development of the *QZSS/Michibiki* second only to the *IGS* system that has garnered so much attention. At first, the system foundered when none of the government ministries could agree to take responsibility for developing it, as the Council on Science and Technology Policy (CSTP) did not approve a mechanism to run the program. Without a definitive government commitment, industry became increasingly reluctant to invest significant sums in a venture that was not likely to make them money in private broadcasting and communication services.

But the push for *QZSS/Michibiki* did not die, and government interest in positioning technologies picked up pace in the aftermath of the first Gulf War as well as the *IGS* saga in which Melco had already played a critical role. By around 2001, NASDA was openly promoting the idea of a quasi-geostationary satellite system, involving three satellites in the service of next-generation mobile communications.[83] In 2003 the METI-related USEF had started the *Advanced Satellite Engineering Research Project (ASER)*, the results of which are expected to be applied to *QZSS/Michibiki*. By 2004 the CSTP had endorsed the idea of an autonomous and GPS-complementary system. By 2006 the Positioning and Geographic Information System Council had released its Basic Policy on the Promotion of the *QZSS/Michibiki* Project. The key legislative move finally came in 2006 when Japan's Diet passed the Basic Act on the Advancement of Utilizing Geospatial Information (AUGI), which was enacted in 2007.[84]

This law would guide the development, distribution, and use of the National Spatial Data Infrastructure (NSDI) as it combined the synergies of technologies from geographic information systems (GIS) with space-based PNT systems derived from the *QZSS/Michibiki*. The law ended a four-year dispute between MEXT, the Ministry of Land, Infrastructure and Transport (MLIT), the Ministry of Internal Affairs and Communications (MIC), METI, and a consortium of private companies as to who would pay for what for the system through the public-private partnership (PPP) approach stressed by Keidanren.[85] The law also helped commit four government ministries led by MEXT to pay ¥75 billion (around US$616 million) for JAXA and Melco to develop an initial satellite based on a DS2000 satellite frame. The government is now responsible for the launch of the *QZSS/Michibiki* system as well as the development and integration of the first satellite system.[86]

To our minds, this suggests that the commercial case for *QZSS/Michibiki*—that is, the need for commercial GPS services and non-positioning services—was difficult to state positively even at the inception. Many of the services trumpeted as important by Japanese actors were already available or becoming available via mobile phones, obviating the need for a complex satellite system at the level of the projected *QZSS/Michibiki*. Specifically, on the GPS side, navigation services with GPS devices using various ground-augmented technologies were and are already widespread and popular in mobile phones and other devices. On the business side, Japan's domestic market has already proven too small to support two commercial satellite service providers (witness the former J-SAT buying the former SCC) and one digital satellite broadcasting platform. In March 2006, after a year and a half of haggling with industry, the Committee on GIS and Positioning Information, a director general-level committee of the four ministries responsible for the government side dealing with *QZSS/Michibiki*, agreed that the government would fund the development and launch of one satellite, leading to the law discussed above. As part of the decision, the communication and broadcasting services were cut entirely from the program.[87] As USEF suggests, from the perspective of the four responsible ministries—namely, MIC, MEXT, METI, and MLIT—the *QZSS/Michibiki* is primarily about significantly improving the accuracy of positioning capabilities for Japan through the near-zenith satellites. According to some members of the private consortium developing it, rather than being merely an augmentation of the GPS, the *QZSS/Michibiki* already has limited accuracy positioning on its own.[88] In reality, then, along with the national security shift toward the publicized *IGS*, the *QZSS*, which is a new-generation GPS space augmentation system, was far more quietly sanctioned because independent positioning technology was also seen as necessary for national security.

Improved Spy Satellites

The timetable for projected spy satellite launches can be seen in Figure 5.1, and here we highlight only a few things in terms of future spy satellite capabilities and directions. There is little question that Japan's existing spy satellites have drawn attention, nowhere more so than among its neighbors.[89] In January 2005, the Japanese government was reported to have committed $44 million to developing a "fifth generation" of *IGS* with improved performance and to begin development of 0.5-meter resolution as well as the ability to switch quickly between targets. China, however, pegged the new satellites as "fourth generation" to be launched by April 2011. According to Xinhua, the current Japanese *IGS* weigh about 2 tons, are "difficult to point," and can take photos of the Democratic People's Republic of Korea (DPRK) once every two days for only several minutes.

Currently, the Japanese government's main policy for the *IGS* system is to maintain a constellation of four satellites available in the form of two optical and two radar satellites at any given time, while improving all the critical technologies for each new generation. In pursuit of this, the original timetable of the Cabinet Satellite Intelligence Center (CSIC) was to launch a second generation of *IGS* around 2008 to replace the first generation, although what improvements the second-generation craft would have remains unclear. The CSIC planned to have six *IGS* in orbit in 2006 and two second-generation satellites with improved imaging by 2009. The November 2003 failure ended this goal. As plans now stand, the CSIC has publicly disclosed the following facts.[90] First, unsurprisingly, the Japanese government is planning to spend in the region of ¥66 billion per year on the *IGS* program through the end of the decade, with one-third of that amount allocated for running costs but the remaining two-thirds to be spent on continually improving the next generations of *IGS*. Meanwhile, the March 2007 failure of the constellation's first radar satellite, originally launched in 2003, may lead to an acceleration (initial discussions were for a three-month acceleration) in the development and launch of its higher-powered replacement.

The next generation of optical *IGS* will be substantially similar to the first generation, but a third generation launched in 2009 included better pointing accuracy and greater overall information-gathering capability. These elements will be further improved for another set of optical and radar satellites in 2011–2012.[91] The improved optical *IGS* has "Quickbird-level" resolution, which suggests 50- to 60-centimeter class resolution. The CSIC remains tight-lipped on improvements to the radar satellites but it is expected to yield higher-resolution images. There are also plans for a slimmed down fourth-generation *IGS*, although no details are available as yet. With the publication of the Basic

Plan in June 2009, the national security role of the *IGS* looks as if it will be handily supplemented, or possibly even superseded, by the launch of the *ASNARO/Sasuke*-based constellation discussed below.

Reentry Technologies

To be clear, Japan's reentry technologies are not exactly emerging. In fact, as discussed in bits and pieces in earlier chapters, they have long been in the works. Not only has Japan acquired sophisticated missile technologies, but it has also developed a range of reentry technologies that can be militarily useful for successfully allowing warheads to enter the Earth's atmosphere. Here we bring together the hitherto scattered strands to provide a more complete overall picture of Japan's reentry technologies at present. The programs conducted to date have involved completely different and entirely legitimate objectives such as, for example, the need for data for developing a space shuttle or to engage in sample-return science experiments. The breadth of and variety of the discrete experiments, discussed briefly below, suggests that detailed understanding and acquisition of reentry technology has always been and remains an important underlying agenda for Japan.[92]

The 1990s saw three major reentry experiments, and, as a preface, it is important to know that there have been mishaps and outright failures along the way. In 2002 the *Demonstrator of Atmospheric Reentry System with Hyper Velocity* (*DASH*) was flown on an *H-IIA* as a piggyback satellite.[93] A wiring mistake meant the separation signal was never received by the reentry vehicle, which thus failed to separate from the fairing. *DASH* was to have separated in GTO orbit, collect thermal data on high-speed atmospheric reentry after three days in orbit, and steer itself down to a landing in Mauritania, West Africa. The element of *DASH* that garnered the most attention was the high-speed return of the ballistic reentry capsule at an estimated rate of 11.6 kilometers per second (km/sec). It was explicitly cited as a demonstrator that would pave the way for the asteroid explorer *Mu Space Engineering Spacecraft* (*MUSES-C/Hayabusa*) discussed below. The *Experiment Reentry Space System* (*EXPRESS*) was also a ballistic reentry vehicle, with a service and reentry module.[94] It was developed both for conducting space-environment-related experiments as well as developing reentry technology. It was launched on an *M3S-II* rocket in January 1995. However, due to the second-stage rocket malfunction, it was lost during its ascent, and the reentry capsule was found later in Ghana. It reportedly provided useful data for reentry tests, such as the capsule's thermal-resistance performance and flight results.

There are also other known technology demonstrations that deserve attention for their relevance to warhead reentry design. The *Orbital Reentry Experiment* (*OREX/Ryūsei)* was conducted in 1994 as part of the research to advance the *H-II Orbiting Plane* (*HOPE*) space plane.[95] The experimental vehicle was a capsule-type reentry vehicle with a blunt-cone shape and a diameter of 3.4 meters. It was the first Japanese vehicle to achieve atmospheric reentry from orbital speed, and its major goal was to obtain data and information under those conditions for future use. This meant evaluating autonomous de-orbit capabilities, testing thermal protection systems and materials for high-speed reentry, and advancing GPS navigation data acquisition during orbit and reentry. The vehicle successfully completed one 450-kilometer orbit, carried out a de-orbit by firing its retro-rockets for atmospheric reentry, and then deployed a parachute before splashing down in the Central Pacific Ocean. *OREX/Ryūsei* succeeded in carrying out an autonomous de-orbit using its navigation system and control systems to measure attitude and velocity. During atmospheric reentry, the vehicle was exposed to extreme aerodynamic heating (around 1,570°C), but the thermal protection material (carbon/carbon and ceramic tiles) on the nose cone survived successfully. These realities have been noted by observers in the United States. In particular, even the very first *OREX/Ryūsei* reentry vehicle is thought to have demonstrated Japanese mastery of reentry technologies in the service of ballistic missiles that may well allow Japan to move from a "countervalue" strategy (targeting of areas and cities) to a more precise "counterforce" strategy (targeting of missile silos or other hard military targets).[96]

A subsequent experiment was the *HYFLEX* in February 1996, also the first and only mission launch of the *J-I*.[97] The roughly 1,000-kilogram vehicle was launched to a maximum altitude of 110 kilometers and was released from the *J-I* while traveling at a speed of approximately 3.9 km/sec. The vehicle flew at a maximum speed of Mach 15 and did a gliding right turn around Chichi-Jima Island in the Ogasawara Islands group. With the aid of a parachute, it finally splashed down in the Pacific northeast of Chichi-Jima. The mission was deemed a failure because the *HYFLEX*'s flotation device failed and the vehicle itself sank in the Pacific Ocean. But, like those before it, Japan's first hypersonic flight was judged to have led to valuable data and experience for reentry technology, such as the thermal protection system and flight data. It also proved invaluable in confirming autonomous guidance and trajectory technology—with the vehicle operating entirely independently of outside control via its on-board computer and Inertial Measurement Unit (IMU). Importantly, the prototype space plane managed to splash down only around 3 kilometers from the planned point.

The 2000s have seen continuing reentry experiment programs.[98] Most notable has been one of several undertaken by USEF, that of the *Unmanned Space Experiment Recovery System* (*USERS*), which was successfully launched in September 2002. *USERS* built on the experience gained through the *Experiment Reentry Space System* (*EXPRESS)* project as well as the earlier *Space Flyer Unit* (SFU) project (for which the prime contractor was Melco). The stated objective for *USERS* was to provide a relatively low-cost experiment platform to test commercial-off-the-shelf (COTS) components and to conduct scientific experiments on-orbit as an alternative to using, for example, the U.S. shuttle or the ISS. The principal contractor for the system was Melco, although other subcontractors were also involved. The unique *USERS* system consisted of both a large service module (SEM) carrying the actual experiments and a large blunted cone-type reentry module (REM) with an abrasion-type heat shield. After about eight and a half months, the two modules separated, and REM returned the experiments to Earth while its SEM remainder continued to orbit. Whatever the probable usefulness of the science experiments, there were no problems with *USERS*'s demonstration of the reentry capability of the bullet-shaped REM in May 2003. It worked exactly as planned after reentry, deployed a parachute before splashing down on target in the ocean off the Japanese Ogasawara Island, and was recovered by ship. All in all, *USERS* was deemed a very ambitious project with impressive results that could be applied to benefit future research programs.

Other experiments and directions are also ongoing that, irrespective of their effectiveness or eventual outcomes, touch on some of the same issues of reentry technology demonstrations. *MUSES-C/Hayabusa* was launched in May 2003 and approached the asteroid Itokawa in September 2005.[99] It had four key technologies for testing, namely, interplanetary cruise via an ion engine as primary propulsion, autonomous navigation and guidance using optical measurement (as highlighted in Chapter 5), sample collection under microgravity, and, of interest here, direct reentry technology. *MUSES-C/Hayabusa* was certainly both commendable and notable for its historic effort to collect samples from an actual asteroid and to bring them back to Earth. It only succeeded in observation, because it remains unclear whether asteroid dust and pebble samples were actually collected. Its return is potentially of interest. *MUSES-C/Hayabusa* also contains a reentry capsule, with a mass of 20 kilograms, a diameter of 40 centimeters, and a convex nose with a 3-centimeter-thick ablative heat shield that is designed to protect the returning samples from extreme heat when the capsule reenters Earth's atmosphere at extremely high speeds of 12 to 13 km/sec and lands somewhere in the Woomera desert in Australia—which, in all, also amounts to testing an extremely high-speed ballistic reentry technology. *MUSES-C/Hayabusa*'s actual return remains uncertain but is estimated to be in 2010.

Missile Launch Early Warning Systems

As Japan's BMD system has become operational, the government has taken steps to ugrade its radars and to boost its tracking and communication capabilities in order to make the overall system function better.[100] As discussed in Chapter 5, moves to equip Japan with independent, or at least its own, military image intelligence date back at least a decade; these moves took on added urgency after the 1998 *Taepodong* launch, when Japan's SDF leaders chafed at the country's limited and one-sided access to U.S. imagery intelligence. The only piece that was missing up until June 2009—Japan's own early warning capability plugged into the system—is now slated for development as a matter of national policy.[101]

As with other technologies, the development of any such sensor grows out of Japan's existing space-based capabilities. JAXA and MEXT officials early on confirmed Japan's intent and capability to build a space-based early warning system.[102] More than anything else, these were based on prior technical feasibility. JAXA has developed a number of designs that are suitable for an early warning satellite, including the *Engineering Test Satellite* (*ETS-VIII/KIKU-8*), the *Wideband InterNetworking Engineering Test and Demonstration Satellite* (*WINDS/Kizuna*), and the *Data Relay Test Satellite* (*DRTS/Kodama*). Although not approved, JAXA also already had a plan to launch a series of LEO pairs of optical and radar observation satellites backed by geostationary observation satellites (although the geostationary observation satellites were not designed to have early warning sensors). A similar story is found at MEXT.[103] It too was conducting internal debates about how best to cope with a request—most likely to come from then JDA in the form of new satellites—to develop an early warning system.

Although the Japanese government's plan suggests that its early warning sensors can be multifunctional, such as in the detection of forest fires, this contradicts existing understanding on the matter. Early warning historically remained controversial precisely because the actual hardware for early-warning satellites (such as the U.S. Defense Support Program, or DSP) is uniquely military.[104] In other words, there are no civilian uses for such satellites' infrared sensors' capabilities for actually detecting the infrared radiation emitted by missile launch plumes and also distinguishing such emissions from fires, explosions, and also Earth's other natural background phenomena. Controversies aside, however, the reality is that the discussion to equip Japan with missile detection/early warning capabilities has been part and parcel of Japan's BMD cooperation from the start, as suggested by Melco's Pitch at the beginning of the IGS saga (in Chapter 5. It is now a critical element of Japan's new space policy. In fact, from Japan's perspective, a sophisticated early-warning system is imperative to an effective BMD response, given that it has little time to respond due to the geographical proximity of North Korean and Chinese launch sites and that it is far more proportionately vulnerable than the United States, with its greater population density.

Advanced Satellite with New System Architecture for Observation (ASNARO) and Space on Demand (SOD)

USEF and NEC are to play a major role in developing Japan's next-generation Earth Observation program through the *ASNARO/Sasuke* project.[105] This involves using the *ASR/Epsilon* to launch between two and six (two test and up to four operational) high-resolution optical and radar satellites, starting in 2011. *ASNARO/Sasuke* is of particular interest for a number of reasons—from the choice of contractors to the high level of the technology used and onto the underlying purposes and sophistication of the program as well as its speed of development and budget.[106]

As with the evolution of other space technologies, the *ASNARO/Sasuke* project is leveraging NEC's know-how through the *Advanced Land Observing Satellite* (*ALOS/Daichi*), the Panchromatic Remote-sensing Instrument for Stereo Mapping (PRISM), and the *IGS* optical sensors (working with spy satellite optics specialist Goodrich Corp.).[107] The goal is to deliver a series of 450-kilogram test satellites (developed via NEC's work with ISAS) with 0.5-meter resolution optical and sub-one meter resolution radar satellites with rapid revisiting times. NEC and USEF are teaming up with Tokyo-based geospatial information provider Pasco Corp. to disseminate the data, aiming at making images available from the highly agile 500-kilometer LEO orbiting satellites to customers within 30 minutes of receipt. As the constellation comes under the national security oriented pan-Asian observation program mandated by Japan's Basic Space Plan, the primary customer for this part-commercial enterprise will be the Japanese government.

The effort is part of METI's SOD program formulated in late 2007 to promote lower-cost satellites with shorter development times and improved functionality for all space systems (utilizing the COTS idea).[108] Effectively, the program seeks to churn out highly functional satellites in twenty-four to thirty-six months at $30 million, a tenth of the price of their JAXA developed antecedents. The SOD concept includes, for example, quick-reaction launch vehicles (such as the *ASR/Epsilon* discussed earlier) or others that could launch from aircraft or submarines, as well as satellites (such as *ASNARO/Sasueke*) that are reprogrammable in orbit. The similarity of the program to ORS overall, including for example the USAF's TacSat program, is striking—a fact noted by METI itself. Although compared to ORS the aims of SOD—primarily to support the industrialization of space without explicitly being a military customer—are certainly different, the dual-use nature of *ASNARO/Sasuke* renders such distinctions meaningless.

It is unclear at the time of this writing whether MOD will focus its reconnaissance program on *ASNARO/Sasuke* and/or *IGS*, or perhaps move to building its own dedicated system. It is clear, however, that MOD has come a very long way from an era in which its predecessor was embroiled in controversy over the use of commercial satellites to one in which it can steadily and legitimately advance a military space program without any blowback. In terms of planning and strategy, MOD's defensive military needs are likely to promote the development of both dedicated military and experimental dual-use technologies from 2010 onward. This is because like other ministries, MOD too has taken the enactment of the Basic Space Law as a departure point for thinking more systematically about the uses of space.[109] It is also going to be doing so explicitly in the context of national security with the establishment in August 2008 of its own Committee for the Promotion of Outer Space Development and Use (CPSDU). In January 2009 this high-level committee conducted a comprehensive audit of its upcoming defensive military space needs, identifying a significant number of space-related technologies and programs as being critical to the development of an integrated Command, Control,

Communications, Computers, Intelligence, Surveillance, Reconnaissance (C4ISR) infrastructure: more and higher-resolution imaging satellites, a dedicated military communications satellite, a missile early warning satellite, small and low-cost satellites that can be launched on short notice, a signals intelligence satellite, independent navigation and positioning capability, satellite protection, and space situational awareness (SSA) capabilities. Apart from noting how much MOD's military space capabilities coincide with other civilian programs, the satellite protection and space situational awareness capabilities are extremely interesting, as they show strategic concerns about the possibility of conflict in space. To allay these concerns may well require the development of extremely small, capable, maneuverable, and reprogrammable satellites, to which we turn next.

Small Satellites: Micro-, Nano-, and Picosatellites

As we discussed in previous chapters, there are strong implications for the militarization (and even weaponization) of space as laboratories around the world, both civilian and military, rush to develop smaller and smaller satellite technologies that take steps beyond SSA. Although there is absolutely no indication that the laboratories involved willfully use the technologies for military applications, the purpose of this brief section is to draw out Japan's participation in this latest direction and to put its development of micro-, nano-, and picosatellites in context. Throughout this book, we have attempted to demonstrate the maturity and potential of Japan's space systems for conversion to (potential and, as we have seen in some cases, actual) military use in their historical development. In doing so, our focus has been mainly on what can best be described as big system technologies.

However, as discussed in Chapter 5, the wide availability of steadily advancing electronics and semiconductor technologies have led to new and profound implications for military space operations, including ORS and SSA. In turn, there have been major and dramatic shifts in our understanding of potential ASAT and defensive and offensive counterspace technologies. The combination of such changes has allowed for the development of complex and capable missions on ever smaller, cheaper, and faster-built platforms—the reality of which is shown, for example, by the fact that ASAT-potential technology could be built on buses as light as 10 kilograms. The changes from satellite programs typical of the 1990s to those of today are emblematic of this evolutionary shift: first a shift from high cost (hundreds of millions of dollars) and long lead times (five years) to low cost (several millions of dollars) and short lead times (under a year); and from there toward on-time assembly of intelligent multi-role (ASAT, situational awareness, communication, refuel and repair missions) and proposed launch on demand systems at present.

Where does Japan stand in these emerging trends?[110] In the case of Japan, there has been a strong shift away by both planners and engineers from "battleship" satellites to smaller, lower-cost, less risky satellites. Japan's shift can, in part, be attributed more specifically to the loss of the near $800 million, 3.6-ton *Advanced Earth Observing Satellite* (*ADEOS/Midori*) in 1997. The collapse of an experimental Toshiba solar array design ruined the international mission that involved both U.S. and European participation. That incident seriously caused both NASDA and SAC to review the practice of building extremely expensive, huge, and very complex satellites with multiple missions all, literally, in one basket. The push toward smaller satellites remains a work in progress.

In addition to the *SmartSat* series mentioned in Chapter 5, one of the major new trends in Japanese satellite building activity that has gained momentum since 2000, and that now figures in Japan's Basic Space Plan in 2009, has been the surge of small satellite projects at Japanese universities that are supported by industry, including NEC and MHI, in particular. The programs are potentially advantageous to Japan's future space development because building such satellites (and spacecraft, lunar probes, and so on) gives the next generation of engineers hands-on experience in building working space systems in projects that typically cost only a fraction or so of government space programs and involve extremely rapid development cycles. Thus future generations of engineers receive practical training at a fraction of the cost of recruiting and employing them on, for example, JAXA programs, and they also do so in an extremely cost-effective way.

Below we briefly outline some of the major research and development programs under way at universities, corporations, and also the relevant government agencies. The widening scale and scope of Japan's burgeoning microsatellite activities means it is difficult to cover the entire range of activities, but an appreciation of the nature and vibrancy of such activities at universities can be obtained by looking at the members of the University Space Engineering Consortium (UNISEC).[111] This NPO, which merged the activities of two organizations focused on satellites and hybrid rockets, took off in 2003. As of

July 2009, it has nearly fifty satellite (or related) or rocket development teams from thirty-four universities.

Although UNISEC supports development of hybrid rocket activities, it has thus far been most active with respect to pushing micro-, nano-, and picosatellite development among university students and may well turn out to be a vehicle to push for government funding for the dozens of proposals made by members. In fact, a wide range of experts and leaders in the field, as well as teams of younger engineers in what was then ISAS and NASDA, all provide critical support for the NPO. Its activities include sharing of information and collaboration, helping students to use ground test facilities of national laboratories, consulting on political and legal matters, coordinating joint development of equipment and projects, and bridging university activities as well as the needs and interests of related persons and organizations. In a competitive civil and military market, the ultimate goal of these ventures, whether led by UNISEC, other teams, or individuals, is the miniaturization of satellite technology, as well as independent capability in small satellites and micro-electro mechanical systems technology.[112]

Already there are a number of present and also future small satellite programs in development across Japan. Table 6.2 sets out our best estimates about the key players, small satellites, and their main missions as of August 2009. As the table makes clear, the research and work on smaller satellite technologies began in the 1990s and are now spreading out from the government and big corporations to include a number of universities and laboratories. Using the table, we briefly outline some of the more salient features below as they show the continuing importance of universities in the process of satellite design, innovation, and evolution. We then provide an examination of government and corporate activity on this front.

Chronologically Japan's first university-built microsatellite was the *Whale Ecology Observation Satellite System* (*WEOS*), developed by Chiba Institute of Technology. However, the actual *CubeSats* (basically, nanosatellites measuring 10 centimeters on each side with a weight of 1 kilogram) were developed by the Tokyo Institute of Technology and the University of Tokyo, beginning with launches of *CubeSat Cute-1* and *CubeSat XI-IV*, respectively, in 2003. Since then, these two institutions have continued to be involved in launching successor *CubeSats*, as indicated in Table 6.2. They are no longer the only players, however. Other *CubeSat* projects that are being carried out include those by Hokkaido Institute of Technology, Nihon University, Kagawa

Table 6.2. Estimated Range of Japan's Small Satellite Programs, 2002–2009

Year		*Launcher*	*Designation*	*Mass*	*Main Objectives*	*Principal Architect*
2002	Feb	*H-IIA*	*DASH*	70 kg[1]	Testing of reentry technology	ISAS/NASDA (subsequently JAXA)
	Dec[2]		*μLabSat; Micro LabSat 1*	54 kg	Testing of miniaturization technologies, small satellite bus technologies, earthly sensors, SELENE relay separating mechanism, remote-controlled inspection technologies	NASDA (subsequently JAXA)
			WEOS Kanta-Kun	50 kg	Testing of observation probe attached to whales to obtain data on position, water pressure, temperature for study of migration patterns and diving depths, tracking, and GPS technologies	Chiba Institute of Technology
2003	Jun	*Rokot*	*(CubeSat) Cute-1*	1 kg	Testing of communication, attitude, and temperature sensing, deployment of antennas and solar paddle	Tokyo Institute of Technology
			CubeSat XI-IV	1 kg	Testing of ultra-small bus technology, transmission of images to ground, education	University of Tokyo
2005	Aug	*R-36M*	*INDEX/Reimei*	70 kg	Testing technologies for future small satellites, simultaneous observing of aurora emission structure and plasma particles	JAXA, universities (Graduate University for Advanced Studies, University of Tokyo, Musashi Institute of Technology, Tokyo University of Science, Soka University, Rikkyo University, Tokyo Metropolitan Institute of Technology, and Tokyo Denki University)[3]

	Oct	*Kosmos*	*Cubesat XI-V*	1kg	Testing of ultra-small bus technology, transmission of images to ground	University of Tokyo
2006	Jan	*S-310-36*	*Furoshiki Sat Test*	–	Testing feasibility of deployment/ extension of mesh/net structure and large-phased array RF performance	University of Tokyo, Kobe University
	Feb	*M-V*	*Cute-1.7+APD*	3.5kg	Testing of attitude and control, APD (Avalanched Photo Diode) sensor for observation, communications[4]	Tokyo Institute of Technology
	Sep	*M-V*	*SSSAT*	6kg	Testing solar power sail extension, attitude control, solar sensor	JAXA
			HITSAT	3kg	Testing small satellite bus, magnetic torque-based attitude control	Hokkaido Institute of Technology, Hokkaido University
2008	Apr	*PSLV*	*Cute-1.7+APD-II*	3kg	Testing of attitude and control, sensor, scientific observation, imaging, communication	Tokyo Institute of Technology
			SEEDS-II	1kg	Testing communications[5]	Nihon University
2009	Jan	*H-IIA*	*SDS-1*	100kg	Testing bus technologies for future small satellites, integrated transponders, network-type data-processing technology, and so on	JAXA
			Kagayaki	20kg	Testing autonomous control system, inflatable progress boom, debris collection, aurora electric current observation, reducing residual magnetism, educational	Sorun Corporation (with Tokai University)
			STARS-I	Parent 3.6kg Child 3.4kg	Testing attitude control and imaging of parent-child satellites during tether deployment	Kagawa University

(continued)

Table 6.2. *(continued)*

Year	*Launcher*	*Designation*	*Mass*	*Main Objectives*	*Principal Architect*
		KKS-1	3 kg	Testing of micro thrusters, three-axis attitude control functions, land imaging	Tokyo Metropolitan College of Industrial Technology
		PRISM	5 kg	Testing ground-image acquisition with expandable refracting telescope, ultra-small/nanosatellite bus	University of Tokyo
		SPRITE-SAT	47 kg	Observing natural phenomena (atmospheric lightning, terrestrial Gamma-ray flashes, very low-frequency electric waves)	Tohoku University
		SOHLA-1	50 kg	Testing of small satellite bus technology , new technologies such as VHF lightning impulse measurement, satellite laser ranging (SLR) for calibration of micro GPS-based satellite positioning	Astro Technology Sohla[6]

SOURCES: JAXA, "Jikken Mokuteki wa Kōsoku Saitotsunyū Toki no Deeta Shutoku" [Experimental Purpose to Obtain Data at Time of High-Speed Reentry], available online from JAXA Space Information Center at www.jaxa.jp (accessed 30 August 2009); JAXA, "Small Satellite Research Activity in JAXA" presentation at the Information Exchange Meeting for Small Satellite Development at the Asia-Pacific Regional Space Agency Forum (AP-RSAF-11), Canberra, Australia, 2 November 2004, pp. 1–33; National Space Development Agency of Japan (NASDA), "ADEOS-II/H-IIA.F4 no Uchiage Keikaku ni Tsuite (Kogata Fukueisei: RedSat, WEOS, μ-LabSat) [Concerning the Launch Schedule for ADEOS-II/H-IIA.F4 9 (Piggyback Satellites: RedSat, WEOS, μ-LabSat)]," press release, 2 October 2002, available online at www.jaxa.jp (accessed 30 August 2009); T. Hayashi et al., "Whale Ecology Observation Satellite (Kanta-kun) System," *Space Japan Review* 32, Dec 2003/Jan 2004, pp. 1–4; Kyoichi Ui et al., "Tokyo Tech's CubeSat CUTE-1: Operation Reports and Analyses of Pico-Satellite Dynamics" ISAS Proceedings of 13th Workshop on Astrodynamics and Flight Mechanics, 2003; abstract only via JAXA at www.jaxa.jp (accessed 30 August 2009); University of Tokyo CubeSat Team, "Cubesat to Wa?" [What is CubeSat?], available online at www.place.t.u-tokyo.ac.jp (accessed 30 July 2009); Interview, Shinichi Nakasuka, "Chōko Gata Eisei ni yoru Uchū Kaihatsu e no Chōsen" [The Challenge of Ultra-Small Satellites for Space Development], 28 August 2006, available online at www.jaxa.jp (accessed 30 August 2009); Shinichi Nakasuka and Nobuyuki Kaya, "Quick Release on Experiment Results of Mesh Deployment and

Phased Array Antenna by S-310-36," *The Forefront of Space Science*, 13 April 2006, pp. 1–3, online version at www.jaxa.jp (accessed 30 August 2009); Shinichi Nakasuka et al., "Sounding Rocket Flight Experiment for Demonstrating 'Furoshiki Satellite' for Large Phased Array Antenna," *Acta Astronautica* 59(1–5) 2006, pp. 200–205; JAXA, "M-V Roketto 8 Gō Ki Tōsai Subupeiroodo no Jikken Kekka ni Tsuite" [Concerning the Experiment Results of the Sub-payload of M-V-8], press release, 3 March 2006, available online at www.jaxa.jp (accessed 30 August 2009); Katsutoshi Imai et al., "Tokyo Tech Small Satellite Development Projects—Cute-1.7 and Tsubame," *Institute of Electronics, Information, and Communication Engineers (IEIC) Technical Reports* 104(698), 2005, pp. 35–40 (abstract only), available online at sciencelinks.jp (accessed 20 August 2008); Ken Fujiwara et al., "Tokyo Tech Nano-Satellite Cute-1.7 + APD Flight Operation Results and the Succeeding Satellite" paper presented at the 17th IFAC Symposium on Automatic Control in Aerospace, Toulouse, France, 25–29 June 2007, pp. 1–6, available online at iss.mes.titech.ac.jp (accessed 29 August 2008); Hiroki Ashida et al., "Design of Tokyo Tech Nano-Satellite Cute 1.7 + APD II and its Operation" paper, The 59th International Astronautical Congress, Glasgow, 29 Sept–3 Oct. 2008, pp. 1–10, available online at iss.mes.titech.ac.jp (accessed 29 August 2008); Takanao Saiki, "Attitude Control System of Spinning Solar Sail Sub-Payload Satellite," *Proceedings of 16th Workshop on JAXA Astrodynamics and Flight Mechanics*, March 2007), pp. 237–241, abstract only at www.jaxa.jp (accessed 30 August 2009); Yuichi Tsuda, "M-V Roketto Tōsai no Subupeirodo" [The Subpayload of the M-V Rocket], *JAXA Special Reference*, February 2008, at www.jaxa.jp (accessed 30 August 2009); Nihon University, "Report of the Launch of SEEDS II from Nihon University," 28 April 2008, Nihon University Official Cubesat Project Official Web Site (for SEEDS), available online at cubesat.aero.cst.nihon-u.ac.jp (accessed 27 July 2009); JAXA, "Kogata Kagaku Eisei 'Reimei'" [Innovative Technology Demonstration Experiment] pp. 1–2, at www.jaxa.jp (accessed 30 August 2009); Hirobumi Saito and Seisuke Fukuda, "Development of Small Highly Functional Science Satellite, INDEX," *The Forefront of Space Science*, 18 November 2004, pp. 1–3, online version at at www.jaxa.jp (accessed 30 August 2009); Masafumi Hirahara et al., "Small Scientific Satellite REIMEI and Auroral Observation," 4 April 2006, pp. 1–3, online version at www .jaxa.jp (accessed 30 August 2009); JAXA, "Kogata Jisshō Eisei 1 Gata" [SDS-1: Small Demonstration Satellite 1], pp. 1–2, at www.jaxa.jp (accessed 30 August 2009); Asia-Pacific Regional Space Agency Forum (APRSAF), "The 6 Small Satellites Will Be Launched Onboard the H-IIA Rocket in Summer 2008, Japan," *Feature Stories*, available online at www.aprsaf.org (accessed 20 July 2009); JAXA, "H-IIA Roketto 15 Gō Ki ni yoru Onshitu Kōka Gasu Kansoku Gijutsu Eisei 'Ibuki' Oyobi Kogata Fukueisei [o]no Uchiage Kekka ni Tsuite (Sokuhō)" [Concerning the Launch of Ibuki and Piggyback Satellites (Announcement)], 28 February 2009), p. 4, at www.jaxa.jp (accessed 30 August 2009); JAXA, "Ibuki Special Site: Piggyback Payload" 9 February 2009), including the linked home Web sites of the principal architects identified, at www.jaxa.jp (accessed 30 August 2009).

1 With the heat shield, the total *DASH* weight was 86 kilograms.

2 FedSAT, a 58-kilogram satellite developed by the Australian Cooperative Research Center for Satellite Systems (CRCSS), was also launched on the same manifest in December 2002, seeking to test magnetic observation mechanisms, Ka/UHF band communications, GPS, and so on.

3 According to *Encyclopedia Astronautica*, Melco was also involved as a private contractor. See www.astronautix.com for information on Japan (accessed 27 July 2009).

4 *Cute-1.7 + APD* stopped functioning after a few months and its missions/experiments were carried over to its successor.

5 *SEEDS* was lost in a rocket failure in 2006, and *SEEDS-II* essentially continued the mission.

6 This is an Osaka-based cooperative undertaking projects sponsored by NEDO since 2003. The name SOHLA refers also to the Space-Oriented Higashiosaka Leading Association, with which JAXA signed a cooperative agreement in May 2004 and agreed to provide technical information and assistance for the development of small satellites *SOHLA-1* and *SOHLA-2*. See Japan Aerospace Exploration Agency, "Concluding a Cooperative Agreement between Japan Aerospace Exploration Agency (JAXA) and Space-Oriented Higashiosaka Leading Association (SOHLA), press release, 20 May 2004, available online at www.jaxa.jp (accessed 27 July 2008).

University, and so on. In all university cases, there is a strong emphasis on nurturing and training the next generation of engineers and satellite builders. Nihon University, for example, started a nanosatellite (*CubeSat*) project as part of student education around 2001, which led to the development of *SEEDS* (*Space Engineering EDucation Satellite*) and its more fortunate descendant *SEEDS-II*.

As the headquarters for UNISEC, the University of Tokyo is pivotal in this new R&D direction. Since launching *PRISM*, it has focused on the METI-funded *Panel Extension Satellite* (*PETSAT*). It seeks to block-build components of the satellite that can then come together in various combinations for on-demand customized satellite missions. The *PETSAT* project is also collaborating with a consortium of small companies in Osaka called the Space-Oriented Higashioka Leading Association (SOHLA), itself working in conjunction with Osaka University, to build a series of spin-stabilized microsatellites with multipurpose panels to follow from *SOHLA-1*. There are also ambitious efforts toward developing systems that can perform modular autonomous operations in space with robotic technology. One line of research at the Research Center for Advanced Science and Technology (RCAST) at the University of Tokyo is working on robot-friendly modularized satellites that could perform a number of in-orbit operations, including satellite capture with an autonomous robotic maintenance vehicle. It is also working on a project called *CellSat* (*Cellular Satellite*), which may lead to the development of technologies to enable functionally segmented cells of a satellite, allowing it to reconfigure itself if necessary. This work has also spun off into a laboratory with the *Robot Oriented Space Evolution Technology Task Force* (*ROSETTA*), focusing more closely on intelligent space robotic functions for a range of on-orbit autonomous operations. The Laboratory for Space Systems (LSS) at the Tokyo Institute of Technology is also advancing novel and modular space robotic systems, such as the *Reconfigurable Brachiating Robot* (*RBR*) system to be tested on *JEM/Kibō* and on robot satellite cluster systems.[113]

On the government side, JAXA, as well as other government actors like METI, are pushing ahead with a range of programs that are relevant here.[114] One of the most important overall tasks of satellite development is the construction of standard bus systems (frames). Since 1998, JAXA has been seriously studying small satellite bus technology at its Space Technology Demonstration Research Center (STDRC). *μLabSat/Micro LabSat 1* launched in 2002 demonstrated the operation of a 50-kilogram bus and an onboard computer.[115]

The *Innovative Technology Demonstration Satellite* (*INDEX*), renamed *Reimei*, was launched in 2005 and was useful above all in showing that satellite development cost and time cycles could be changed into high-speed low-cost ones in order to test next-generation small satellite components. The STDRC's goals are to develop standard bus technology for small satellites and one example is developing a 50-kilogram class bus capable of inter-satellite communications, autonomous orbit maneuvering, and rendezvous technology. Another example is a standard 100-kilogram class bus, the first example being the *Small Demonstration Satellite* (*SDS-1*) launched in early 2009. The emphasis is on the low-cost, high-speed construction of highly functional satellites that can be used for disaster monitoring (or early warning), weather observation, and mobile communication.

There are also other projects that highlight the dual-use complexities described earlier. ISAS/JAXA, for example, would like to launch the *SCOPE* (*cross Scale COupling in the Plasma universE*) mission to investigate the Earth's magnetotail.[116] The mission would feature formation-flying satellites, including a mothership and probably four daughterships able to communicate with each other up to a maximum distance of 5,000 kilometers. JAXA has also reached out to the various skills and knowledge bases emerging from Japan's university-based satellite design efforts, largely in an effort to boost the agency's ability to deliver on microsatellite technology. One avenue in which this has moved forward is JAXA's interest in an active space debris removal system.[117] It is working along with Kyushu University on related international experiments, such as microsatellite impact testing with NASA's Orbital Debris Program Office. JAXA has also promoted the idea of a small spacecraft that captures large debris objects in useful orbits and transfers them to disposal orbits. Put simply, the capture/removal system uses electrodynamic tethers (EDT), which are essentially long "wires" that unfurl beneath the target and "drag" them lower and lower until they de-orbit.

Finally, there is also some related corporate activity on the small satellite front that has surfaced publicly.[118] MHI, for example, has done work on orbit approach rendezvous technology in order to cope with elliptic orbits and with formation-flying control involving a target and chaser satellite—a sort of super *ETS-VII/Kiku-7* experiment. Toshiba's name has also surfaced prominently in connection with the design and development of the *μLabSat/Micro LabSat 1*. Another intriguing program is a conceptual plan called the Micro Satellite Launch System (MSLS) that was first put forward in public and then retracted as being a voluntary activity not recognized officially by KHI.[119] In any event, the idea behind this was that a dedicated small, two-stage, solid-fuel rocket satellite launcher (launching a 40-kilogram payload into a 200-kilometer orbit) would offer a quick fix for microsatellite launches and would also involve minimal investment and new technology. Whatever the actual state of MSLS, or something like it, the fact is that Japan could rapidly design and produce a launch-on-demand small satellite booster if needed.

Although the major established rocket and satellite players do not appear to be heavily involved at this stage, they are no doubt keenly aware of the small-satellite trend, which may well become bigger with backing from the government. This is because rocket makers such as MHI have launched the smaller satellites on the *H-IIA*, and more of such business would be welcomed, especially if foreign launchers are somehow disqualified from launching. Satellite makers, such as Melco and NEC, also have an interest in pursuing such research and it is difficult to imagine, given their long experience, that they would not be able to press the talents of the existing cadres of young engineers into service. It should also be remembered that the *Pencil*, which launched Japan's rocket program, started out with engineers and visionaries who were just as determined in the early postwar period as those at the forefront of small satellite programs in Japanese universities today. For both rocket and satellite makers, all this would make even more profitable sense if, in the current environment, the smaller satellites themselves lent themselves more explicitly to military applications. Thus although corporate players are small at this stage, they are not to be taken lightly. One known private firm, Sorun Corporation, has already launched *Kagayaki*, which is billed as the first small satellite built by private companies.

The above overview barely scratches the range of research being carried out in the microsatellite field in Japan. This is a field which deserves the utmost attention in our view, especially because the government has emphasized the need for close-knit public-private relations in the interest of securing space-related human resources in Article 21 of the Basic Space Law (see Appendix II here). As this brief account indicates, there are some important research directions in Japan in the next-generation satellite and spacecraft technologies. As their proponents have suggested, these smaller satellite projects will certainly remain useful for attempting to cut development costs and time, testing new technologies and components, and delivering full science missions on extremely tiny physical frames. As with programs around the world, they are to be lauded. However, we believe that this trend toward ever more sophisticated miniaturization of satellites and spacecraft can also serve military purposes. As discussed in Chapter 5, the very same technologies that can reenter, fly apart, dock, communicate, de-orbit others, and so on can also be deployed to work as ASAT weapons that can take out enemy satellite and spacecraft. Although the budget, schedule, and on-orbit demonstrations of these new systems remains to be seen, this new research needs to be borne in mind when evaluating Japanese military space capabilities in the future.

Unmanned Aerial Vehicles (UAVs)

A final area that deserves monitoring is UAVs that may go on to be even more integrated with spaceborne surveillance assets in the future given, for example, their dependence on GPS to aid navigation. Although UAVs have been around since World War I, they really began to draw attention after their battle use in Operation Desert Storm in 1991 and were further legitimated after 9/11.[120] UAVs essentially make airborne surveillance possible, and have gained ground around the world as intelligence, surveillance, and reconnaissance (ISR) systems. UAVs have begun to make a big dent in many Asia-Pacific countries concerned with monitoring archipelagos, long coastlines, and sea lanes. For Japan, these high-altitude reconnaissance platforms, usually equipped with infrared, electronic surveillance, and maritime radar sensors (such as Northrop Grumman's Global Hawk), could identify intruding ships, spot cruise missile launches, track small spy ships, and even monitor missile launches. From there, some research concepts by Raytheon suggest that long-endurance UAVs (roughly the size of General Atomic Aeronautical Systems' Predator B) could carry adequate missiles and fly high enough to set up launch-area denial spheres over enemy missiles while orbiting in international waters over, say, the Sea of Japan. In all, the popularity of UAVs has risen rapidly, and U.S. companies hold about 60 percent of the market in an industry whose sum total is expected to be worth about $16 billion through 2015.

Japan has also expressed an interest in the development of a domestic UAV.[121] With some modifications (such as "sense and avoid" capability to reduce the risk of collisions in national or civilian airspace with airplanes), UAVs may also certainly make a mark in a range of commercial or civilian markets, with varying degrees of government involvement. These markets include, for example, the observation of disaster zones and dangerous areas, border patrol missions, protection of civil airports, search-and-rescue missions, crop spraying, and so on.

But the military angle is all important, as some developments make clear. One ample demonstration comes in the Yamaha Motor Company debacle from 2006 onward.[122] Yamaha develops unmanned helicopters, such as the *RMAX* helicopter with autonomous flight capability, for disaster-area observation and surveillance, as well as agricultural crop spraying. In 2005 Yamaha helped METI actually craft stricter export controls on unmanned helicopters; however, in a stunning revelation in 2006, Yamaha was discovered to have illegally exported remote-controlled helicopters to Chinese companies (Poly Technologies, Beijing BE Technology) under the direct supervision of or connections with the Chinese Army. According to Yamaha, the helicopters were to be used for spraying pesticides, but according to the Japanese government, they could also be diverted for military uses such as spraying hazardous chemical or biological agents. Because an unlicensed export of sensitive items constituted a violation of the Foreign Exchange and Foreign Trade Law, Yamaha was barred from exporting its remote-controlled helicopters and related components for nine months from 18 May 2007 and was further subject to hefty fines in summary courts in Iwata, Shizuoka Prefecture, where it is based.

Another pertinent example is that UAVs continue to draw the attention of the Japanese defense establishment and may well go on to become a sizable element in integrated airborne- and spaceborne ISR for Japan.[123] The Japan Air Self-Defense Force (JASDF) has expressed interest in using UAVs to collect tactical information about the size and capabilities of the enemy and to conduct battle damage assessments largely with a view to operations focused on the air defense of Japan. Although it is not looking to use weapon-carrying UAVs for attack operations at this stage, the acquisition of that capability on its own, irrespective of costs, cannot be ruled out in the future. MOD has been seeking to build a long-endurance and high-flying surveillance drone, equipped with an AIRBOSS infrared camera which would be used to detect rocket plumes in missile launches and also provide extended surveillance of regional neighbors such as North Korean and China.[124] As in the case of other space technologies, the U.S. has been initially opposed to the project as, in its view, Japanese requirements could be met with U.S. systems. But as also in those other cases, with the lead industrial player in the drone's development likely to be MHI (which was awarded a contract in 2007 to study the concept), Japan is likely to pursue its development.

CONCLUDING ASSESSMENT

Drawing on its existing technological base of rockets and satellites, Japan is engaged in a wide range of research and development that is designed to keep it abreast of next-generation technologies in space. On the rocket side, Japan has moved conceptually well beyond the present heavy-launcher *H-IIA*, onto the ambitious solid-fuel *ASR/Epsilon*. The latter rocket seeks to increase speed of access to space and also to revolutionize launch infrastructures; it may, therefore, be of service in a military context. Component technologies, as well as the integrated infrastructure, of Japan's rocket program in aid of further military goals are also being aided by Japan's continuing deployment of the BMD system in cooperation with the United States. Additionally, there has been a long-standing hope for a Japanese space plane in the future. On this front, Japan is continuing research on RLVs, as well as scramjet engines, as part of the effort of staying abreast of new propulsion technologies and the ultimate goal of achieving hypersonic flight capabilities.

On the satellites and spacecraft side, Japan continues to seek improvements in its *IGS* or spy satellites, which is to be expected given that they have already been deployed. Through *ASNARO/Sasuke*, it is seeking to move toward an additional new set of surveillance satellites. In line with improving its BMD-related infrastructure, Japan is also taking steps to move toward satellites that will allow for an early warning system of its own rather than being dependent on the United States. In the interest of improving and upgrading its ISR system, Japan is taking significant steps in new directions as well. The country now has a legal setup that shows its determination to have the *QZSS/Michibiki*, which is its version of a regional GPS or Galileo system and which can serve also as a counterpoint to China's Compass system in the region. At present, a number of university-, corporate-, and government-based research programs under way on ever more sophisticated, miniaturized, and autonomously operational satellites and spacecraft systems also showcase the way civilian technologies can potentially be transposed in the service of militarized space assets.

As we stated at the outset, these largely unknown research efforts and programs in Japan may not come to fruition or may even evolve differently than expected. However, some of their aspects deserve attention in line with our market-to-military thesis, particularly in the post-Basic Space Law era. As some have correctly noted, for example, it is not clear what the potential overseas or domestic market is for small satellites; rather, as with most countries, Japan is more likely to use its small satellites for reconnaissance or other military purposes.[125] In addition, from a commercial point of view, small satellites of all stripes are less attractive to industry as they produce fewer profits than big ones. Although things may certainly change, much the same argument can be applied to Japan's ambitions for the next generation of rocket launchers, which do not presently stand a chance in global commercial markets given the trends. Indeed, with respect to the *ASR/Epsilon*, the commercial rhetoric has not really come to the fore as it has done for almost all other SLVs in Japan. Put simply, then, the survival, viability, and sheer profitability of some of Japan's emerging satellite and rocket technologies may well depend on securing a military angle.

7 IN DEFENSE OF JAPAN

IN THIS BOOK WE CONCENTRATED on Japan's capabilities in the space industry—the one strategic sector that now underpins civil, commercial, scientific, and security concerns for all great powers. As we detail, there has been an incremental but solid shift from the market-to-the-military in Japan's space policy overall. Practically, this means that Japan's militarization—meaning, in this book, the use of space for military purposes to support, enable, or conduct defensive, and, potentially, even offensive actions, to protect the homeland—has already become reality.

We wrote this book primarily to document specific developments in Japanese space assets spread across a range of public and especially private actors who have been critical players in Japan's space saga: rockets and satellites and, through them, a set of emerging related technologies. Through these players, Japan has acquired capabilities within its civil space program in plain sight of the public, component by painstaking component, and, until very recently, with a stated focus on each capability's technological level, scientific utility, and commercial potential. But as experts point out, space technology is inherently dual-use, with an estimated 95 percent of all space technologies having both civil and military applications.[1] A leading space expert puts it in pragmatic terms:

> Rockets, launchers, and missiles for military use rely on the same basic parameters as space flight for success. If a country can build a missile, it clearly has the technical knowledge to build a launch vehicle. The primary technical differences between the two are trajectory, payload, and guidance systems; the

> primary difference overall is intent of use. Images taken by remote-sensing satellites can be used to maximize crop rotations for increased yield or to target weapons. Navigation satellites, such as those in the GPS, keep civilian airliners properly spaced and on course as well as guide munitions with precision accuracy. Other technologies are more difficult to identify as potentially dual use. . . . Component parts are perhaps even more problematic than large pieces of hardware in determining intent. Military and commercial space systems share components for electronics and computers, optics, propulsion, and sensors.[2]

What we seek to show is that this duality has always been as true for Japan as elsewhere—despite the long-standing government emphasis on commerce and science in the space program, despite little public acknowledgement of the reality of militarized space assets across the postwar period, and, most important of all, despite the country's pacifist constitution and official orientation under the peaceful purposes resolution from 1969. Thus a singular problem-focused issue drove our book: How, when, and why did things come to the overall shift from the market-to-the-military in Japan's space policy?

There are of course long-standing debates about Japan's militarization, namely controversies about realism and constructivism, as well as the legal furor over constitutional change. We, however, focused on process-tracing the space-related activities of the Japanese public and, where possible, private players. The sum total of their actions over the past several decades suggests that we also need to pay attention to the highly important role of concrete corporate interests. As stated at the beginning in Chapter 1, we did not set out to provide any kind of robust theoretical tests, or to make any kind of sweeping proclamations about international relations theory or legal change or even interest-based approaches that are more common to political economy studies. Rather, we aimed to use our specialist evidence to briefly assess the interplay of such approaches to the market-to-military trend in Japan's space technology and policy. Overall, we find that the key ideational, institutional, and legal approaches provide a very rich tapestry for the interest-based narrative in this book. Before getting to a discussion of the principal focus, findings, and implications of our study, which forms the bulk of this chapter, we briefly set out some of our key assessments with respect to the theoretical and constitutional debates in light of the findings.

To the extent that realists, whether of the defensive or offensive stripe, are correct in suggesting that structural changes will elicit more aggressive security policies, we find that in many ways Japan's market-to-military conversion in the space industry has been facilitated by the rising insecurities across borders, both globally (such as in the U.S.-led war on terror) and regionally (such as the fact of nuclear-armed North Korea and a rising China). But as many have already observed, we too assert that it is difficult to predict responses to any structural shifts in the distribution of capabilities (or, for that matter, threats) without an understanding of the domestic ideas, institutions, and especially interests that make such responses possible. This realist emphasis on external uncertainties chimes, as we indicated, very nicely with the interests of Japan's corporations, who have an eye on their economic livelihood in the future both within Japan and abroad. Successfully exploiting the dual-use ambiguities inherent in space technologies in a volatile security environment, these are some of the very agents who have been central in facilitating an actual shift toward more realist-like stances and norms in national space policy. Their actions are consistent with claims that realist pragmatism has historically informed Japan's foreign policy apparatus.[3]

Second, to the extent that constructivist arguments, based on the culture of antimilitarism have relevance for the market-to-military trend, we also find that in many ways the legal and institutional elements that make up Japan's postwar pacifist strictures, particularly the peaceful purposes resolution, were key in framing policy debates about the uses of space.[4] But as space technologies progressed, Japan also successively either worked around them or did away with them. Practically put, even with the peaceful purposes resolution in place, there have already been significant breeches in Japan's postwar security identify of domestic antimilitarism, some of them with particular resonance in Japan's space-based capabilities as we go on to discuss below.[5] There is the Ministry of Defense (MOD) that, like all bureaucratic entities, will continue to expand in influence and may well acquire significant control over space assets in the future; there is the development of an indigenous solid-fuel ballistic missile capability coupled with the potential for nuclear warheads; there are the BMD system technologies, the most promising potential of which lies in technologies targeting incoming ballistic missiles in the boost phase on foreign territory and even potentially other spacecraft; there are rapidly improving reconnaissance satellites and significant counterspace technologies; and there have already been SDF boots on foreign soil. Until recently, especially for the space field, these were scattered dots that did not really seem to cohere into an explicit whole. But, as we will discuss below, the new Basic Space Law changes that. In the process of legally redefining the peaceful purposes resolution for Japan's space activities, it not only augurs an alternate future for Japan's space technologies but potentially also for the country's security identity.

Third, with respect to the dimension of constitutional revision, particularly Article 9, we find that it has never seriously hindered Japan's quest for security.[6] It has been less important for what it says than for what it has been stretched to mean. Because the sincere renunciation of war and forces had to confront the practicalities of homeland defense from the start, Article 9 has turned out to be the fount of contortions that strain legal credulity on almost every front.[7] This is evident in a number of ways:[8] For example, the existence of the Japan Self-Defense Force (JSDF)—whose "war potential" and functions have continued to expand since their formation by law in 1954 in the wake of the Korean War—cannot really be reconciled with the text or spirit of the Article as successive tortuous interpretations by the CLB in 1952 and 1954 show only too well.[9] Additionally, the aversion to collective self-defense—to which the government has the *right* (under Article 51 of the United Nations Charter) but still not the *right to exercise* (under Article 9) according to a 1981 interpretation by the CLB—does not square well with the existing military realities in which Japanese troops are already engaged.[10] Although debate will continue about whether such moves on the part of the JSDF will lead to full-blown military operations abroad all the time, they are even more instructive about the limitations of existing legal interpretations.[11]

While constitutional mandates have been brought to bear on such debates, more on point for the purposes of this book is Japan's peaceful purposes resolution in terms of the use of outer space. As we saw in Chapter 2, since 1969 it was interpreted as being strictly "nonmilitary" in Japan, well beyond its more common international interpretation of merely "nonaggressive" in terms of the 1967 Outer Space Treaty. Even the use of satellite imagery by the JSDF was controversial, and required legal legitimization as Japan moved from acquiring foreign satellite imagery to using the country's own commercial satellite transponders, and onto actually launching its own spy satellites. From a long-term legal perspective, the constitutional interpretations are going to be increasingly under the spotlight with the new Basic Space Law, which is billed as being in the name of the peace constitution but which now overturns this long-standing Japanese interpretation, as we discuss more fully later in this chapter.

Rather than focusing, as others have already done, on theoretical paradigms or constitutional interpretations that help frame our work, we set our sight on specific industry actors—such as MHI, Melco, and IHI: What they wanted, what kind of space technologies they produced, what those technologies mean for Japan's shift from the market-to-the military, and how also the legal and institutional structure began to change in line with their demands. These actors, after all, have the most to gain or lose economically in making the shift from the market to the military a reality in the space sector. What we can also say is that the lobbying by such actors, and the space technologies they are capable of producing, has found a visibly conducive environment as the need for military defense becomes more visible amid a host of much highlighted security concerns for Japan, especially within the region.[12] These include, for example, the 1998 Taepodong trigger, the growing competition posed by China, the chance of a conflict over Taiwan, the competition for undersea gas and other natural resources, and the presence of unresolved territorial issues. Such concerns have all tacitly combined to legitimize boosting Japan's military capability in the eyes not just of the policy elites but also of increasing segments of the general public.

More critically, such concerns have served as near-perfect fodder for Japan's extremely powerful stakeholders, the defense companies that also double over as the key makers of space technologies. They have long been vexed, for example, that without the military to absorb costs, Japan's civilian SLVs are too costly to win international launch contracts.[13] They have also long been irritated at the inability to militarize Japan's satellite and thus reconnaissance capabilities. By this we do not mean that Japan's defense-related companies acted with any kind of cohesive unity in developing space technologies, or that they operated under some grand design. If anything, there has been severe competition among them and technological acquisition was hardly ever on an onward and upward trajectory. Our point from a political economy perspective is far more basic: Like all other concentrated interests, their economic motives are understandably straightforward; and at this stage in the Japanese government's known threat perceptions and assessments their interests and capabilities in the space sector have also fortuitously become intertwined with being in defense of Japan.

With this as a background, the remainder of the chapter turns more specifically to the central focus of this study in two parts. The first part turns more specifically to the findings in this study, underscoring the market-to-military trend in Japan's space saga that reflects the role of Japan's defense-related corporations and the resurrection of their specific economic interests amid newer geopolitical uncertainties. It summarizes the evidence across Japan's space-related technologies, and also briefly examines the relevance of nuclear realities. The second part ends with some broader policy concerns that speak to Japan's future security directions: the impact of the new Basic Space Law on the future of Japan's military space agenda, the issue of whether alternating political parties, such as the newly-elected DPJ, will stay the market-to-military course, and the trends in Japan's relations with other space powers such as the United States and China.

JAPAN'S SPACE-BASED CAPABILITIES IN CONTEXT

Throughout, we have concentrated primarily on understanding how and why Japan is increasing its military capabilities in the space sector. To summarize the principal findings we focus, first, on the conceptual shift in favor of space-based assets in Japan and Japanese space-based capabilities as discussed in the book. We then turn, second, to the technologies themselves, focusing on space launch vehicles as well as satellites and spacecrafts. Although we maintain that, if things stay the course, Japan's space-related technologies will make it an even more important military space power in the medium-term, we make no overblown claims about the power of space assets alone to transform Japan's national security. We need to contextualize those space realities, both by looking back at Japan's conventional forces and by looking forward to its nuclear capabilities. The evidence does suggest that, independent of the United States, Japan has already taken significant steps in terms of the dimensions of the military use of space identified by the United States Joint Chiefs of Staff as well as the United States Air Force Space Command (AFSPC) at the start of the book—namely, space support, space force enhancement, space control, and space force application. More importantly, Japan is also continuing to keep pace with cutting-edge technologies across all of those dimensions.

THE SHIFT FROM THE MARKET-TO-THE-MILITARY IN SPACE POLICY

Japanese thinking about the unique capabilities of space systems, over and above non-space ones, has moved in tandem with concrete shifts in related assessments by U.S. defense planners.[14] There is little question that the highest political levels in the United States have responded to the need for enhancing military space prowess as articulated by a national security space community—remote sensing from space, intelligence collection in Earth orbit, robust bandwidth-on-demand telecommunications, and (graphic real-time) information dissemination infrastructures that directly aid war fighters. In 2005, the president of the United States authorized a new national policy, one that underscored the importance of U.S. space transportation programs and assured access to space for guarding national security and homeland security, as well as a range of civil, scientific, and economic interests. All of these facets, as public and private actors realized, were increasingly dependent on both U.S. government and commercial space assets.

The United States was the biggest user of these assets; it also stood to be the biggest loser if they were targeted and destroyed. The spreading recognition of this dependence was critical, particularly in the aftermath of Operations Desert Storm and Iraqi Freedom, and has suffused the U.S. space policy with the core idea of Operationally Responsive Space (ORS).[15] Simply put, ORS emphasizes the importance of having the capability to promptly, accurately, and decisively position, operate, and, if necessary, protect national military space assets. The core consensus within Japan regarding Space on Demand (SOD) has come thus to emphatically echo the U.S. idea of ORS.[16] All this helped to focus attention on the critical and growing importance of integrated space assets (satellite systems, which are themselves dependent on space launch capability) for safeguarding a wide range of military and intelligence operations around the world: communications, intelligence, surveillance, reconnaissance, early warning, situational awareness, precision targeting, navigation and timing, and meteorology/oceanography.

For Japanese defense planners, this evolutionary view about the importance of space assets to national security took on additional policy hues given the range of concerns in Japan's external environment: a rising China with nuclear capabilities and moves to increase its prowess in space; a nuclear North Korea with a history of lobbing missiles around Japan; foreign air and naval movements around, especially, the Senkaku Islands; and a continuing U.S.-led war on terror with unpredictable blowbacks taking place, not just in the Middle East but across the globe in South Asia, Africa, and Southeast Asia, and even possibly on U.S. bases in Japan. Thus, from Japan's perspective, when North Korea conducts nuclear tests or launches multi-stage missiles, or China rattles its own missiles or anti-satellite (ASAT) weapons, a space-based response becomes a necessity. Whether or not legally or politically defensible, the reality is that Japan's burgeoning space-based assets—such as satellites for the purposes of more accurate information behind diplomacy, or even offensive missiles standing in the background for the purposes of coercive diplomacy—are the fundamental infrastructure that now gird potentially independent military operations in defense of the homeland.

The above two changes—namely, the conceptual shifts in recognizing the importance of space assets to military theaters more generally, and external instability in both the international and regional security environment for Japan more specifically—also resonated with the concrete concerns of a specific set of actors: Japan's defense-related corporations who remain key drivers and beneficiaries of the market-to-military trend in Japan's space policy. The fact is that these shifts have taken place as they continue to face difficult prospects for commercial profits. They are also faced with the fact that military space will be a significant profit area going forward and that their investments in space technologies will likely wither in a crowded commercial market. The advancement of Japan's space technologies would be nowhere without the interest and technical capabilities of these private actors. Without their eye on the space-related defense market, we might very well not be looking at the increasingly defense-bound nature of Japan's space production.

Who are these actors? As this book has detailed, there have been a number of key private players on both the rocket and satellite side, the twin pillars of Japan's space program. Chief among them are those from the Mitsubishi group—specifically, Mitsubishi Heavy Industries (MHI) on the rocket side, and Melco for satellites—who after considerable twists and turns and intense competitive pressures with their Japanese rivals in the postwar period have emerged to dominate Japan's space program at present. They may not always dominate, if for no other reason than that the Japanese government would be more secure with a diversified set of suppliers. These are likely to include other established players, such as Ishikawajima-Harima Industries (IHI, now known as IHI Corporation) and Nec Corporation/ NT Space, whose economic fortunes, incidentally, appear to be rising once again under the new military paradigm in space with new sets of rockets and satellites.

Through various production strategies, defense contractors remain pivotal to the ongoing saga of Japan's space-related militarization. Indeed, we would maintain their concern with their own economic dire straits could not have come at a more fortuitous moment in postwar history in which Japan's political parties—whether historically dovish or hawkish in their orientation—

have more or less yielded to the exigencies of the external security environment. Even more fortuitous, external structural shifts and internal generational changes have no doubt further contributed to dampening antagonism toward incremental institutional, legal, and especially technical changes that have taken Japan further down the path to militarized space technologies. Happily for Japan's space-related corporations, this then is the environment in which they now operate.

On the surface, Japan has not sought a full-spectrum independent military space capability. But, thanks to the concerted efforts of key industry actors over the postwar period, who went through twists and turns, as well as failures and setbacks, a basic modern infrastructure is already in place spearheaded by Japan's largely independent and civilian space program. To place Japan's space assets in comparative perspective, in Table 7.1 we compare its civilian SLVs and spacecraft to the known military-related ones of the other major powers. In doing so we are not in any way suggesting that Japanese dual-use assets are in the same military category as those of the other countries. Our point is, rather, that they share the same underlying technologies, which can be transposed to serve military ends. As the table suggests, Japan has an impressive arsenal of technologies developed and in development in the space industry. Keeping these comparisons in mind, below we also briefly summarize the more specific evidence for space launch vehicles and spacecraft.

Space Launch Vehicles (SLVs)

With respect to SLV technology, Japan's program did not start from scratch in the early 1950s. The program, which actually originated during World War II, was bolstered by the acquisition of German technology. In the postwar period, Japan has come a very long way from its humble origins with the *Pencil*. Although the Japanese media is fond of excoriating Japan's record of failures, the reality is that Japan has a superb record of success in launch vehicle development, and its learning curve has been rapid. The *Pencil* effort was driven by engineers related to the wartime military programs. From that point on the country moved determinedly in search of independent SLV capability on both the solid- and liquid-fuel fronts. Despite setbacks, technological progress continued on both fronts, and the varied problems were rightly attributed more to quality and control issues than to the functioning of specific rocket technologies.[17]

Table 7.1. Japan's Space-Related Capabilities in Comparative Perspective[1]

				Military				Civilian and Dual-Use
Service	*Measurements*	*United States*	*South Korea*	*North Korea*	*China*	*Russia*	*India*	*Japan**
Space	Ballistic Missiles (including ICBMs, IRBMs, SRBMs)	LGM-30G/ Minuteman III LGM-118/MX/ Peacekeeper UGM-96/Trident I C-4 UGM-133/ Trident II D-5	(NHK-I/II Hyonmu)[2]	(Nodong)[2]	DF-3A DF-4 DF-5A DF-21 DF-21A DF-15/M-9 DF-11/M-11 DF-7 DF-31 JL-1	SS-18 SS-19 SS-24 SS-25 SS-27 SS-N-8 SS-N-18 SS-N-20 SS-N-23 SS-N-30	AGNI-2 AGNI-3 SS-150/ Prithvi SS250/Prithvi	*M-3SII* *M-V* *J-I* **ASR/Epsilon* *Air Launch (proposed)
	Military Communications Satellites	DSCS-3 Milstar SDS UHF Follow-on	0	0	Feng Huo	Molniya Geizer Strela	0	(transponders on commercial satellites) *COMETS/ Kakehashi* *WINDS/Kizuna* *OICETS/Kirari* *ETS-VI/Kiku-6* *ETS-VIII/Kiku-8* *DTRS/Kodama* **DRTS* follow-on **OICETS follow-on* *WINDS* follow-on **QZSS/Michibiki* *SIGINT (proposed)

Navigation, Positioning, and Timing Satellites	GPS	0	0	Beidou	Parus Glonass	0	*WINDS/Kizuna* *OICETS/Kirari* *ETS-VIII/Kiku-8* *DTRS/Kodama* *EGS/Ajisai* **QZSS/Michibiki*
Meteorology, Oceanography, and Earth Observation Satellites	DMSP GFO Wide-Area Survey Follow-on	0	0	0	US-PU Okean-1 Globus/ Raduga	0	*ALOS/Daichi* *EOS/Aqua* *TRMM* *INDEX/Reimei* *EXOS-D/Akebono* *GEOTAIL* **GOSAT* **GPM* **GCOM* **EARTHCARE*
Intelligence, Surveillance, and Reconnaissance	EIS KH-12 Lacrosse/Onyx Mercury New SIGINT Trumpet Advanced/Orion	0	0	ZY-2	Tselina-2 Cosmos	0	*IGS-1* *IGS-2* *IGS-3* *IGS-4* *Future *IGS* **ASNARO*-1-4/ *Sasuke* 1–4 *SIGINT (proposed)
Nuclear Early Warning Satellites	DSP	0	0	0	Oko	0	Planned for 2016

SOURCES: Information for each country other than Japan is for representative purposes only and is not an indication of actual capability or quantity. All information other than Japan is based on that found in *The Military Balance* 103(1), 2003, p. 228, and *The Military Balance* 107(1), 2007, as follows by country: for the United States, pp. 28–29; for South Korea, pp. 359–361; for North Korea, pp. 357–359; for China, pp. 346–348; for Russia, p. 195; and for India, p. 315. Information on military satellites is from *The Military Balance* 103(1), 2003, p. 230, which also lists the Japanese *IGS*. Information on Japan is based on findings in this book.

(*continued*)

Table 7.1. *(continued)*

1. Japan's capabilities are not directly comparable to the actual weapons and military capabilities of the other countries noted here, and the chart categorizes Japanese civilian and dual-use assets along the same dimensions based only on assessments about potential similarities in use and functions in the underlying technology as discussed in the book. Under ballistic missiles, our estimates include Japan's SLVs that are discontinued or are in design stage (indicated by *). Under Japan's satellites and spacecraft, our estimates are based on all those that are currently in operation, discontinued, or are in development (indicated by *); in addition, as based on official representations by JAXA, our estimates also suggest that some satellites go across the standard categories as indicated.
2. For both South Korea and North Korea, the only given information is listed under SSMs (surface-to-surface missiles).

The rubric behind Japan's SLV development was first "catch-up," then conversion from the laboratory to the market, and later if not the market then the military, In this book, we focused on understanding less the market, which never really emerged as far as Japan was concerned, and more the military side of affairs, which is playing an ever more important role. As explained in detail, SLVs and ballistic missiles share the same underlying technologies. However, we did not focus as much on whether Japan has acquired a ballistic missile force structure, which remains a matter of political will. Rather, we concentrated on assessing the acquisition of component technologies that are considered to be critical in the manufacture, launch, and, from there, possibly construction of a force structure of ballistic missiles—specifically, propulsion systems, structures, staging ability or large boosters, sophisticated guidance and control systems, reentry vehicles, and flight operation skills, as well as the technical and institutional infrastructure necessary for integration.[18]

Not only have each of these elements matured in Japan's case, but their historical progression leaves little doubt that Japan has the technical experience and capability for assembling sophisticated state-of-the-art intercontinental ballistic missiles (ICBMs). On the solid-fuel side, which is of more relevance to modern ballistic missile arsenals, the *Baby*, *Kappa*, *Lambda*, and, especially, upgrades in the *M* series allowed Japan to acquire and integrate technologies to such an expert level that they were immediately noted for ballistic missile conversion. The *M-3SII* was marked as a sophisticated Intermediate Range Ballistic Missile (IRBM) by the Americans; the *M-V*, famous as one of the world's best multi-stage solid rockets, was marked as an ICBM. On the liquid-fuel side, based on limited technology transfers from the United States, there was the progression from the *N* rockets to the increasingly and then wholly indigenous *H* series. The continued evolution of the latter series in particular also garnered world, and especially U.S. attention, not so much for ballistic missile conversion as for the sophistication of its technologies and total systems integration skills. The combination of solid-fuel and liquid-fuel technologies that came together in the service of the short-lived *J-I* was also noted widely; the *J-I* itself was marked for potential ICBM conversion by the United States.

With this as a background, it is important to remember that Japan's rocket technology, whether in part or in whole, has not reached some plateau. So advanced is Japan's rocket technology that it seems a relatively trivial exercise now to convert the SRB-AII technology of the *H-II* series for an ORS-capable, quick-launch *Advanced Solid Rocket* (*ASR/Epsilon*). Air launch LVs are also being planned. Apart from the highly visible rockets, there are discrete elements that deserve greater attention. These range from reentry vehicle experiments (vital to technologies for functioning ballistic missile warheads) to the direct-ascent interceptors in a sophisticated BMD system (vital to getting technologies at the cutting-edge of a highly controversial military application that can double over as ASAT weapons). All in all, we believe that irrespective of commercial prospects Japan's long trajectory of SLV development shows that the country will continue to secure independent access to space through the efforts of private makers of space technology.

By extension of the underlying SLV technology, there is little question that Japan has ballistic missile capability. But as is widely recognized, any kind of ballistic missile capability needs also to be assessed in the context of warhead capability. As we studiously avoided this issue earlier, we now turn to an assessment below focusing only on Japan's nuclear one. As in our discussion of SLVs, we are not assessing whether Japan has any particular warhead; rather below we are only assessing Japan's capabilities with respect to building them.

Nuclear Assessment for Ballistic Missile Capability

Japan is of course a non-nuclear weapons state.[19] In addition, its principled commitment to non-nuclearization, as well as nuclear disarmament and nonproliferation around the world, is not in doubt.[20] In fact, it is specifically evident in a set of formal and informal instruments.[21] However, on closer examination, nuclear politics is also where an application of Japan's strategic hedging may well be most brilliantly on display.[22] Nuclear weapons have in fact been an integral part of the discourse over Japanese security in the postwar period. They were declared to be constitutionally acceptable as early as 1957 in line with the interpretive emphasis by the Cabinet Legislation Bureau (CLB) on a minimum level of military potential required to exercise self-defense. Since then political leaders have raised the possibility of Japan having the industrial and scientific capability for developing nuclear weapons, stirring much controversy, and government feasibility assessments have also made their way to the public.[23]

According to most accounts, the most formidable technical, industrial, and financial obstacle for countries interested in building nuclear weapons is also the first step—the acquisition of fissile material in sufficient quantity.[24] Given the presence of fissile materials in civil nuclear power programs for stable sources of energy, Japan's nuclear power industry has consequently come under intense public scrutiny.[25] Some estimates of plutonium put it at 145,000 kilograms by 2020 (and perhaps more, as Japan admitted in January 2003 to have "lost" at least 280 kilograms over the past decade or so from reprocessing plants right under the watchful eye of the International Atomic Energy Agency [IAEA]).[26] In addition, Japan has also made moves to amass supergrade plutonium through fast-breeder reactors, which actually produce more plutonium than they consume.[27] Other developments, such as the new Rokkasho-mura reprocessing plant in Aomori prefecture, as well as controversial plutonium disposal methods such as fabrication into mixed-oxide (MOX) fuels in commercial reactors, only continue to add to the existing stockpiles.[28]

All this too has generated much controversy.[29] As the IAEA was alerted in early 2006, the fear is that if things stay the course and no more concrete plans emerge to consume the official 43 tons of plutonium Japan already has, then Japanese surplus plutonium may be about 78 tons by 2012—a figure comparatively more in line with the U.S. military inventory of 99.5 tons of separated plutonium and the UK military and civilian inventory of 77.8 tons.[30] Whatever the accuracy of the estimates of the stockpiles, to achieve an explosive yield of just 1 kiloton, Japan would need a mere 1 to 3 kilograms of weapons-grade plutonium and between 2 and 7 kilograms of highly enriched uranium, depending on the technological sophistication (low, medium, high) of the design.[31] Thus the Japanese government's peaceful nuclear energy drive over the postwar period has allowed it to amass the potential for (at least) a few hundred 1-kiloton-yield nuclear bombs, should it choose to make them.

The other key point here is that the arguments often made about reactor-grade plutonium being unsuitable for nuclear weapons are no longer credible. As early as 1962, the United States conducted a nuclear test using reactor-grade plutonium that successfully resulted in a nuclear yield (a fact not declassified until 1994), and by 1967 it admitted that reactor-grade plutonium could be successfully used to make nuclear bombs.[32] In 1976, thinking it prudent in terms of the goals of nuclear non-proliferation, the U.S. Department of Energy briefed all active nuclear powers of the utility of reactor-grade plutonium in the making of nuclear bombs. By 1990, even the IAEA had reversed the agency's position on reactor-grade plutonium and concluded that virtually all isotopes of plutonium, including those that comprise reactor-grade plutonium, can be used to construct nuclear weapons. Even if reactor-grade plutonium has some uncertainty in explosive yield (and there are said to be advanced techniques that make it possible to offset its generally high rate of spontaneous fission and thus susceptibility to pre-detonation), some scientists assert that it will still produce a destruction radius that is roughly one-third of the Hiroshima bomb (*Little Boy*).[33] It should also be kept in mind that Japan is not limited to the reactor-grade plutonium route to a nuclear weapons program, as there are several plausible options available to obtaining plutonium and weapons-grade uranium.[34]

There is also the issue of Japan's scientific capability to actually produce nuclear weapons—an issue that was raised in the United States as early as 1957.[35] Any non-nuclear weapons state with even remote nuclear ambitions today is advantaged by the wealth of information on nuclear weapons design that has been made public over the past sixty years, information on which it can improve.[36] No competent power in global politics today can bill itself as being innocent of this mass of technical knowledge. The United States is also reported to be willing to supply and share some fundamental technology, such as refining bomb-grade plutonium from breeder reactors.[37] Some estimates of Japanese capabilities to deal with problems related especially to the use of reactor-grade plutonium in nuclear weapons are also suggestive. Knowledge of Japanese research in areas such as high-explosive technology, inertial fusion, and production and handling of hydrogen isotopes has led to assertions that Japanese scientists may well be able to deal with related problems on point, specifically pre-detonation.[38] By around 2000, the technical competence of Japanese scientists was thought to be at the intermediate-to-high end of the nuclear weapons capability spectrum. Japan, in short, is widely recognized by experts to be capable of constructing nuclear weapons in a short time.[39] According to government reports, the Japanese government has also come to the same conclusion several times since at least the early 1980s.

In light of Japan's infrastructure, resources, and technologies already in place, it is little surprise that leading Japanese politicians and government agencies can fan the reality of links between Japan's plutonium stockpiles and its ability to convert them to nuclear weapons.[40] However, we believe there is still the highly important matter of the Japanese public, whose deep sentiments on the nuclear issue should not be under- or overestimated. Japan has a long-standing anti-nuclear peace movement, and whatever the sentiments regarding its usefulness, its continued presence is evidence of popular sentiments against nuclearization.[41] Here there are some sobering realities to keep in mind. First, the passage of U.S. nuclear armaments and warships from the early postwar period has cast some doubt on the three nuclear principles.[42] Second, war and nuclear memories are fading in Japan just as external nuclear realities, regional power shifts, and nationalist trends impress themselves on the younger generations.[43] In addition, the anti-nuclear sentiments of the Japanese public have had the luxury of extended nuclear deterrence under the United States; only when it is absent can some stronger conclusions emerge about ideational and legal constraints.[44] Finally, from our perspective, the incremental developments in the space industry suggest that norms, principles, and laws can turn in the face of technological changes and geopolitical realities.[45] Such creeping change in the defense arena all around—what was once not permissible too soon becomes reality—is surely noteworthy. This is especially instructive in the nuclear saga because, echoing the developments in the space sector, even back in 1969 Japan had stated its resolve to maintain the economic and technological potential to manufacture a nuclear device so as not to be restrained by others.

Though the Japanese government has only done so indirectly, the fact is that even a virtual or latent nuclear deterrence posture against rivals gives Japan leverage in reinforcing its diplomacy.[46] As the country's foreign policy elites recognized long ago, the acquisition of such capabilities is undoubtedly the most important insurance Japan can provide in the face of future uncertainties in both international and regional politics. From our perspective, Japan's nuclear posture is all the more credible precisely because of Japan's long experience with rocket and satellite systems as traced in this book. In short, it is not fiction to state that Japan has an independent and indigenous state-of-the-art capability to build solid-fuel ICBMs, acquired through its long-standing civil space program, and to arm them with nuclear warheads.[47]

To be clear, of course, we cannot be sure about confidential elite planning for a functional nuclear force structure, any more than a ballistic missile arsenal. We cannot also say how and whether Japan's public will ultimately influence the course of Japan's nuclear or military space politics. But we can say with some confidence that were Japan to acquire nuclear warheads it certainly also has the capability to deliver them worldwide through its long-standing rocket program—elements that, perhaps like the spy satellite saga, may well come to the fore and be legitimized in the domestic defense discourse with a trigger security event.

Satellite and Spacecraft Technology

With respect to satellite and spacecraft technology, Japan has also come a very long way from the launch of its first test satellite, the *Ohsumi*. Our case for the ever-increasing market-to-military trend in Japan's space assets has been made easier by Japan's deployment of an actual *Information Gathering Satellites* (*IGS*) structure, which today is a major fixed component of Japan's space budget overall. A more obvious measure of a militarized space program could not be found. The Japanese defense establishment now more gathered and concentrated than ever before in postwar history in the shape of the Ministry of Defense (MOD), has always had access to foreign commercial satellite imagery and U.S. military intelligence, as well as transponders on the country's own commercial satellites. But as the defense establishment, backed by the country's formidable defense contractors, has long agitated, none of these has provided for the type of control and access that comes from having national assets. This is true irrespective of whether other ministries and agencies also have access to information gathered from such satellites, a feature that itself may change.

Although the 1998 Taepodong incident pushed Japan to officially reorient the space program toward the full-fledged acquisition of basic satellite reconnaissance capability, there is a far more basic reality that needs to be kept more firmly in mind. The reality is that Japan has always been in the business of observing the Earth through its extensive Earth observation (EO) program, focusing on environment and meteorology, resources and natural disaster surveillance, precision mapping, and more generally land, ocean, and coastal topography. The information-gathering aspects of the spy satellites, in fact, thus have a clear technological lineage in the country's long-standing EO program. The very remote-sensing technologies that made ever more precise observations of land, sea, and air phenomena possible, specifically in the *Advanced Land Observing Satellite* (*ALOS/Daichi*), were transposed into the technological makeup of the spy satellites. In our view, it is the underlying technology, and more specifically its makers, that have been the true drivers of the militarization of Japan's satellite and spacecraft programs.

In line also with the RMA's emphasis on secure satellite-based communication networks, Japan has taken considerable steps toward the next frontier in the form of research and development on laser or optical inter-satellite links, which allow for extremely high bandwidth communications that cannot be jammed. In some ways, by the mid-1990s, its technology programs—such as that involving the *Engineering Test Satellite* (*ETS-VI/Kiku-6*), which conducted the world's first known experiment with lasers to provide two-way communication with space—serve to underscore the advanced nature of Japan's achievements. Similarly, the country's *Optical Inter-orbit Communications Engineering Test Satellite* (*OICETS/Kirari*) is also the world's first known satellite to achieve bi-directional optical inter-satellite links. The failure (partial or complete) of any one satellite program is less important than the ways it allows the country to continue to improve technologies in the service of next-generation spacecraft. Evolving from these earlier programs, Japan's hopes for an operational space communication infrastructure is pinned on improving the *Data Relay Test Satellite* (*DRTS/Kodama*) to far more advanced versions in the near future. Japan's communication satellite infrastructure also has a regional focus, specifically in the Asia-Pacific. This is a region in which Japan's satellites, such as the *ETS-VIII/Kiku-8* as well as *Wideband InterNetworking Engineering Test and Demonstration Satellite* (*WINDS/Kizuna*), are being configured to speed up satellite-positioning and communication technologies.

Just as with changes in SLV developments today, Japan's satellite and spacecraft paradigm is now shifting toward smaller satellites. This move was fueled by the evolutionary shift toward smaller and smaller satellites worldwide. In Japan, the first phase at the beginning of the 2000s aimed this to train a new workforce to take advantage of this technological evolution. The present phase a decade later is now explicitly marked by the Ministry of Economy, Trade, and Industry (METI) for national security satellites and by MOD for space situational awareness (SSA) and defensive counterspace purposes. Indeed Japan's abilities for on-orbit reprogrammability and maneuvering were demonstrated, ahead of the curve worldwide, with *ETS-VII/Kiku-7*. Of course, *Orihime* and *Hikoboshi* were designed to help give Japan the ability to dock the *H-II Transfer Vehicle* (*HTV*) at the International Space Station (ISS). But within a few years Japanese scientists were looking at the *Micro OMS Light Inspection Vehicle* (*Micro LabSat/Micro-OLIVe*) and even the fledgling *SmartSat* venture to demonstrate improved capabilities for autonomous proximity operations in space.

The continued militarization of satellite technology in the near future looks set, especially given the realities of the new Basic Space Law discussed at the end in this chapter—in the *IGS*, in the *Advanced Satellite with New system Architecture for Observation* (*ASNARO/Sasuke*), in the Ballistic Missile Defense- (BMD) related infrastructure, and now even moves toward early warning satellites. In the interest of improving and upgrading its Command, Control, Communications, Computers, Intelligence, Surveillance, Reconnaissance (C4ISR) system, Japan is also taking significant steps in new directions. The country now has a legal and institutional setup that shows its determination to have the *Quasi-Zenith Satellite System* (*QZSS/Michibiki*), which is its version of a regional Global Positional System (GPS) or Galileo system and which can serve as a counterpoint to the expected one by China's Compass system in the region. In addition, a number of university-, corporate-, and government-based research programs under way on ever more sophisticated, miniaturized, and autonomously operational satellite and spacecraft systems also showcase how civilian technologies hold the potential to some day be transformed into even more potent space-based ASAT weapons.

In summary, based on the evidence overall across SLV and satellite technologies, it is helpful to see where Japan stands in term of military dimensions of the use of space. We believe the discrete programs for different civil purposes add up to an impressive stock of dual-use technologies when taken as a whole. The presence of these slowly acquired and discrete elements in Japan's space program demands very careful attention; taken together, they stand to have a significant impact on the contents and direction of Japan's space policy and, from that basis, the country's security directions in the near future. Table 7.2 shows what capabilities Japan has or is developing in each of the key mission areas identified by the U.S. military. It is instructive in showing the broad range of dual-use technologies already held by the country. It is thus, we believe, not an exaggeration to call Japan a military space power.

LOOKING AHEAD

Our findings on the market-to-military shift in Japan's space developments speak to broader debates about the directions of Japan's security policies, and also for Japan's standing in world politics. Here, then, we also assess the fronts we consider important: the impact of the Basic Space Law, the role of alternating political parties, and relations with other space powers like the United States and China.

Table 7.2. Assessment of Japanese Military Capabilities by Space Mission Areas

Mission Area	*Operational Objectives*	*Main Capabilities to Achieve Stated Objectives*	*Japanese Capabilities*
Space Support	To deploy and sustain military and intelligence systems in space	Spacelift Operations (capability for deploying payloads, including operationally responsive [ORS] launch capabilities, through SLVs)	Yes (ORS capability in development)
		Satellite Operations (capability to maneuver, configure, operate, and sustain on-orbit satellites and spacecraft)	Yes
		Rendezvous and Proximity Operations (capability to intentionally bring two resident space objects operationally close together)	Tested
		Reconstitution of Space Forces (capability to replenish lost or diminished space assets and augment with civil and commercial capabilities as needed)	Possible
Space Force Enhancement	To increase joint force effectiveness of military forces	ISR	Yes
		Missile Warning	Proposed
		Environmental Monitoring	Yes
		Satellite Communications	Yes
		Space-based Positioning, Navigation, and Timing	In Development
Space Control	To ensure freedom of action in space for friendly forces; when directed, to deny freedom of action to adversary	Space Situational Awareness (SSA) (fundamental enabler for all other space control tasks requires complete common operational picture of home, neutral, ally, and adversary space capabilities within terrestrial and space domain, as well as space debris tracking	In Development

(*continued*)

Table 7.2. *(continued)*

Mission Area	*Operational Objectives*	*Main Capabilities to Achieve Stated Objectives*	*Japanese Capabilities*
		[capability includes components of space force enhancement, such as ISR and environmental monitoring])	
		Defensive Space Control (DSC) (capability to protect home, neutral, and ally space capabilities using protective and defensive prevention measures)[1]	Yes
		Offensive Space Control (OSC) (destructive and nondestructive capability to deceive, disrupt, deny, degrade, or destroy an adversary's space systems and services, including ground, data and information links, user, and/or space segments)	Potentially Yes
Space Force Application	To conduct combat in, through, and from space to influence course and outcome of conflict by holding terrestrial targets at risk	Ballistic Missile Defense[2]	Potentially Yes
		Force Projection	Potentially Yes

SOURCE: For fundamentals of military space operations, and identification of general mission areas, objectives, and capabilities in columns 1 to 3, see United States Joint Chiefs of Staff, *Space Operations*, Joint Publication 3-14, 6 January 2009, pp. ix–xi, II.1–II.10 available online via the Defense Technical Information Center (DTIC) at www.dtic.mil (accessed 28 August 2009). Column 4 is based on the authors' assessment as discussed in the book and refers to the presence of such capabilities rather than any particular level or number. See also Paul Kallender-Umezu, "Amid Shift in Power Japan Seeks Space Budget Hike," *Space News*, 7 September 2009, available online at www.spacenews.com (accessed 18 September 2009).

1. Means of protection include ground facility protection (for example, security, concealment, mobility, hardening, and so on), spare satellites, link encryption, increased signal strength, satellite radiation hardening, signal monitoring (for example, electromagnetic interference [EMI]), space debris protection measures, and so on. Means of defensive prevention include diplomatic, informational, and economic measures (for example, sanctions, and reputational pressures) as appropriate.

2. As the *Space Operations* report does not list specific capabilities under this mission area, they are identified from the *USDOD Dictionary of Military Terms* (for term "Space Force Application"), available online at www.dtic.mil (accessed 28 August 2009).

Impact of the Basic Space Law

To our minds, the clearest barometer of change in Japan's space saga is in the shape of law, which at long last provides a legal foundation for developments in the space industry. The Basic Space Law that was passed by the 169th session of the Diet in late May 2008 was approved by the upper and lower houses with great speed and with meager substantive opposition from the DPJ, which swept into power just as this book was being finalized.[48] As our translation in Appendix II reveals, although this law seeks to promote space development in Japan consistently with international treaties and agreements, as well as the pacifist orientations of the Japanese constitution, in actuality its specific provisions leave little doubt that it also specifies autonomous space-based capabilities for safety, security, and military purposes. Of particular interest is Article 2 (Peaceful Uses of Outer Space) of the Basic Space Law, which mandates that Japanese space development and utilization shall henceforth follow the 1967 Outer Space Treaty, at long last in line with international interpretations of "peaceful uses" as being nonaggressive rather than nonmilitary.

As a legal provision, Article 2 has practical consequences. At the governmental level, the Japanese government no longer has to justify its space ventures in the name of commercialization or scientific exploration, but can actively do so in the name of national military security. Japan can also now openly continue to refine its dual-policy technology acquisition in all areas pertinent to developing strategic space military abilities, and can move toward their use and deployment probably led by MOD and/or METI. At the private corporate level, the provision also legalizes space technology developments by private corporations that can serve military purposes. The change in the legal orientation was a move notably welcomed at Nippon Keidanren, for example, whose players now foresee the dawn of a new Japanese space era with a framework both facilitating Japan's space utilization and also integrating policies across ministries into a national strategy under strong leadership.[49] Overall, if the political will and budgetary flexibility exists, both government and corporate players can now expand the scope and contents of Japan's space technologies in ways that were not openly possible before, as long of course as they are henceforth presented as nonaggressive like those of the other major space players.

Altogether, with the Basic Space Law in place, we could not have hoped for a more tangible validation of our market-to-military thesis.[50] Japan's space saga has entered an altogether different era. We are now no longer alone in so claiming outright. As other observers also immediately noted, this law is a significant departure from the past, which stressed the peaceful purposes resolution from 1969 onwards. That resolution—which did not prevent the militarization of Japan's space assets as we suggested through various ways in this book, and to which even illustrious players in Japan's space policy are willing to draw attention to once again after having suffered threats of condemnation in earlier times—is now more clearly behind us than ever before.[51] Japan does have a genuine, full-fledged, and impressive space science program with an international reputation and it is possible for Japan to play a role in commercial space in some ways.[52] Through this book, however, we show that Japan's space program can also be viewed from another angle—a more military-oriented one—that, until the Basic Space Law came along, was not acknowledged officially.

Where, then, is Japan headed? In the space sector, if the present law continues to exist without controversy and continues to shape space policy, we believe that it could move, ever more openly, from the market to the military. Of course, this is not guaranteed. By the conditional tense we acknowledge only that alternative futures are still possible, though in our assessment, barring some domestic upset or reversal, those futures are unlikely. As we know too well, there have been twists and turns, progress and setbacks in Japan's space ventures—all of which will continue. We cannot say whether Japan's space-related militarization will be successful in warding off attacks or allowing Japan to defend its interests far from home—if theories and constitutions cannot do that, neither can the militarization of space technology alone. But we maintain that, given a growing military institutional structure, sliding anti-militarist domestic sentiments, and unreliable international structures, space technology will take Japan much further in its quest for national defense.

It should also be remembered that what the country has achieved incrementally—certainly from space probes to planned lunar expeditions, but also from the *Pencil* to the *ASR/Epsilon*, and from EO to reconnaissance satellites and onto even potential counterspace capabilities—has mostly thus far been under a pacifist constitution and the peaceful purposes resolution. As far as Japanese space policy is concerned under this new law, the market may well be saved by the military direction in the foreseeable future. As the country's BMD infrastructure continues to expand and reinforce itself, MOD stands to become probably the most important direct customer of space-based systems necessary for homeland defense, either purchasing them from contractors or working through what is likely to be a revamped Japan Aerospace Exploration Agency (JAXA). The institutional infrastructure that allows command, control, and communication to be integrated at home and with allies is of course critical too. Some proposals, as yet not made public, suggest an independent military space organization, either as part of MOD or related to it, needs to be set up to more effectively run national security programs. In this sense we believe that a new, and from a historical perspective, a far more transparent, saga in Japan's space policy is about to unfold—a good thing perhaps for both Japan's allies and rivals who seek to assess where Japan stands in the provision of its own defense.

The Role of Alternating Political Parties

With the historic election of the DPJ in 2009, and the fact that it has publicly opposed many initiatives undertaken by its predecessor, the Liberal Democratic Party (LDP), there is concern as to whether the new party leadership will stay the market-to-military course. Importantly, just as this book was being finalized, the indications are that the newly installed DPJ administration will continue to support the Basic Space Law broadly, as well as the budgets requested for the plethora of military and national security space programs.[53] There are several reasons we can expect the DPJ, or for that matter any other party that comes into power, to stay the market-to-military course.

The first, and pragmatically the most important, is that like other advanced industrial great powers of the day, Japan is highly dependent on space assets in the commercial and economic realms. Like the LDP, the DPJ also does not have the luxury to turn away from the necessity of protecting the very same space assets that also fuel Japan's economic and industrial growth. These assets continue to be vulnerable to both man-made and natural disruptions, and militarized space assets can be an answer to dealing with both.

Second, also like the LDP, the DPJ faces the same external constraints and uncertainties both globally and regionally. Although it can certainly choose to exercise a different direction, external threats that have been highly publicized in the domestic discourse over defense, such as the economic and military rise of China and the erratic actions of North Korea, have not dissipated. Both of these also weigh ever more on the public's mind. Their very presence demands a prudent and security-conscious response from a DPJ in power as compared to a DPJ posturing to get into power. Certainly the DPJ's leadership has historically engaged in exactly such prudent responses. Already in 1998, in the aftermath of the Taepodong incident, then deputy head of the DPJ Yukio Hatoyama, who is now prime minister, had pragmatically stated that the peaceful purposes resolution could well be stretched to allow defense-related missile monitoring sensors aboard a satellite.[54]

Third, as discussed in this book, the Basic Space Law was put into motion at a time of publicized threats and a future of geopolitical uncertainties, all of which only serve to advantage those actors who have and make the actual space technologies to secure Japan. The very existence of the peaceful purposes resolution long frustrated certain elements within the LDP, and especially space-related actors in the corporate world. Nippon Keidanren, Japan's powerful industrial and business lobbying organization, spent well over a decade trying to convince the government to focus more cohesively on space activities. In many ways, it was the Basic Space Law that caught up with the technological capabilities and political maneuverings of some of Japan's most formidable defense contractors, not the other way around.

The proposed centralization of Japan's space policymaking, under the direct control of the prime minister in the Cabinet, speaks to their long-standing concerns that Japan needs strong national leadership and strategy in space—now recognized as perhaps the most critical theater of military operations that underpins all other land, sea, and air theaters. The appointment of Japan's first minister of space development in June 2008 was no doubt a harbinger of policy and institutional moves that seek to alleviate exactly such concerns in the future.[55] Thus it is important to know that private corporations involved in the making of space technology are powerful stakeholders in the market-to-military trend. They have an incentive to push the status quo, whether with the DPJ or any other party in power. Moreover, as they can now also articulate, the new Basic Space Law mandates that the government focus on and support their concrete needs quite clearly (see Article 4, Article 16, and Article 17 in Appendix II at the end of the book).

Relations with Other Space Powers

There is little question that what Japan does and the ways in which it sets about deciding the future course of its national security is an issue of seismic proportions for other countries. Dominant powers, such as the United States and even China, may or may not care for Japan's course, which itself may or may not be advantageous from their economic, political, and strategic perspectives in the present and future.[56]

In terms of its conventional forces, as we saw in Chapter 1, Japan is hardly a military pygmy, and the Self-Defense Force (SDF) may well be far more advantaged through space-based capabilities vis-à-vis rivals in the region than by a focus on amassing conventional forces. This is because the sobering reality is that even with the protection of its powerful ally, there are some serious constraints on Japan's ability to achieve its diplomatic goals and military objectives. The space-based emphasis is perhaps all the more important in light of some of the concerns expressed for the operations of the United States in the East Asian region.[57] Whether its actions are actually judged to be effective or not, China is widely seen as beefing up its capabilities to pursue an area-denial strategy regarding the United States. As U.S. military leaders noted starting in the mid-1990s, with a foe determined enough to have an asymmetric strategy to oppose the deployment and movement of U.S. military forces in a region, there is a strong possibility that the United States could suffer prohibitive losses in even projecting its forces into a disputed theater, much less carrying out operations. Both anti-access strategies to prevent U.S. forces entry into a military theater and, especially, area-denial operations to prevent their actions within a narrow battlespace are critical realities in present-day East Asia—realities to which U.S. military forces may well have to craft a proactive response in concert with others, such as a far more space-capable ally like Japan.

The United States, which has had a competitive relationship with Japan in space technologies in the past but is now focused on a more cooperative one stemming from the technical interoperability of BMD, needs to consider proactive solutions with its closest ally on a far more equal footing than ever before. Given the historical trends, the integration of space technologies into Japan's national defense will continue forward, and this area offers an avenue for cooperation in terms of intelligence, surveillance, and reconnaissance (ISR).[58] The U.S. needs to factor in the possibility that, as Japan reinforces its military operations across the key mission areas as in Table 7.2, it can afford to be more assertive politically and it may well seek more autonomy within the alliance. The DPJ appears to be particularly keen on such types of moves as it had already declared its interest in constructing a more autonomous (from the United States) foreign policy in its pre-election manifesto.[59]

China, which has both civil and military space ambitions of its own, also needs to consider whether and how far it can push Japan in the space arena. As we saw, Japan has many cutting-edge space technologies by working with but also beyond allies like the United States, which has sometimes tried to block certain technical aspects of Japan's SLV progress. Japan may also do the same with neighbors like China, which has already engaged in manned space and counterspace activities of its own. There is little doubt that China's destruction of its own weather satellite in 2007 only further alerted Japan's then newly-minted MOD to the vulnerability of Japanese space assets. Perhaps more consequential, within the domestic political arena in Japan, China's actions will also have the long term effect of legitimizing Japanese moves to equip itself technologically to protect its own space assets. As it is, through its cooperation on Ballistic Missile Defense (BMD) with the U.S., Japan has access to the very same technology namely the modified SM-3, with which the U.S. shot down one of its own satellites after China's test. Additionally, Japan is in some ways at the cutting edge of autonomous proximity operations, which are potentially key for both commercial and military service of satellites and spacecraft.

Japan has space interests, both civilian as has long been known, and now explicitly military. The steps Japan takes and what specifically it chooses to do in terms of the provision of its own defense are currently in the hands of the DPJ. Even as the DPJ has moved to embrace the vision of an East Asia Community including China, its leaders harbor serious concerns about neighboring countries' continued modernization of military capabilities, including no doubt those in military space.[60] Although in comparison to the push under the LDP, the DPJ may certainly be less willing to broach the subject of the militarization of space, it is nevertheless more important to see what it actually does.[61] Already, the DPJ, as noted above, appears disposed to continue the military space programs under the Basic Space Law, which is key to Japan's future space directions in many ways.[62]

In light of such technological and political realities, the warning by China's air force chief in 2009 about the inevitability of a military space race need to be heeded by the leadership in both countries.[63] China needs to carefully consider what has already happened before the Basic Space Law went into place: In the field of space technology for SLVs, for example, Japanese development has proceeded in steps, that is taking liquid U.S. technology (Delta, Atlas) and combining it with increasingly advanced refinements *after* the nation had already put in place basic IRBM and ICBM technologies with its solid rocket program. Similarly, the conversion of satellite technology to military purposes, whether for reconnaissance efforts or for autonomous proximity operations, has already been actualized. Japan, in short, put in a Basic Space Law precisely because it draws attention to itself and what it affects.

To conclude, Japan has traveled quite the market-to-military distance in its civil space program, and this route equips it go further more solidly. We are often asked, and so would like to be clear in the end: We are focused on exposition in this book. We do not take a position on whether what Japan seeks to do with the militarization of its space technologies or with changes in the related legal and institutional infrastructure is good or bad, effective or not. Clearly, Japan needs to clarify its security postures in line with its space-based capabilities and to do so in the context of its conventional and nuclear realities as suggested in this work.

In the meantime, it is not an exaggeration to say that it is the Basic Space Law which actually makes possible newer directions in Japanese security policy and visions as a whole. To be sure, the Basic Space Law is only one step. But from a legal and policy standpoint, an extremely important one that will reverberate outward from mere space technologies to the whole gamut of security and political dimensions in Japanese society.[64] It legitimizes technology pathways that were conceived as improbable and rejected rigidly by the original peaceful purposes resolution, which was consciously wedged in the path of military space development as Japan's civil space program got under way. Given the fortunate timing of the bill's passage, the MOD will have enough time to deal with such concerns more concretely by incorporating space programs into its next five-year Midterm Defense Plan slated to begin in 2010. Space, put simply, is going to be used in defense of Japan. No longer do individuals, such as the new minister of space; corporations like MHI, Melco, IHI, and NT Space, or government entities like MOD, METI, and SHSP need to couch strategic militarily-oriented space plans in niceties to pacify the world.

As the law at last makes explicit, the future of Japan's space saga is one in which Japanese space development will be promoted vigorously to the benefit of its industrial makers, will continue to be geared toward autonomous capabilities while engaging international cooperation, and will ultimately be designed on many fronts to contribute to the national security of Japan. Seen from the side out in, then, at no time in Japan's postwar history has it been more necessary for Japan's leadership to clarify the country's defense and security postures. Just as the basis of any security policy must well be its own efforts, so Japan's security postures are ultimately also a thing for the Japanese to contest and shape.[65] Given its standing as a military space power, Japanese choices on these fronts will matter for Japan's relative position in the world, for its concrete relationships with allies and rivals, and for the broader course of diplomacy and cooperation in military space politics worldwide.

APPENDIX I

Timeline of Principal Launches by or Involving the Japanese Space Program, 1955–2009[1]

Appendix 1.

Launch Date (dd/mm/year)[2]	*Flight Name or SLV*	*Mission*	*Launch Site*
11/09/2009	H-IIB-TF1	H-II Transfer Vehicle (HTV) demonstration flight of unmanned cargo transporter to International Space Station (ISS)	Tanegashima Space Center
23/01/2009	H-IIA-F15	Ibuki (GOSAT) greenhouse gas monitoring SDS-1 (Small Demonstration Satellite-1) Piggy-back satellites: KAGAYAKI, STARS, KKS-1, PRISM, SOHLA-1, SPRITE-SAT	Tanegashima Space Center
16/06/2009	U.S. Space Shuttle Endeavour	JEM/Kibō's Exposed Facility to ISS	Kennedy Space Center
01/06/2008	U.S. Space Shuttle Discovery	Japanese Experiment Module/Kibō's (JEM/Kibō) Pressurized Module and Remote Manipulator System to ISS	Kennedy Space Center
11/03/2008	U.S. Space Shuttle Discovery	JEM/Kibō's Experiment Logistics Module Pressurized Section to ISS	Kennedy Space Center
28/04/2008	PSLV	Cute-1.7+APD-II &, SEEDS-II microsatellites	Sriharikota
23/02/2008	H-IIA-F14	Kizuna (WINDS) communications satellite	Tanegashima Space Center
14/09/2007	H-IIA-F13	Kaguya (SELENE) lunar probe	Tanegashima Space Center
24/02/2007	H-IIA-F12	IGS-2 (surveillance); IGS-3V (surveillance)	Tanegashima Space Center
18/12/2006	H-IIA-F11	Kiku-8 ETS-VIII (Engineering Test Satellite-VIII)	Tanegashima Space Center
3/10/2006	Ariane 5	Piggy-back satellite: LDREX-2	Kourou
23/09/2006	M-V-7	Hinode SOLAR-B Piggy-back satellite: SSSat—picosatellite, Hitsat—picosatellite	Uchinoura Space Center
11/09/2006	H-IIA-F10	IGS-3A Next Generation IGS (Information Gathering Satellites) #1	Tanegashima Space Center
22/02/2006	M-V-8	Akari ASTRO-F, 21st scientific satellite Piggy-back satellite: Cute-1.7 picosatellite	Uchinoura Space Center

18/02/2006	H-IIA·F9	MTSAT-2 (Multi-functional Transport Satellite 2)	Tanegashima Space Center
24/01/2006	H-IIA·F8	Daichi (ALOS) (Advanced Land Observing Satellite)	Tanegashima Space Center
22/01/2006	S-310-36	Furoshiki Technology Test	Kagoshima Space Center
10/10/2005	NAL-735	NEXST-1 scaled experiment SST	Woomera
27/10/2005	Kosmos 3	Cubesat XI-V	Plesetsk
24/08/2005	Dnepr	Kirari OICETS (Optical Inter-orbit Communications Engineering Test Satellite); Reimei INDEX (Innovative Technology Demonstration Experiment Satellite)	Baikonur
10/07/2005	M-V-6	Suzaku ASTRO-EII, 23rd scientific satellite	Uchinoura Space Center
26/02/2005	H-IIA·F7	MTSAT-1R (Multi-Functional Transport Satellite-1 Replacement) Third Ignition Experiment of H-IIA Upper Stage Engine	Tanegashima Space Center
09/08/2004	S-310-34	Solar Sails Technology Mission	Kagoshima Launch Center
29/11/2003	H-IIA·F6	Information Gathering Satellites IGS-2a, IGS-2b launch failure after SRB-A malfunction	Tanegashima Space Center
30/10/2003	Rokot	SERVIS-1 (Space Environment Reliability Verification of Integrated System) for USEF	Plesetsk
30/06/2003	Rokot	Piggy-back satellites: Cute-1 and CubeSat XI-IV picosatellites	Plesetsk
09/05/2003	M-V-5	Hayabusa MUSES-C, asteroid approach and landing mission: research on new engineering technology such as electric propulsion, autonomous navigation, and Earth reentry (sample return)	Uchinoura Space Center
28/03/2003	H-IIA·F5	IGS-1a and IGS-1b, Information Gathering Satellites	Tanegashima Space Center
14/12/2002	H-IIA·F4	Midori-2 ADEOS-II (Advanced Earth Observing Satellite II) Piggy-back payloads (Micro LabSat, WEOS, Fed Sat)	Tanegashima Space Center
10/09/2002	H-IIA·F3	Kodama DRTS (Data Relay Test Satellite); USERS (Unmanned Space Experiment Recovery System)	Tanegashima Space Center
14/07/2002	NAL-735	National Experimental Supersonic Transport NEXST-1 test failure due to loss of stability	Woomera

(*continued*)

Appendix 1. *(continued)*

Launch Date (dd/mm/year)[2]	*Flight Name or SLV*	*Mission*	*Launch Site*
04/02/2002	H-IIA-F2	MDS-1 (Mission Demonstration Test Satellite-1/Tsubasa); VEP-3 (H-IIAVehicle Evaluation Payload #3); DASH (Demonstrator of Atmospheric Reentry System with Hyper Velocity)	Tanegashima Space Center
29/08/2001	H-IIA-F1	Flight Demonstration Test LRE (Laser Range Experiment VEP-2 [H-IIA] Vehicle Evaluation Payload #2)	Tanegashima Space Center
20/12//2000	Ariane 5	LDREX antenna experiment for ETS-VIII	Kourou
11/10/2000	U..S. Space Shuttle Discovery	Astronaut Wakata's shuttle flight (STS-92)	Edwards Air Force Base
11/02/2000	U.S. Space Shuttle Endeavour	Astronaut Mohri's second shuttle flight (STS-99)	Kennedy Space Center
10/02/2000	M-V-4	ASTRO-E X-ray astronomy satellite, launch failure	Uchinoura Space Center
15/11/1999	H-II No. 8	MTSAT (Multi-functional Transport Satellite) launch failure, first-stage malfunction	Tanegashima Space Center
19/11/1998	TR-1A-7	Microgravity Experiments	Tanegashima Space Center
29/10/1998	U.S. Space Shuttle Discovery	Astronaut Mukai's second shuttle flight (STS-95)	Kennedy Space Center
04/07/1998	M-V-3	Nozomi PLANET-B Mars orbiter	Uchinoura Space Center
02/06/1998	U.S. Space Shuttle Discovery	Space Radiation Environment Program (STS-91/9th Shuttle/Mir Mission)	Kennedy Space Center

17/04/1998	U.S. Space Shuttle Columbia	Neurolab Program (STS-90)	Kennedy Space Center
21/02/1998	H-II No.5	Kakehashi COMETS (Communications and Broadcasting Engineering Test Satellite)	Tanegashima Space Center
22/01/1998	U.S. Space Shuttle Endeavour	Space Radiation Environment Measurement Program (STS-89)	Kennedy Space Center
28/11/1997	H-II No. 6	Orihime/Hikoboshi ETS-VII, TRMM (Tropical Rainfall Measuring Mission)	Tanegashima Space Center
19/11/1997	U.S. Space Shuttle Columbia	Astronaut Takao Doi performed Japan's first spacewalk (STS-87)	Kennedy Space Center
07/08/1997	U.S. Space Shuttle Discovery	Manipulator Flight Demonstration (MFD) (STS-85)	Kennedy Space Center
05/07/1997	PROGRESS-M35 Cargo Supply Craft	MIR Utilization Space Experiment	Baikonur cosmodrome
01/07/1997	U.S. Space Shuttle Columbia	First Microgravity Science Laboratory (MSL-1R) (STS-94)	Kennedy Space Center
15/05/1997	U.S. Space Shuttle Atlantis	6th Shuttle/Mir Mission (STS-84)	Kennedy Space Center
04/04/1997	U.S. Space Shuttle Columbia	First Microgravity Science Laboratory (MSL-1) (STS-83)	Kennedy Space Center
12/02/1997	M-V-1	Halca MUSES-B, Very Long Baseline Interferometry (VLBI) for radio astronomy	Uchinoura Space Center
17/08/1996	H-II No. 4	Midori ADEOS (Advanced Earth Observing Satellite) Piggy-back JAS-2 amateur radio satellite	Tanegashima Space Center
06/07/1996	(NA)	ALFLEX (Automatic Landing Flight Experiment)	Woomera
12/02/1996	J-I No.1	HYFLEX (Hypersonic Flight Experiment)	Tanegashima Space Center
11/01/1996	U.S. Space Shuttle Endeavour	Space Experiments/SFU Recovery (Astronaut Koichi Wakata on board) (STS-72)	Kennedy Space Center

(*continued*)

Appendix 1. *(continued)*

Launch Date (dd/mm/year)[2]	*Flight Name or SLV*	*Mission*	*Launch Site*
18/03/1995	H-II No. 3 (test vehicle)	GMS-5 & SFU (Space Flyer Unit) launch	Tanegashima Space Center
15/01/1995	M-3 SII-8	EXPRESS 1 (launch failure, payload too heavy)	Uchinoura Space Center
28/08/1994	H-II No. 2 (test vehicle)	Kiku-6 ETS-VI launch (satellite kick motor failed)	Tanegashima Space Center
08/07/1994	U.S. Space Shuttle Columbia	IML-2 (Astronaut Mukai on board) (STS-65)	Kennedy Space Center
04/02/1994	H-II No. 1 (test vehicle)	OREX (Orbital Reentry Experiment), Myojo Vehicle Evaluation Payload	Tanegashima Space Center
20/02/1993	M-3 SII-7	Asuka ASTRO-D satellite for X-ray observation of high-energy sources	Uchinoura Space Center
12/09/1992	U.S. Space Shuttle Endeavour	Fuwatto '92/Spacelab-J (Astronaut Mohri on board) (STS-47)	Kennedy Space Center
22/01/1992	U.S. Space Shuttle Discovery	IML-1 (STS-42)	Edwards Air Force Base
25/08/1991	H 23F (H-I 8F)	BS-3b launch	Tanegashima Space Center
30/08/1991	M-3 SII-6	Yohkoh SOLAR-A satellite for high-resolution imaging of solar flares	Uchinoura Space Center
28/08/1990	H 22F (H-I 7F)	BS-3a launch	Tanegashima Space Center
07/02/1990	H-21F (H-I 6F)	MOS-1b, JAS-1b launch	Tanegashima Space Center
24/01/1990	M-3 SII-5	Hiten MUSES-A satellite for lunar swing-by orbiting demonstration Hagoromo lunar orbiter	Uchinoura Space Center

06/09/1989	H 20F (H-I 5F)	GMS-4 launch	Tanegashima Space Center
20/08/1989	TR-I-3	Technical data acquisition for H-II development	Tanegashima Space Center
21/02/1989	M-3 SII-4	Akebono EXOS-D satellite for active dynamic observation of aurora	Uchinoura Space Center
27/01/1988	TR-I-2	Technical data acquisition for H-II development	Tanegashima Space Center
16/09/1988	H 19F (H-I 4F)	CS-3b launch	Tanegashima Space Center
06/09/1988	TR-I-1	Technical data acquisition for H-II development	Tanegashima Space Center
19/02/1988	H 18F	CS-3a launch	Tanegashima Space Center
27/08/1987	H 17 (H-I test vehicle)	Kiku-5 ETS-V launch/three-stage rocket test	Tanegashima Space Center
19/02/1987	N16 (N-II 8F)	MOS-1 launch	Tanegashima Space Center
05/02/1987	M-3 SII-3	Ginga ASTRO-C satellite for X-ray observation of active galactic nuclei	Uchinoura Space Center
13/08/1986	H 15 (H-I test vehicle)	EGS, JAS-1 & MABES launch	Tanegashima Space Center
12/02/1986	N14 (N-II 7F)	BS-2 launch	Tanegashima Space Center
18/08/1985	M-3 SII-2	Suisei PLANET-A for observation of Halley's comet	Uchinoura Space Center
07/01/1985	M-3 SII-1	Sakigake MS-T5 first interplanetary probe	Uchinoura Space Center
03/08/1984	N13 (N-II 6F)	GMS-3 launch	Tanegashima Space Center
23/01/1984	N12 (N-II 5F)	BS-2a launch	Tanegashima Space Center
14/02/1984	M-3 S-4	Onzora EXOS-C satellite for upper atmosphere observation	Uchinoura Space Center
19/08/1983	TT-500A 13F	Material processing experiments using small rocket	Tanegashima Space Center
06/08/1983	N11 (N-II 4F)	CS-2b launch	Tanegashima Space Center
04/02/1983	N10 (N-II F3)	CS-2a launch	Tanegashima Space Center
27/01/1983	TT-500A F12	Material experiments in microgravity environment: semiconductor experiments and performance test of halogen lamp	Tanegashima Space Center
20/02/1983	M-3S-3	Tenma ASTRO-B for observation of X-ray stars and gamma bursts	Uchinoura Space Center
05/09/1982	MT-135P T33	Meteorological observation	Tanegashima Space Center

(*continued*)

Appendix 1. *(continued)*

Launch Date (dd/mm/year)[2]	*Flight Name or SLV*	*Mission*	*Launch Site*
03/09/1982	N9 (N-I F7)	Kiku-4 ETS-III launch	Tanegashima Space Center
16/08/1982	TT-500A F11	Performance tests of radar tracking system and material experiment recovery system	Tanegashima Space Center
03/02/1982	MT-135P T32	Meteorological observation	Tanegashima Space Center
02/02/1982	MT-135P T31	Meteorological observation	Tanegashima Space Center
11/08/1981	N8 (N-II F2)	GMS-2 launch	Tanegashima Space Center
02/08/1981	TT-500A F10	Performance test: radar tracking system and ground systems; material experiments in microgravity environment	Tanegashima Space Center
12/02/1981	MT-135P T30	Meteorological observation	Tanegashima Space Center
11/02/1981	N7 (N-II F1)	Kiku-3 ETS-IV launch	Tanegashima Space Center
15/01/1981	TT-500A F9	Performance test: radar tracking system and ground systems; material experiments in microgravity environment	Tanegashima Space Center
21/02/1981	M-3S-2	Hinotori ASTRO-A satellite for solar flare imaging by hard X ray	Uchinoura Space Center
15/09/1980	MT-135P T29	Meteorological observation	Tanegashima Space Center
14/09/1980	TT-500A F8	Performance test: radar tracking system and ground systems; material experiments in microgravity environment	Tanegashima Space Center
13/09/1980	MT-135P T28	Meteorological observation	Tanegashima Space Center
23/02/1980	MT-135P T27	Meteorological observation	Tanegashima Space Center

22/02/1980	N6	ECS-b launch (failed to reach geostationary orbit)	Tanegashima Space Center
28/01/1980	TT-500 F7	Performance test: radar tracking system and ground systems	Tanegashima Space Center
17/02/1980	M-3S-1	Tansei-4 MS-T4 satellite for launch verification and engineering.	Uchinoura Space Center
27/08/1979	MT-135P T26	Meteorological observation	Tanegashima Space Center
27/08/1979	MT-135P T25	Meteorological observation	Tanegashima Space Center
25/08/1979	TT-500 F6	Performance test: radar tracking system and ground systems	Tanegashima Space Center
07/02/1979	MT-135P T24	Meteorological observation	Tanegashima Space Center
06/02/1979	N 5	ECS launch (failure to reach geostationary orbit)	Tanegashima Space Center
27/01/1979	TT-500 F5	Performance test: radar tracking system and ground systems	Tanegashima Space Center
21/02/1979	M-3C-4	Hakucho CORSA-b satellite for X-ray observation of high-energy sources in the universe	Uchinoura Space Center
26/08/1978	MT-135P T23	Meteorological observation	Tanegashima Space Center
24/08/1978	MT-135P T22	Meteorological observation	Tanegashima Space Center
25/08/1978	TT-500 F4	Performance test: radar tracking system and ground systems	Tanegashima Space Center
07/04/1978	DELTA F140 (TYPE 2914)	BSE communications satellite launch	Kennedy Space Center
17/02/1978	MT-135P T21	Meteorological observation	Tanegashima Space Center
16/01/1978	TT-500 F3	Performance test: radar tracking system and ground systems	Tanegashima Space Center
16/02/1978	N4	ISS-b launch	Tanegashima Space Center
16/09/1978	M-3H-3	Jikiken EXOS-B satellite for aurora observation	Uchinoura Space Center
04/02/1978	M-3H-2	Kyokko EXOS-A satellite for geomagnetic field exploration	Uchinoura Space Center
14/12/1977	DELTA F137 (TYPE 2914)	CS communications satellite launch	Kennedy Space Center
26/08/1977	MT-135P T20	Meteorological observation	Tanegashima Space Center
23/08/1977	MT-135P T19	Meteorological observation	Tanegashima Space Center
25/08/1977	TT-500 F2	Performance test: radar tracking system and ground systems	Tanegashima Space Center

(continued)

Appendix 1. *(continued)*

Launch Date (dd/mm/year)[2]	*Flight Name or SLV*	*Mission*	*Launch Site*
14/07/1977	DELTA F132 (TYPE 2914)	GMS 1 launch	Kennedy Space Center
24/02/1977	MT-135P T18	Meteorological observation	Tanegashima Space Center
25/01/1977	TT-500 F1	Performance test: radar tracking system and ground systems	Tanegashima Space Center
23/02/1977	N3	Kiku-2 ETS-II launch	Tanegashima Space Center
19/02/1977	M-3H-1	Tansei 3 satellite for launch verification and engineering	Uchinoura Space Center
25/09/1976	MT-135P T17	Meteorological observation	Tanegashima Space Center
24/09/1976	TT-210 F3	Performance test: radar tracking system	Tanegashima Space Center
23/09/1976	MT-135P T16	Meteorological observation	Tanegashima Space Center
27/01/1976	TT-210 F2	Performance test: radar tracking system	Tanegashima Space Center
29/02/1976	N-2	ISSa launch	Tanegashima Space Center
04/02/1976	M-3C-3	Corsa-A failed satellite launch	Uchinoura Space Center
10/09/1975	MT-135P T15	Meteorological observation	Tanegashima Space Center
17/08/1975	TT-210 F1	Performance test: radar tracking system	Tanegashima Space Center
09/09/1975	N-1	Kiku-1 ETS-I launch	Tanegashima Space Center
07/02/1975	MT-135P T14	Meteorological observation	Tanegashima Space Center
30/01/1975	MT-135P T13	Meteorological observation	Tanegashima Space Center

05/02/1975	ETV-2	N rocket engineering test: two-stage propulsion system, control system, ground facilities, and so on	Tanegashima Space Center
24/02/1975	M-3C-2	Taiyo satellite for observation of solar X rays and solar UV radiation	Uchinoura Space Center
03/09/1974	MT-135P T12	Meteorological observation	Tanegashima Space Center
03/09/1974	MT-135P T11	Meteorological observation	Tanegashima Space Center
02/09/1974	ETV-1	N rocket engineering test: two-stage gas jet system, attitude control system, ground facilities, and so on	Tanegashima Space Center
10/02/1974	MT-135P T10	Meteorological observation	Tanegashima Space Center
02/02/1974	MT-135P T9	Meteorological observation	Tanegashima Space Center
09/02/1974	LS-C7	N rocket engineering test: two-stage LE-3 liquid engine	Tanegashima Space Center
01/02/1974	JCR-10	Electrical linkage check between control devices and ground facilities; attitude control test; destruction command system test	Tanegashima Space Center
16/02/1974	M-3C-1	Tansei 2 satellite for launch verification and satellite engineering	Uchinoura Space Center
05/09/1973	MT-135P T8	Meteorological observation	Tanegashima Space Center
05/09/1973	MT-135P T7	Meteorological observation	Tanegashima Space Center
07/09/1973	JCR-9	Electrical linkage check between control devices and ground facilities; attitude control test; destruction command system test; gas jet control system performance test	Tanegashima Space Center
08/02/1973	MT-135P T6	Meteorological observation	Tanegashima Space Center
05/02/1973	MT-135P T5	Meteorological observation	Tanegashima Space Center
07/02/1973	JCR-8	Performance test: gas jet control system; PCM-PM telemeter devices; S-band command receiver; C-band radar transponder	Tanegashima Space Center
25/09/1972	LS-C6	Performance test: Gimbal control system; gas jet control system	Tanegashima Space Center
30/08/1972	MT-135P T4	Meteorological observation	Tanegashima Space Center
30/08/1972	MT-135P T3	Meteorological observation	Tanegashima Space Center
06/02/1972	JCR-7	Performance test: gas jet control system; PCM-PM telemeter devices	Tanegashima Space Center
07/02/1972	MT-135P T2	Meteorological observation	Tanegashima Space Center

(continued)

Appendix 1. *(continued)*

Launch Date (dd/mm/year)[2]	*Flight Name or SLV*	*Mission*	*Launch Site*
02/02/1972	MT-135P T1	Meteorological observation	Tanegashima Space Center
19/08/1972	M-4S-4	Denpa REXS satellite for measurement of plasma waves, plasma density, electron flux, and so on	Uchinoura Space Center
17/09/1971	JCR-6	Performance test: gas jet control system	Tanegashima Space Center
10/09/1971	LS-C5	Performance test: Gimbal control system; gyroscope; gas jet control system	Tanegashima Space Center
11/09/1971	SB-III A15	Meteorological observation	Tanegashima Space Center
06/09/1971	SB-III A14	Meteorological observation	Tanegashima Space Center
03/02/1971	SB-III A13	Meteorological observation	Tanegashima Space Center
03/02/1971	SB-III A12	Meteorological observation	Tanegashima Space Center
01/02/1971	JCR-5	Performance test: gas jet control system; disturbance measurement	Tanegashima Space Center
28/09/1971	M-4S-3	Shinsei satellite for observation of ionosphere, cosmic rays, and so on	Uchinoura Space Center
16/02/1971	M-4S-2	Tansei-1 satellite for environment measurement and engineering test	Uchinoura Space Center
07/09/1970	NAL-7BS9	Aerodynamics stability study	Tanegashima Space Center
09/09/1970	LS-C4	Performance test: Gimbal control system; gyroscope; gas jet control system	Tanegashima Space Center
03/02/1970	LS-C3	Performance test: engine with tubewall chamber; Gimbal control system	Tanegashima Space Center
25/09/1970	M-4S-1	SS failed satellite launch for magnetosphere observation	Uchinoura Space Center
11/02/1970	L-4S-5	Ohsumi 5: Japan's first satellite launch	Uchinoura Space Center

01/02/1970	JCR-3	Performance test: flight characteristic, separation mechanism, command system	Tanegashima Space Center
20/09/1969	NAL-7 F7	Aerodynamics study, vibration and deflection test	Tanegashima Space Center
20/09/1969	NAL-16H F2	Performance test: rocket engine; aerodynamics separation mechanism; lightning of the body	Tanegashima Space Center
16/09/1969	JCR-2	Attitude control around the roll axis	Tanegashima Space Center
15/09/1969	JCR-1	Performance test of gas jet control systems	Tanegashima Space Center
10/09/1969	LS-C2	Performance test: liquid propellant engine; flight performance	Tanegashima Space Center
09/09/1969	SB-III F11	Meteorological observation; flight route measurement	Tanegashima Space Center
01/02/1969	NAL-25*31	Gas jet operation test	Tanegashima Space Center
30/01/1969	NAL-16*31D	Performance test: solid rocket; separation mechanism Measurement test: aerodynamics characteristics; base drag	Tanegashima Space Center
08/02/1969	SC-4	Flight performance test; meteorological observation	Tanegashima Space Center
07/02/1969	SC-3	Flight performance test; meteorological observation	Tanegashima Space Center
08/02/1969	SC-1	Flight performance test; meteorological observation	Tanegashima Space Center
06/02/1969	LS-CI	Liquid propellant engine performance test	Tanegashima Space Center
22/09/1969	L-4S-4	Ohsumi 4 failed satellite launch	Uchinoura Space Center
19/09/1968	NAL-16H F1	Flight performance test	Tanegashima Space Center
19/09/1968	LS-C-D	1st stage SRB performance test; first- and second-stage separation mechanism test	Tanegashima Space Center
17/09/1968	SA-II A9	Meteorological observation; flight route measurement	Tanegashima Space Center
13/04/1967	L-4S-4	Ohsumi 3 failed satellite launch	Uchinoura Space Center
1967	NAL-16TR	Rehearsal only	Tanegashima Space Center
20/12/1966	L-4S-2	Ohsumi 2 failed satellite launch	Uchinoura Space Center
26/09/1966	L-4S-1	Ohsumi 1 failed satellite launch	Uchinoura Space Center
17/11/1965	NAL-16TR	Flight performance test	Niijima

(continued)

Appendix 1. *(continued)*

Launch Date (dd/mm/year)[2]	*Flight Name or SLV*	*Mission*	*Launch Site*
22/11/1965	LS-A3	Flight performance test with the liquid propellant engine	Niijima
16/11/1965	ST-I F1	Safety check: AP propellant combustion/load	Niijima
19/11/1965	SB-II F7	Performance test of rocket engine; operation check of onboard equipment	Niijima
18/11/1965	SB-II F6	Performance test of rocket engine/operation check of onboard equipment	Niijima
18/06/1965	HM-16-IT	Flight performance test of the main body; meteorological observation	Niijima
17/06/1965	HM-16-D	Flight performance test of the main body	Niijima
16/06/1965	S-B5	Performance test of rocket engine; operation check of onboard equipment	Niijima
15/06/1965	S-B4	Performance test of rocket engine; operation check of onboard equipment	Niijima
22/07/1964	LS-A2	Liquid propellant engine performance test	Niijima
22/07/1964	LS-A1	Liquid propellant engine performance test	Niijima
23/07/1964	S-B3	Flight test; nose cone separation check; meteorological observation	Niijima
19/07/1964	S-B2	Flight test; nose cone separation check; meteorological observation	Niijima
17/07/1964	S-B1	Flight test; nose cone separation check; meteorological observation	Niijima
10/08/1963	LS-A	Performance test: liquid propellant engine; flight performance; operation check of onboard equipment; nose cone separation check and its recovery	Niijima
10/08/1963	S-A3	Performance characteristics check: engine combustion; spin stability analysis; nose cone separation test/parachute test; falling velocity measurement; meteorological observation	Niijima

10/08/1963	S-A2	Performance characteristics check: engine combustion; spin stability analysis; nose cone separation test/parachute test; falling velocity measurement; meteorological observation	Niijima
10/08/1963	S-A1	Performance characteristics check: engine combustion; spin stability analysis; nose cone separation test/parachute test; falling velocity measurement; meteorological observation	Niijima
20/05/1963	Kappa 9M	Ionosphere mission	Kagoshima
18/12/1962	Kappa 8	Cosmic rays mission	Kagoshima
23/08/1962	Kappa 8L	Test mission	Kagoshima
01/08/1962	K150	Test mission	Kagoshima
25/05/1962	Kappa 8	Ionosphere mission failure	Akita
26/12/1961	Kappa 9L	Test mission	Akita
30/10/1961	Kappa 8	Ionosphere mission	Akita
24/10/1961	Kappa 8	Ionosphere mission	Akita
21/07/1961	Kappa 8	Aeronomy/Ionosphere mission	Akita
18/06/1961	Kappa 4	Sigma 4 Rockoon chemical release mission	Obachi
18/04/1961	Kappa 8	Ionosphere mission	Akita
01/04/1961	Kappa 9L	Test mission	Akita
27/03/1961	Kappa 8	Ionosphere mission	Akita
29/11/1960	Kappa 6H	K-6H-1 Aeronomy mission	Akita
26/09/1960	Kappa 8	Ionosphere mission	Akita
22/09/1960	Kappa 8	Ionosphere mission	Akita
17/07/1960	Kappa 8	Test mission	Akita
11/07/1960	Kappa 8	Test mission	Akita
28/03/1960	Kappa 8	Test mission	Akita
18/11/1959	Kappa 7	Test mission	Akita

(*continued*)

Appendix 1. *(continued)*

Launch Date (dd/mm/year)[2]	*Flight Name or SLV*	*Mission*	*Launch Site*
01/07/1958	Kappa 6tw	30-mile altitude test flight	Michikawa Rocket Center
16/06/1958	Kappa 6	Test mission	Akita
14/06/1958	K245	Test mission	Akita
08/04/1958	K150	Test mission	Akita
20/09/1957	Kappa 4	Chemical release mission	Akita
26/07/1957	Kappa3	Test mission	Akita
30/04/1957	Kappa 2	Test mission	Akita
13/12/1956	Kappa 1	K-I-T-4 Test mission	Akita
24/09/1956	Kappa 1	K-I-S-1 Test mission	Akita
1955	PENCIL	Horizontal test flight	Kokubunji

SOURCES: In addition to various sources as cited in text, data and information on SLVs and satellites comes from JAXA, ISAS, and NASDA, all available online at www.jaxa.jp (accessed 14 March 2010); as well as that on Japan available online through the *Encyclopedia Astronautica* at www.astronautix.com (accessed 14 March 2010). Data and information concerning the Delta launch vehicle is available at the official Boeing Web site at www.boeing.com/defense-space/space/bls/missions/index.html (accessed 14 March 2010).

1. Focus is only on principal launches to clarify the main trajectory of events, and not all missions, repeat tests, commercial broadcasting and communications satellites, other payloads, etc. are included.
2. Launch dates reflect the date of the country in which the launch center is located.

APPENDIX II

Basic Space Law 2008

LAW NO. 43 (28 MAY 2008)[1]

Table of Contents

CHAPTER 1: GENERAL PROVISIONS

(Purpose)

Article 1: Accompanying the scientific and technological advances as well as changes in internal and external conditions, taking into account the increasing importance of space development and utilization, (hereafter space-related), being based on the principle of pacifism in the Japanese constitution, giving due consideration to harmony with the environment, this law is established to expand the role played by space development and utilization in our country.

[1] The text represents the authors' translation, and in the event of any divergences in interpretive translation, only the authentic Japanese text of the Basic Space Law prevails. For the law itself, see "Uchyū Kihon Hōan" [Basic Space Law], available online through the Strategic Headquarters for Space Policy at www.kantei.go.jp (accessed 16 April 2009).

In relation to space development and utilization, the purpose of this law is to establish the basic doctrines and the basic particulars to actualize them, to clarify the government's responsibilities, to also formulate a basic space plan, and to establish such things as the Strategic Headquarters for Space Policy. Accordingly, the goal is to advance comprehensive and systematic measures related to space development and utilization, improve the livelihood of citizens, and contribute to the development of the economic system, world peace, and human welfare.

(Peaceful Uses of Outer Space)
Article 2: The development and utilization of space will follow the Treaty on Principles Governing the Activities of States in the Exploration and Use of Outer Space, including the Moon and Other Celestial Objects; treaties related to space development and utilization, as well as other international agreements; and conform to the principles of pacifism in the Constitution of Japan.

(Improvement of Citizens' Livelihood)
Article 3: The development and utilization of space must improve the livelihood of citizens, create a safe and secure society for living, eliminate numerous threats to human existence and life such as disasters and poverty, and contribute to the peace and security of international society as well as the national security of our country.

(Promotion of Industry)
Article 4: The development and utilization of space must be promoted actively and systematically, and the results of research and development will be commercialized without delay. Accordingly, these must contribute to the promotion of our industry, and to the enhancement of the technology and international competitiveness of our space industry and related industries.

(Development of Human Society)
Article 5: The development and utilization of space must contribute to the realization of humankind's dream of space and the development of human society. Keeping in mind that the accumulation of space-related knowledge is vital for the intellectual property of human beings, we must promote advanced space development and utilization, and space science.

(International Cooperation etc.)

Article 6: The development and utilization of space must actively advance space-related international cooperation and diplomacy. Accordingly, we must actively contribute to the global role played by our country and increase our own national interests in international society.

(Environmental Considerations)

Article 7: We must give due consideration to the impact of space development and utilization on the environment.

(Obligations of the Government)

Article 8: The government bears the obligation for enacting and enforcing comprehensive measures related to space development and utilization, and also for conforming to the basic principles (hereafter basic principles) related to space development and utilization from Article 2 to the previous Article.

(Duty of Efforts by Local Public Entities)

Article 9: With regards to space development and utilization, local public entities must conform to the basic principles and have an appropriate division of labor between themselves and the national government. They must make efforts to enact and enforce independent measures that capitalize on their distinct regional characteristics.

(Strengthening Coordination)

Article 10: The government will devise reciprocal coordination with local public entities, universities, private companies, and so on, to have cooperation. Keeping in mind the plans for effective advancement of space development and utilization, the government will take necessary measures to strengthen coordination between them.

(Legal Measures etc.)

Article 11: In order to enforce policies related to space development and utilization, the government will take necessary legal, monetary, tax, financial, and other measures.

(Upkeep of Administrative Organizations)

Article 12: As the government takes measures related to space development and utilization, it will strive accordingly to ameliorate the upkeep and management of the administrative organizations.

CHAPTER 2: BASIC POLICIES

(Use of Satellites to Contribute to the Improvement of Citizens' Livelihood etc.)

Article 13: In order to improve the livelihood of citizens, to create a safe and secure society for living, and to contribute to the elimination of threats to human existence and life such as disasters and poverty, the government will promote the maintenance and will also take other necessary measures for such things as a stable information and communications network using satellites, an information system related to observation, and an information system related to positioning.

(Ensuring the Peace and Safety of International Society as Well as Our National Security)

Article 14: In order to promote the contribution of space development and utilization to ensuring international peace and safety, as well as our national security, the government will take necessary measures.

(Independent Launch of Satellites etc.)

Article 15: Keeping in mind the importance of our country's ability to have independent development, launch, tracking, and operation of satellites, the government will take necessary measures such as the acquisition of indispensable equipment (including spare parts thereof), the promotion and upkeep of technological research and development, the maintenance of facilities and institutions and so on, and the securing of frequencies used by our country's space development and utilization.

(Promotion of Space Development and Utilization by Private Enterprise)

Article 16: Considering the importance of the role played by the private sector in space development and utilization, the government will promote space-related enterprise activity (including research and development). In striving to strengthen the technological strength and the international competitiveness of our space industry, as well as related industries, when the government takes operational initiative related to space development and utilization, it will harness the strengths of the private sector. In addition the government will give systematic consideration to the procurement of goods and services, and to the maintenance of launch facilities (place where rocket launches take place), as well as testing, research, and other facilities and establishments. The government will have tax and financial measures as well as other necessary policies to expedite the transfer of space-related R&D results to the private enterprises; to promote the privatization of civilian space-related R&D results; and to facilitate investments in space-related enterprises by the private sector.

(Support and Improvement of Reliability)

Article 17: Considering the importance of striving to support and improve the reliability of technology related to space development and utilization, the government will take necessary measures and promote fundamental space-related research and basic technology R&D.

(Promotion of Advanced Space Development and Utilization etc.)

Article 18: The government will take necessary measures to promote academic research and so on related to space development and utilization for such things as space exploration as well as the space sciences.

(Promotion of International Cooperation etc.)

Article 19: In the field of space development and utilization, the government will actively fulfill our country's role in international society. In addition to advancing our space-related national interest in international society, the government will take necessary measures for promoting international alliances for R&D, international technology cooperation, and other international cooperation, as well as deepening foreign countries' appreciation of our country's space development and utilization.

(Environmental Conservation)

Article 20: The government will take necessary measures to promote space development and utilization giving due consideration to harmony with the environment.

20.2 The government will make an effort to secure international alliances in order to preserve the environment of space.

(Securing Human Resources etc.)

Article 21: To promote space development and utilization, the government will take necessary measures to improve the securing, training, and quality of space-related human resources while striving for close-knit coordination and cooperation between universities, private enterprises, and so on.

(Promotion of Education and Learning etc.)

Article 22: To strengthen the understanding and interest of the public related to space development and utilization widely, the government will promote space-related education and learning, enhance public relations, and take other necessary measures.

(Management of Information Related to Space Development and Utilization)

Article 23: Considering the special characteristics of space development and utilization, the government will take necessary measures for the appropriate management of space-related information.

CHAPTER 3: BASIC SPACE PLAN

Article 24: In striving to promote comprehensive and systematic measures related to space development and utilization, the Strategic Headquarters for Space Policy must prepare a Basic Space Plan (hereafter Basic Space Plan).

24.2 The Basic Space Plan is determined by the following stipulations.

1. Basic principles related to the promotion of space development and utilization.
2. Policies related to space development and utilization that the government should implement comprehensively and systematically.
3. In addition to the previous point, other necessary provisions for the government in order to promote policies related to space development and utilization comprehensively and systematically.

24.3 As for the policies set out in the Basic Space Plan, there should be, in principle, concrete goals and achievement timetables for the relevant policies.

24.4 As provided for in Clause 1, the Strategic Headquarters for Space Policy must publicize the Basic Space Plan at the time of completion through the internet and other appropriate means without delay.

24.5 As provided in Clause 3, at the appropriate time, the Strategic Headquarters for Space Policy must investigate the actual level of achievement of the objectives, and publicize the results through the internet and other appropriate means.

24.6 Considering the conditions of progress of space development and utilization, and the effectiveness of the government's space-related policies, the Strategic Headquarters for Space Policy can, at its discretion, also investigate the Basic Space Plan. If it is found to be necessary, the Strategic Headquarters for Space Policy must revise the Basic Space Plan. In that case it will apply the provisions of Clause 4, *mutatis mutandis.*

24.7 In striving to secure the necessary funds for the operating expenses of the Basic Space Plan, the government must, financial conditions permitting, make an effort every year to appropriate the budget and so on, and take necessary measures for smooth implementation.

CHAPTER 4: THE STRATEGIC HEADQUARTERS FOR SPACE POLICY

(Establishment)

Article 25: In order to promote policies related to space development and utilization comprehensively and systematically, the Strategic Headquarters for Space Policy (hereafter Headquarters) will be established within the Cabinet.

(Administrative Tasks)

Article 26: The Headquarters will undertake the following administrative tasks.

26.1 Prepare the Basic Space Plan and promote its implementation.

26.2 In addition to those articulated in the previous clause, undertake investigations and reviews of other significant plans related to space development and utilization policies, as well as the promotion and overall adjustment of those policies.

(Organization)

Article 27: The Headquarters consists of the Director, the Deputy Director, and other staff.

(Director of the Strategic Headquarters for Space Policy)

Article 28: The Directorship of the Strategic Headquarters for Space Policy (hereafter Director) will be assumed by the Prime Minister.

28.2 The Director will undertake overall administration of the Headquarters, and will direct and supervise its staff.

(Deputy Directors of the Strategic Headquarters for Space Policy)

Article 29: The Headquarters will have Deputy Directors of the Strategic Headquarters for Space Policy (hereafter Deputy Directors). The Chief Cabinet

Secretary and the Minister of Space Development and Utilization (a Minister of State who takes orders from the Prime Minister, and assists the Prime Minister in space-related matters) will assume the position of Deputy Directors.

29.2 The Deputy Directors will assist the work of the Director.

(Staff of the Strategic Headquarters for Space Policy)

Article 30: The Headquarters will have the Strategic Headquarters for Space Policy Staff (hereafter Staff).

30.2 With the exception of the Director and the Deputy Directors, the Staff will be composed of all Ministers of State.

(Submission of Documents and Other Cooperation)

Article 31: If especially necessary to the execution of its administrative tasks, the Headquarters can seek submission of documents, statements of opinions and explanations, as well as any other necessary cooperation from related government agencies, local public entities, heads of independent administrative agencies (as stipulated in Article 2.1 of the Act on General Rules for Incorporated Administrative Agency [Law No. 103, Heisei 11]), as well as representatives of semi-governmental corporations (legal persons established either by a law directly or by a special set up act of a special law, as applied under the provisions of Article 4.15 of the Act for Establishment of the Ministry of Internal Affairs and Communications [Act No. 91, Heisei 11]).

31.2 If necessary to the execution of its administrative tasks, the Headquarters can also request necessary cooperation from persons other than those stipulated in the previous clause.

(Administrative Matters)

Article 32: Administrative matters pertaining to Headquarters will be dealt with in the Cabinet Secretariat, and be directed through orders given to the Assistant Deputy Chief Cabinet Secretaries.

(Minister in Charge)

Article 33: With respect to matters pertaining to Headquarters, the Minister in Charge as stated in the Cabinet Act (Act No. 5, Showa 22) will be the Prime Minister.

(Delegating to Cabinet Orders)

Article 34: In addition to the things determined in this law, other matters pertaining to Headquarters will be determined by Cabinet orders.

CHAPTER 5: UPKEEP OF LEGAL SYSTEM RELATED TO SPACE ACTIVITIES

Article 35: In order to enforce regulations having to do with space activities, other treaties related to space development and utilization, as well as other international agreements, the government must improve the legal system for matters necessary for enforcement comprehensively, systematically, and speedily.

35.2 The improvement of the legal system in the previous clause must be done to advance our country's interest in international society and to contribute to increasing the public use of space development and utilization.

SUPPLEMENTARY PROVISIONS

(Dates of Enforcement)

Article 1: Within three months from the date of promulgation, this law will be put into force from the day of the Cabinet order onwards.

(Upkeep of the Legal System etc. to Allow the Cabinet Office to Deal with Administrative Matters Related to Headquarters)

Article 2: Using one year after the enforcement of this law as a benchmark, the government must improve the necessary legal system and take other measures in order to allow the Cabinet Office to deal with administrative matters related to Headquarters.

(Examination Related to [Independent Administrative Agency] Japan Aerospace Exploration Agency [JAXA] etc.)

Article 3: Using one year after the enforcement of this law as a benchmark, the government will examine and review the goals, functions, operational range, ideal organizational structure, and the administrative supervision of appropriate agencies and so on related to [independent administrative agency] Japan Aerospace Exploration Agency [JAXA], as well as any other agencies related to space development and utilization.

(Examination of Ideal Administrative Organization etc. to Advance Policies Related to Space Development and Utilization Comprehensively and Seamlessly)

Article 4: In order to advance the policies related to space development and utilization comprehensively and seamlessly, the government will examine the ideal administrative organization and so on, and based on that result will take necessary measures.

Signed by the Prime Minister and Ministers of:
Internal Affairs and Communications
Justice
Foreign Affairs
Finance
Education, Culture, Sports, Science, and Technology
Health, Labor, and Welfare
Agriculture, Forestry, and Fisheries
Economy, Trade, and Industry
Land, Infrastructure, Transport, and Tourism
Environment
Defense

NOTES

Chapter 1

1. Earlier work on the Japanese space industry has been carried by journalists such as Paul Kallender-Umezu and Eiichiro Sekigawa, both of whose work is cited extensively throughout the book, as well as a few academics: Setsuko Aoki, "Military Uses of Outer Space: Law and Policy in Japan" (International Symposium on Space Technology and Science Paper 2004-r-32, 2004), pp. 1–6; Kazuto Suzuki, "Administrative Reforms and the Policy Logics of Japanese Space Policy," *Space Policy* 21(1), 2005, pp. 11–19; Steven Berner, "Japan's Space Program: A Fork in the Road?" RAND National Security Research Division, Santa Monica, CA, 2005, available online at www.rand.org (accessed 22 December 2006), pp. 1–37; Andrew L. Oros, *Normalizing Japan: Politics, Identity and the Evolution of Security Practice* (Stanford, CA: Stanford University Press, 2008), pp. 122–148; Andrew L. Oros, "The New Politics of Antimilitarism: Explaining Japan's Development of Surveillance Satellite Capabilities" (paper presented at the annual meeting of the International Studies Association, Honolulu, Hawaii, 1–5 March 2005), pp. 1–36; Saadia M. Pekkanen, *Picking Winners? From Technology Catch-up to the Space Race in Japan* (Stanford, CA: Stanford University Press, 2003), esp. pp. 161–190; Seungjoo Lee, "Autonomy or International Cooperation? The Japanese Space Industry Responds to U.S. Pressure," *Business and Politics* 2(2), 2000, pp. 225–250; and William D. Wray, "Japanese Space Enterprise: The Problem of Autonomous Development," *Pacific Affairs* 64(4), 1991, pp. 463–488.

2. For a view on how Japanese defense production has been embedded in the larger commercial economy and thus allowed the fusion of "technology-conscious industrial policy with national security policy," see Richard J. Samuels, *"Rich Nation, Strong Army": National Security and the Technological Transformation of Japan* (Ithaca, NY: Cornell University Press, 1996), esp. pp. 154–197.

3. Kazuto Suzuki, "Space: Japan's New Security Agenda," Research Institute for Peace and Security (RIPS) Policy Perspective, No. 5, Tokyo, October 2007, esp. pp. 1–5.

4. For an introduction on dual-use, see especially Joan Johnson-Freese, *Space as a Strategic Asset* (New York: Columbia University Press, 2007), pp. 6–7. This multifaceted story resonates well with most analysts' emphasis that space policy is generally propelled by some combination of three key drivers—namely, science, commerce, and security. See Kurt M. Campbell, Christian Beckner, and Yuki Tatsumi, "U.S.–Japan Space Policy: A Framework for 21st Century Cooperation" (Washington, DC: Center for Strategic and International Studies, July 2003), pp. 3–4.

5. On dual-use as well as the militarization-weaponization divide on space assets, see especially Johnson-Freese, *Space as a Strategic Asset*, pp. 2–6, 27–50, 82–140. See also Michael E. O'Hanlon, *Neither Star Wars Nor Sanctuary: Constraining the Military Uses of Space* (Washington, DC: Brookings Institution Press, 2004), pp. 1–28; and for the controversies over defining space weapons, see James Clay Moltz, *The Politics of Space Security: Strategic Restraint and the Pursuit of National Interests* (Stanford, CA: Stanford University Press, 2008), esp. p. 43.

6. For fundamentals of military space operations and identification of general mission areas, as well as clear and practical explanations of military space concepts, see United States Joint Chiefs of Staff, *Space Operations*, Joint Publication 3-14, 6 January 2009, pp. ix–xi, II.1–II.10, available online via the Defense Technical Information Center (DTIC) at www.dtic.mil (accessed 28 August 2009); and also U.S. Air Force Space Command (AFSPC), *Strategic Master Plan FY06 and Beyond* (Peterson AFB, CO: Air Force Space Command, 1 October 2003), pp. 2, 17–33.

7. See, under the chairmanship of Donald Rumsfeld, *Report of the Commission to Assess United States National Security Space Management and Organization*, Pursuant to Public Law 106-65, 11 January 2001, available online at www.fas.org (accessed 1 June 2009), esp. pp. 18, 22. The report warned explicitly that because the United States was more dependent on space than any other nation, it was an attractive candidate for a "Space Pearl Harbor." Concern especially with threats to satellites has led to innovative ways to shield them from offensive attacks, such as *Starfire*, the ground-based laser system being developed at an unclassified lab in New Mexico. It is a category of ground-based anti-satellite (ASAT) weapons that are not prohibited by law or treaty. The trend clearly is toward the development of a next generation of defensive and offensive space weapons that are potentially of interest to other countries like Japan. See James Kitfield, "The Permanent Frontier," *National Journal*, 17 March 2001, available online at www.globalsecurity.org (accessed 25 October 2006); Charles V. Peña and Edward L. Hudgins, "Should the United States 'Weaponize' Space? Military and Commercial Implications," *Policy Analysis* 427, 18 March 2002; and William J. Broad, "Administration Researches Laser Weapon," *New York Times*, 3 May 2006.

8. See Campbell, Beckner, and Tatsumi, "U.S.–Japan Space Policy," esp. pp. 19–28.

9. O'Hanlon, *Neither Star Wars Nor Sanctuary*, esp. pp. 22–23.

10. See some of the key academic and policy debates in Jennifer M. Lind, "Pacifism or Passing the Buck? Testing Theories of Japanese Security Policy," *International Security* 29(1) 2004, pp. 92–121; Michael J. Green, *Japan's Reluctant Realism: Foreign Policy Challenges in an Era of Uncertain Power* (New York: Palgrave, 2001), pp. 11–34; Mike M. Mochizuki, "Terms of Engagement: The U.S.–Japan Alliance and the Rise of China," in *Beyond Bilateralism: U.S.–Japan Relations in the New Asia-Pacific*, edited by Ellis S. Krauss and T. J. Pempel (Stanford, CA: Stanford University Press, 2004), pp. 87–114; Paul Midford, *Japanese Public Opinion and the War on Terrorism: Implications for Japan's Security Strategy*, Policy Studies 27 (Washington, DC: East-West Center Washington, 2006); Christopher W. Hughes, *Japan's Re-emergence as a "Normal" Military Power* (Oxford: Oxford University Press for the International Institute for Strategic Studies, 2004); Andrew L. Oros, *Normalizing Japan*, 2008; Christopher W. Hughes, *Japan's Remilitarisation* (London: Routledge for the International Institute for Strategic Studies, 2009); Richard J. Samuels, *Securing Japan: Tokyo's Grand Strategy and the Future of East Asia* (Ithaca, NY: Cornell University Press, 2007); Kenneth B. Pyle, *Japan Rising: The Resurgence of Japanese Power and Purpose* (New York: PublicAffairs, 2007); and also the commentary by T. J. Pempel, Mike M. Mochizuki, Ming Wan, Christopher W. Hughes, Richard J. Samuels, and Kenneth B. Pyle in "Book Review Roundtable: Kenneth B. Pyle's *Japan Rising* and Richard J. Samuels, *Securing Japan*," *Asia Policy* 4, July 2007, pp. 187–211.

11. "LDP Policy Chief Calls for Debate on Nuke Option," *Japan Times*, 16 October 2006. The remark, made during a television show by the chairman of the Liberal Democratic Party's (LDP) Policy Affairs Research Council, caused much controversy. See also David Pilling, "Abe Fails to Quell Allies' Call for Debate on Going Nuclear," *Financial Times*, 9 November 2006. The possibility of a nuclear option for Japan under changing circumstances had also been raised much earlier in the 2000s by Chief Cabinet Secretary Yasuo Fukuda and by Shinzo Abe, then deputy cabinet secretary. See also Dan Plesch, "Without the UN Safety Net, Even Japan May Go Nuclear," *Guardian*, 28 April 2003, available online at www.guardian.co.uk (accessed 24 October 2006).

12. Samuels, *Securing Japan*, p. 5.

13. See the defense chronology in Ministry of Defense (MOD), *Nihon no Boeishō 2008* [Defense of Japan 2008], available online at www.mod.go.jp (accessed 1 June 2009), pp. 454–465; and also in Samuels, *Securing Japan*, esp. p. 93 (table 1).

14. Samuels, *Securing Japan*; and Robert Pekkanen and Ellis S. Krauss, "Japan's 'Coalition of the Willing' on Security Policies," *Orbis* 49(3), 2005, pp. 429–444. See also the view from inside the Japanese military that has very nuanced implications for Japan's "militarization debates in theory and practice in Sabine Frünstück, *Uneasy Warriors: Gender, Memory, and Popular Culture in the Japanese Army* (Berkeley and Los Angeles: University of California Press, 2007).

15. Hughes, *Japan's Remilitarisation*, pp. 19–20, 35–52, 91–98, 99–138; and Oros, *Normalizing Japan*, esp. pp. 1–9.

16. For the latest update, see Christopher W. Hughes, *Japan's Remilitarisation*, pp. 11–20.

17. See Lind, "Pacifism or Passing the Buck?" pp. 92–121; and Hughes, *Japan's Re-emergence*, pp. 67–96. Both offer the most cohesive evidence on existing Japanese military power in air and on sea and land, and for correcting images of Japan's military weaknesses. For earlier overviews of indigenous defense production, its high-profile nature as well as attendant costs and benefits, see Michael J. Green, *Arming Japan: Defense Production, Alliance Politics, and the Postwar Search for Autonomy* (New York: Columbia University Press, 1995); and also Samuels, *"Rich Nation, Strong Army,"* pp. 154–269. See more recently also Richard J. Samuels, "'New Fighting Power!' Japan's Growing Maritime Capabilities and East Asian Security," *International Security* 32(3), 2007–2008, pp. 84–112.

18. We set aside considerations of how government expenditures over dual-use technology-related programs (which are of special interest especially for what some have called Japan's techno-nationalism) are or are not accounted for in reported overall defense spending.

19. Hughes, *Japan's Remilitarisation*, pp. 37–40. Defense expenditures should be approached with some caution, as there are other realities to consider. In Japan's case, although defense budgets may be declining in relative importance to social and economic concerns, they have managed to hover at around 6 percent of total government spending from the mid-1980s to 2008. While this puts downward pressure on equipment acquisition (23 percent in 1988 to 17 percent in 2008 of the defense budget), the reality is that quantitative restrictions have not put a brake on the qualitative expansion of military capabilities because of budgetary flexibility and creativity. The practice of deferred payments since the mid-1970s (to spread the costs of weapons systems over a number of years), for example, is estimated to run upward of 60 percent of defense expenditures—a fact that shows formal breaches in the 1 percent ceiling. Of course the 1 percent ceiling was more visibly breached to much fanfare in 1987, and as a formality it may also come under pressure to increase as China's defense expenditures grow.

20. For far more extensive coverage, see especially Lind, "Pacifism or Passing the Buck?" pp. 94–101; and Hughes, *Japan's Re-emergence*, pp. 67–96.

21. Eiichiro Sekigawa, "Revisiting the Threat; Japanese Defense Spending Faces Shakeup as Government Cites Terrorism, Missile Buildup," *Aviation Week & Space Technology* (hereafter *AWST*), 18 August 2003, p. 30. Although any kind of a direct amphibious invasion of Japan has lessened in threat perceptions, the GSDF is fully capable of defending against a large-scale landing invasion or, possibly, even a special-ops one. This is largely because the GSDF has geo-positional advantages (natural

moat, mountainous terrain, familiarity with home territory), and can, if necessary, rely on the protection of the United States as well as its own air force.

22. Norimitsu Onishi, "Bomb by Bomb, Japan Sheds Military Restraints," *New York Times*, 23 July 2007.

23. Aerial refueling capabilities, which extend the range of aircraft for both counter-air and strike missions, raise concerns that extending the range of ASDF fighters will no longer be strictly limited to self-defense and will thus violate the pacifist constitution. See "Air Tankers Refused for Fiscal 2000," *Japan Times*, 17 December 1999; "Extending Fighters' Range: ASDF's First KC-767 Aerial Refueling Aircraft Arrives in Gifu," *Japan Times*, 21 February 2008; Boeing, "Boeing Delivers First KC-767 Tanker to Japan," News Release, 19 February 2008, available online at www.boeing.com (accessed 26 February 2008). On the modernization of its fleet, the ASDF had earlier set its eyes on the controversial fifth-generation stealth fighter (F-22) and stressed the possibility of producing some at home. Only 190 (150 F-15Cs and 40 F-2s) of ASDF's fighter aircraft are modern in comparison to, for example, China which has a combined modern total of 299 fighters (116 Su-27SKs and 62 J-10s), strike aircraft (73 Su-30MKK), and naval strike aircraft (48 Su-30MK2). The discussion on F-22s is from "Jiki Sentōki—F22 Dōnyū—Takai Kabe" [High Barriers to Introduction of Next Generation Fighter F-22], *Asahi Shinbun*, 4 June 2007; Demetri Sevastopulo, "U.S. Urges Japan to Consider F-35 Jets," *Financial Times*, 14 May 2007; Demetri Sevastopulo and David Pilling, "Stealth Debate Precedes Abe U.S. Visit," *Financial Times*, 22 April 2007; Chisaki Watanabe, "Japan to Develop Prototype Fighter Jet," *Associated Press*, 24 July 2007; and David A. Fulghum and Douglas Barrie, "F-22 Tops Japan's Military Wish List," *Aviation Week*, 23 April 2007, available online at www.aviationweek.com (accessed 4 October 2007).

24. Michiyo Nakamoto, "Japan Plans to Speed Up Anti-Missile Programme," *Financial Times*, 26 October 2006; and Hughes, *Remilitarising Japan*, p. 46.

25. Like the other Japanese forces, the MSDF also has some relative imbalances. One weakness is that MSDF ships and submarines do not carry land-attack cruise missiles like the U.S. Tomahawk, meaning that they cannot attack land targets. The MSDF also remains potentially vulnerable to Chinese (and Russian) state-of-the-art anti-ship cruise missiles (ASCMs). As China's space-based C4ISR (Command, Control, Communication, Computers, Intelligence, Surveillance, Reconnaissance) and target acquisition abilities continue to grow, these vulnerabilities may become even more acute over time. Richard Fisher, Jr., "Growing Asymmetries in the China-Japan Naval Balance," especially section on anti-ship missiles, International Assessment and Strategy Center, 22 November 2005, available online at www.strategycenter.net (accessed 6 October 2007), pp. 1–13 (online version). Vulnerability to attack may become even more acute over time as China's space-based C4ISR (Command, Control, Communication, Computers, Intelligence, Surveillance, Reconnaissance) and target acquisition abilities continue to grow. On this see Office of the Secretary of Defense, *Annual*

Report to Congress: Military Power of the People's Republic of China 2006 (Washington, DC: United States Department of Defense, 2006), esp. pp. 11, 31.

26. See particularly Hughes, *Japan's Re-emergence*, pp. 83–88.

27. See the comments by Michael J. Green, "Seven Questions: Reshaping Japan's Security," available online at www.foreignpolicy.com, August 2006 (accessed 14 June 2007).

28. A long and rich tradition of realism is summarized by Sean M. Lynn Jones and Steven E. Miller, "Preface," in *The Perils of Anarchy: Contemporary Realism and International Security—An International Security Reader*, edited by Michael Brown, Sean M. Lynn-Jones, and Steven E. Miller (Cambridge, MA: The MIT Press, 1995), pp. ix–xxi; Benjamin Frankel, "Restating the Realist Case: An Introduction," in *Realism: Restatements and Renewal*, edited by Benjamin Frankel (New York: Frank Cass & Co., 1996), pp. ix–xx; and Jeffrey W. Taliaferro, "Security Seeking Under Anarchy: Defensive Realism Revisited," *International Security* 25(3), 2000, esp. pp. 128–135. See also John J. Mearsheimer, "Back to the Future: Instability in Europe After the Cold War," *International Security* 15(1), Summer 1990, pp. 5–56; and Joseph M. Grieco, "Anarchy and the Limits of Cooperation: A Realist Critique of the Newest Liberal Institutionalism," *International Organization* 42(3), 1988, pp. 485–507.

29. For the latest cogent restatement on the divisions and similarities, see also Jack Snyder, "Correspondence: Defensive Realism and the 'New' History of World War I," *International Security* 33(1), 2008, esp. pp. 181–185.

30. Samuels, *"Rich Nation, Strong Army,"* pp. x, 32–67, esp. p. 78; and also Peter J. Katzenstein, *Cultural Norms and National Security: Police and Military in Postwar Japan* (Ithaca, NY: Cornell University Press, 1996), p. 151. See also Wray, "Japanese Space Enterprise," pp. 463–488; and Green, *Arming Japan*, pp. 7–30, 53. For a more general emphasis on the costs of closure (or autarky) in weapons production and the necessity of international production through multinational corporations (MNCs) under conditions of globalization today, see Stephen G. Brooks, *Producing Security: Multinational Corporations, Globalization, and the Changing Calculus of Conflict* (Princeton, NJ: Princeton University Press, 2005), esp. pp. 76–78.

31. The Council on Security and Defense Capabilities (CSDC), *The Council on Security and Defense Capabilities Report: Japan's Visions for Future Security and Defense Capabilities* (Tokyo: CSDC, October 2004), pp. 21–22, states: "A truly efficient, competitive defense production and technological base can be maintained only by making a clear distinction between what domestic capabilities should be retained and what needs should be outsourced to other nations, and by retaining the state-of-the-art production capabilities in core weapon systems."

32. For a summary of this vast literature, see Jack S. Levy, "The Offensive/Defensive Balance of Military Technology: A Theoretical and Historical Analysis," *International Studies Quarterly* 28(2), 1984, pp. 219–238; Keir A. Lieber, "Grasping the Techno-

logical Peace: The Offense-Defense Balance and International Security," *International Security* 25(1), 2000, pp. 71–104; and Keir A. Lieber, *War and the Engineers: The Primacy of Politics Over Technology* (Ithaca, NY: Cornell University Press, 2005). Beyond a narrow focus on technology, see also broader conceptualizations and hypotheses in Robert Jervis, "Cooperation Under the Security Dilemma," *World Politics* 30(2), 1978, pp. 167–214; Stephen Van Evera, "Offense, Defense, and the Causes of War," *International Security* 22(4), 1998, pp. 5–43; and Charles L. Glaser and Chaim Kaufman, "What Is the Offense-Defense Balance and Can We Measure It?" *International Security* 22(4), 1998, pp. 44–82.

33. The discussion on constructivism generally draws on Alexander Wendt, "Anarchy Is What States Make of It: The Social Construction of Power Politics," *International Organization* 46(2), 1992, pp. 391–425; Alexander Wendt, "Constructing International Politics: A Response to Mearsheimer," *International Security* 20(1), 1995, pp. 71–81; Martha Finnemore and Kathryn Sikkink, "International Norm Dynamics and Political Change," *International Organization* 52(4), 1998, esp. pp. 888–889, 909–915; and Peter J. Katzenstein, "Introduction: Alternative Perspectives on National Security," in *The Culture of National Security: Norms and Identity in World Politics*, edited by Peter Katzenstein (New York: Columbia University Press, 1996), pp. 1–32.

34. Thomas U. Berger, "From Sword to Chrysanthemum: Japan's Culture of Antimilitarism," *International Security* 17(4), Spring 1993, esp. p. 120; Thomas U. Berger, "Norms, Identity, and National Security in Germany and Japan," in Katzenstein, *Culture of National Security*, pp. 317–356; Katzenstein, *Cultural Norms and National Security*, pp. 33–58, 115–130, 191–209; Peter J. Katzenstein and Nobuo Okawara, "Japan's National Security: Structure, Norms, and Policies," *International Security* 17(4), Spring 1993, pp. 84–118; Yoshihide Soeya, "Japan: Normative Constraints versus Structural Imperatives," in *Asian Security Practice: Material and Ideational Influences*, edited by Muthiah Alagappa (Stanford, CA: Stanford University Press, 1998), pp. 198–233; Paul Midford, "Japan, Germany, and the 'War on Terrorism': Culturalism, Defensive and Offensive Realism" (paper presented at the annual meeting of the International Studies Association, Honolulu, Hawaii, 1–5 March 2005), pp. 1–26; Akitoshi Miyashita, "Where Do Norms Come From? Foundations of Japan's Postwar Pacifism," *International Relations of the Asia-Pacific* 7(1), 2007, pp. 99–120; and Oros, *Normalizing Japan*.

35. Although in Japan's case there is a distinction to be made between anti-militarism (specific opposition to military aggrandizement) and pacifism (transcendental opposition to violence), following Miyashita, "Where Do Norms Come From?" p. 102 n3, we use the word *pacifism* interchangeably with *anti-militarism*. See also Midford, "Japan, Germany, and the 'War on Terrorism,'" esp. p. 3, who uses the term *pacifism* in his work while being mindful of its historical origins in anti-militarism with respect to Japan.

36. See, for instance, Masaru Tamamoto, "Ambiguous Japan: Japanese National Identity at Century's End," in *International Relations Theory and the Asia-Pacific*,

edited by G. John Ikenberry and Michael Mastanduno (New York: Columbia University Press, 2003), esp. pp. 206–208; David C. Kang, "Getting Asia Wrong: The Need for New Analytical Frameworks," *International Security* 27(4), 2003, esp. pp. 73–79; and Eugene A. Matthews, "Japan's New Nationalism," *Foreign Affairs* 82(6), 2003, p. 79. For earlier views along the same vein, see Richard K. Betts, "Wealth, Power and Instability: East Asia and the United States After the Cold War," *International Security* 18(3), Winter 1993/1994, p. 60; Denny Roy, "Hegemon on the Horizon? China's Threat to East Asian Security," *International Security* 19(1), 1994, pp. 149–168; and Aaron Freidberg, "Ripe for Rivalry: Prospects for Peace in a Mulitpolar Asia," *International Security* 18(3), 1993, pp. 5–33.

37. Thomas U. Berger, "The Pragmatic Liberalism of an Adaptive State," in *Japan in International Politics: The Foreign Policies of an Adaptive State*, edited by Thomas U. Berger, Mike M. Mochizuki, and Jitsuo Tsuchiyama (Boulder, CO: Lynne Rienner Publishers, 2007), esp. 259–260.

38. Andrew L. Oros, "Explaining Japan's Tortured Course to Surveillance Satellites," *Review of Policy Research* 24(1), 2007, p. 35.

39. See, for example, H. Richard Friman, Peter J. Katzenstein, David Leheny, and Nobuo Okawara, "Immovable Object? Japan's Security Policy in East Asia," in *Beyond Japan: The Dynamics of East Asian Regionalism*, edited by Peter J. Katzenstein and Takashi Shiraishi (Ithaca, NY: Cornell University Press, 2006), pp. 85–107; and also Oros, *Normalizing Japan*.

40. Jefffrey T. Checkel, "The Constructivist Turn in International Relations Theory," *World Politics* 50(2), 1998, esp. pp. 340–342.

41. 1947 Japanese Constitution, Chapter II (Renunciation of War), Article 9.

42. Glenn D. Hook and Gavan McCormack, *Japan's Contested Constitution: Documents and Analysis* (London: Routledge, 2001), pp. 3, 13–17; and see also generally, J. Patrick Boyd and Richard J. Samuels, *Nine Lives? The Politics of Constitutional Reform in Japan*, Policy Studies 19 (Washington, DC: East-West Center, 2005), pp. 1–77.

43. "Shushō 'Ninkichū ni Kaiken' [Prime Minister: 'Constitutional Revision While in Office']," *Asahi Shinbun*, 1 November 2006; "Shūdanteki Jieiken o Kenkyū" [Research on the Right to Collective Self-Defense], *Asahi Shinbun*, 30 September 2006; David Pilling, " 'To Befit the Reality': How Abe Aims to Secure Japan Its Desired World Status," *Financial Times*, 1 November 2006; Reiji Yoshida, "Schieffer's Call for Missile Defense Help Raises Constitution, *Japan Times*, 28 October 2006; "Education, Defense Bills Passed," *Japan Times*, 16 December 2006; and " 'Defense Ministry' Will Bear Watching," *Japan Times*, 17 December 2006.

44. Hook and McCormack, *Japan's Contested Constitution*, pp. 55–176. Proposals for revisions have come from several leading media sources, as well as politicians.

45. For the discussion below, see Ministry of Foreign Affairs (MOFA), "Japan Announces the Implementation Guideline for Humanitarian and Reconstruction Assis-

tance in Iraq," available online at www.mofa.go.jp (accessed 14 October 2004); "Armitage: Article 9 Hinders Japan's Alliance with U.S.," *International Herald Tribune/Asahi Shinbun*, 23 July 2004; and "Kaigai de Buryoku Kōshi Kanō [Possibility of the Use of Force Abroad]," *Ashahi Shinbun*, 30 July 2004.

46. As we discuss in a later chapter, Japanese moves to research, develop, and deploy missile defense (MD) raised questions about the country's right to collective self-defense that has the formal commitment of the government.

47. Paul Midford, "Japan's Response to Terror: Dispatching the SDF to the Arabian Sea," *Asian Survey* 43(2), 2003, pp. 329–351.

48. Green, *Japan's Reluctant Realism*, pp. 25–26. For an overview of the various political party proposals, see Christopher W. Hughes, "Why Japan Could Revise Its Constitution and What It Would Mean for Japanese Security Policy," *Orbis* 50(4), Autumn 2006, pp. 736–741.

49. Pekkanen and Krauss, "Japan's 'Coalition of the Willing' on Security Policies," pp. 431–439.

50. 1947 Japanese Constitution, Chapter IV (The Diet), Article 59. A bill requires passage by both houses. The Upper House can, as Article 59.4 suggests, sit on a bill as a tactic to gain time and refuse to take any action for sixty days, which constitutes rejection of a bill. In addition, in the event that the Upper House makes a different decision on a bill than that of the Lower House, then the bill can still become law if passed a second time a majority of two-thirds or more of the members of the Lower House. Depending on who is in power, this is hardly guaranteed if genuine policy differences on security emerge and continue between the LDP and DPJ.

51. Interview, Kazuto Suzuki, Tokyo, 27 July 2007.

52. Such high-profile maneuvering was revealed at the close of 2007, for example, in the expressed resistance by opposition parties to extend the ATSML, which had allowed the continuation of MSDF refueling activities in the Indian Ocean to support NATO-led anti-terrorist missions. After six years of operation, despite the dominance of the LDP in the lower house, the public stance of the opposition-controlled Upper House (which hitherto had hardly ever been a force in postwar electoral politics) carried the day and the Japanese ships had to be turned back. See "Abe's Pledge to Afghan Mission Raises Opposition's Ire," *Japan Times*, 10 September 2007; "Bracing Against the Opposition," *Japan Times*, 27 September 2007; Yamamoto Daisuke and Yamada Akihiro, "Tero Tokusohō Kigen Kire" [Special Anti-Terrorism Law Expires], *Asahi Shinbun*, 2 November 2007; "MSDF Mission Now Over," *Japan Times*, 2 November 2007; and David Pilling, "Japan's Opposition Rejects PM's Coalition Offer," *Financial Times*, 3–4 November 2007.

53. See Secretariat for Strategic Headquarters for Space Policy (SHSP), "Heisei 22 Nendo Yosan (Seifu Genan) Ni Okeru Uchū Kaihatsu Yosan ni Tsuite (Sokuhōchi) [Space Development Budget in Heisei 22 Budget (Government Original Bill) (Preliminary Figures)], Tokyo, 12 January 2010.

54. Pekkanen and Krauss, "Japan's 'Coalition of the Willing' on Security Policies," pp. 436–437; Richard J. Samuels, "Securing Japan: The Current Discourse," *Journal of Japanese Studies* 33(1), esp. pp. 135–137; and Jacob Brown, "Catalysts, Choices and Cooperation: Japanese Military Normalization and the U.S.–Japan Alliance," *Stanford Journal of East Asian Affairs* 5(2), 2005, esp. pp. 41–42.

55. See, for instance, Masami Ito, "Hawks Expected to Push Fukuda Hard," *Japan Times*, 24 September 2007; and Michael J. Green, "Fukuda, the Other Side of Koizumi's Magic," *Financial Times*, 28 September 2007.

56. The chronology of constitutional revision steps between 2000 and 2006 is taken from the Constitutional Revision Research Project at the Reischauer Institute of Japanese Studies, Harvard University, available online at www.fas.harvard.edu (accessed 24 September 2007). Additional information is from "Kokumin Tōhyō Hōan Kaketsu" [National Referendum Bill Passes], *Asahi Shinbun*, 13 April 2007; "Kokumin Tōhyō Hōan, Sanin e" [National Referendum Bill Heads For Upper House], *Asahi Shinbun*, 14 April 2007; "Kokumin Tōhyō Hōan 14 Nichi Seiritsu" [National Referendum Bill Enacted on the 14th], *Asahi Shinbun*, 12 May 2007; and Nippon Keidanren, "Looking to Japan's Future: Keidanren's Proposal on Constitutional Policy Issues," 18 January 2005, available online at www.keidanren.or.jp (accessed 24 September 2007), ¶3.

57. 1947 Japanese Constitution, Chapter IX (Amendments), Article 96.1. Article 96 of the Japanese Constitution stipulates that any amendments to the constitution require a concurring vote of two-thirds or more of all the members of each House and then ratification by the people through an affirmative majority of all votes cast through a special referendum or election.

58. Kazuaki Nagata, "Both Sides on Constitutional Change Hold Rallies," *Japan Times*, 4 May 2009; "Anti-piracy Law," *Japan Times*, 23 June 2009; Craig Martin, "Piracy and the Constitution," *Japan Times*, 26 March 2009; Jun Hongo, "Defense-Only Posture Needs Reviewing: Panel," *Japan Times*, 5 August 2009; and Kazuaki Nagata and Mizuho Aoki, "Constitution Day Marked with Rallies and Protests," *Japan Times*, 4 May 2010.

59. Boyd and Samuels, *Nine Lives?* esp. pp. 59–61.

60. Green, *Arming Japan*, pp. 3–4.

61. CSDC, *Council on Security and Defense Capabilities Report*, pp. 21–22. For an overview, see also Andrew L. Oros and Yuki Tatsumi, "Japan's Evolving Defense Establishment," in *Japan's New Defense Establishment: Institutions, Capabilities, and Implications*, edited by Yuki Tatsumi and Andrew L. Oros (Washington, DC: The Henry L. Stimson Center, 2007), pp. 9–21.

62. Interview, Gen Nakatani, member of the House of Representatives, Tokyo, 4 December 2004.

63. CSDC, *Council on Security and Defense Capabilities Report*, pp. 3–4.

64. *Heisei 17 Nendo ni Kakaru Bōei Keikaku no Taikō ni Tsuite* [National Defense Program Guideline, FY 2005], available at the official Web site of the Japan Defense Agency, www.jda.go.jp (accessed 8 June 2005), pp. 1–2.

65. Jiyū Minshūto Seimu Chōsakai, Uchū Kaihatsu Tokubetsu Iinkai [Policy Affairs Research Council of the Liberal Democratic Party, Special Committee on Space Development], *Aratana Uchū Kaihatsu Riyō Seido no Kōchiku ni Mukete: Heiwa Kokka Nihon Toshete no Uchū Seisaku (An)* [Toward the Establishment of a New Space Development and Utilization System: Japanese Space Policy as a Peaceful Nation (Proposal)], September 2006, esp. pp. 12–13.

66. "Kitachōsen ga Kakujikken" [North Korea Conducts Nuclear Tests], *Asahi Shinbun*, 11 October 2006; "'Bōeisho' Masu Seijiryoku" ['Ministry of Defense' Gains Political Power], *Asahi Shinbun*, 15 December 2006; and "'Nihonban NSC' Kaigi Kyō Hassoku: Abe Kōsō Jikken e Ippō" [Conference From Today on 'Japanese NSC': One Step Towards the Realization of Abe's Scheme], *Asahi Shinbun*, 22 November 2006.

67. See Green, *Japan's Reluctant Realism*, pp. 77–144; Richard J. Samuels, "Japan's Goldilocks Strategy," *Washington Quarterly*, Autumn 2006, esp. pp. 114, 120–121; Samuels, *Securing Japan*, pp. 135–157; and most recently, Hughes, *Japan's Remilitarization*, pp. 139–143. Also on North Korea, see "Misairu Hakai Meirei Hatsurei e" [Toward a Missile Destruct Order Proclamation], *Asahi Shinbun*, 21 March 2009; Kazuaki Nagata, "North's Launch Spurs Lower House Censure," *Japan Times*, 8 April 2009; "Kitachōsen ga Kakujikken [North Korea Conducts Nuclear Tests]," *Asahi Shinbun*, 26 May 2009; Choe Sang-Hun, "North Korea Says It Tested Nuclear Device," *New York Times*, 25 May 2009; Choe Sang-Hun, "A Defiant North Korea Test-Fires 3 More Missiles," *New York Times*, 27 May 2009; "LDP Defense Panels Pitch Ability to Strike Enemy Missile Sites," *Japan Times*, 10 June 2009; "North Threatens to Shoot Spy Planes," *Japan Times*, 28 June 2009; and on China see Ming Wan, *Sino-Japanese Relations: Interaction, Logic, and Transformation* (Washington, DC: Woodrow Wilson Center Press/Stanford, CA: Stanford University Press, 2006), pp. 142–167, 338–345; Pyle, *Japan Rising*, 331–339; Reinhard Drifte, *Japan's Security Relations with China since 1989: From Balancing to Bandwagoning* (London: RoutledgeCurzon, 2002); and Kent E. Calder, "China and Japan's Simmering Rivalry," *Foreign Affairs* 85(2), 2006, pp. 129–139.

68. Samuels, "Japan's Goldilocks Strategy," esp. pp. 113–116.

69. Samuels, *"Rich Nation, Strong Army,"* esp. pp. 99–100. For example, in the interwar period, some zaibatsu used the war with China after 1937 to consolidate their financial and technological positions more broadly in the Japanese economy. See also a brief treatment on private industry in Oros, "Explaining Japan's Tortured Course to Surveillance Satellites," pp. 33–34, and on military-industrial interests in Hughes, *Japan's Remilitarisation*, p. 48.

70. This is based on a vast literature, which is summarized succinctly by Jeffrey Frieden and Lisa L. Martin, "International Political Economy: Global and Domestic

Interactions," in *Political Science: The State of the Discipline*, edited by Ira Katznelsona and Helen V. Milner (New York: W. W. Norton & Company, 2002), pp. 126–136.

71. Keiko Chino, "Interest Grows in Space Program," *Yomiuri Shinbun*, 3 October 2006, speaks of the growing momentum for security-centric space development.

72. Samuels, *"Rich Nation Strong Army,"* esp. pp. 90–93.

73. Harvey Morris, "UN Set to Tighten N. Korea Sanctions After Rocket Test," *Financial Times*, 13 April 2009.

74. Eric Heginbotham and Richard J. Samuels, "Mercantile Realism and Japanese Foreign Policy," *International Security* 22(4), Spring 1998, pp. 171–203.

75. Eric Heginbotham and Richard J. Samuels, "Japan's Dual Hedge," *Foreign Affairs* 81(5), September/October 2002, p. 111.

76. Green, *Arming Japan*, esp. p. 3.

77. Berner, "Japan's Space Program," pp. 17, 19, suggests not only that Japan has used its civil space program to militarize some of its space assets, such as the reconnaissance program, but that overall Japan's military space program will compete with the civil one for available resources in the future.

Chapter 2

1. See "Waga Kuni ni okeru Uchū no Kaihatsu Oyobi Riyō no Kihon ni kan suru Ketsugi [Resolution Concerning Japan's Basic Development and Utilization of Space," (Plenary Session of the House of Representatives, 5 May 1969); and also " 'Heiwa' to iu Go no Imi ni kan suru Nihonkoku no Kokumu Daijin no Genmei (1968 Nen Dai 61 Kokkai Kagaku Gijutsū Shinkjō Taisaku Tokubetsu Iinkai Giroku)" [Declaration of Japan's Minister of State Concerning the Term 'Peace,' (Minutes of the Special Committee on Science and Technology Promotion Policy, 61st Diet, 1968], both available via the official index on Uchūhō [Space Law] online through the library at Japan Aerospace Exploration Agency (JAXA) at www.jaxa.jp (accessed 30 June 2008).

2. For background, see Setsuko Aoki, "Tekihō na Uchū Gunji Riyō Kettei Kijūn toshite no Kokkai Ketsugi no Yūyōsei" [The Significance of the Diet Resolution in the Legitimate Standards for the Weaponization of Outer Space], (Sōgō Seisakugaku Working Paper Series, No. 68, Keio University, April 2005), esp. pp. 5–6, 16–22; and also Setsuko Aoki, "Military Uses of Outer Space: Law and Policy in Japan" (International Symposium on Space Technology and Science Paper 2004-r-32, 2004), pp. 1–6.

3. Outer Space Treaty, 1967, Article IV, ¶2. See also the brief discussion by Tetsuo Tamama, "Japanese Space Policy Revision and Future of National Security," *DRC Annual Report 2002*, online version available at www.drc-jpn.org (accessed 5 November 2009).

4. "Uchū Kaihatsu Jigyōdan Hō ni Tai Suru Kokkai no Futai Ketsugi" [Supplementary Resolution by the Diet Concerning the National Space Development Agency Law], 13 June 1969, available via the official index on Uchūhō [Space Law] online through the library at JAXA at www.jaxa.jp (accessed 30 June 2008).

5. The following draws on Hiroshi Ikawa, "Review for Japan's Three Principles on Arms Export," *DRC Annual Report 2005*, AR-9E (Tokyo: Defense Research Center, 2005), pp. 7–11, available online at www.drc-jpn.org (accessed 5 November 2009). Notably the earliest efforts at arms control for Japan were external. Japan joined the Coordinating Committee for Multilateral Export Controls (CoCOM) in September 1952, and under it Japanese arms exports were controlled domestically through the Foreign Exchange and Foreign Trade Control Law. During CoCOM's tenure between 1949 and 1994, one of the most egregious examples of its informal embargoes involved a Japanese company, Toshiba Machine Company, a subsidiary of Toshiba Corporation. See Michael Mastanduno, "Trade as a Strategic Weapon: American and Alliance Export Control Policy in the Early Postwar Period," *International Organization* 42(1), 1988, esp. pp. 132–139; George E. Shambaugh IV, "Dominance, Dependence, and Political Power: Tethering Technology in the 1980s and Today," *International Studies Quarterly* 40(4), 1996, esp. pp. 581–582; and Michael Lipson, "The Reincarnation of CoCOM: Explaining Post-Cold War Export Controls," *Nonproliferation Review*, Winter 1999, pp. 33–51.

6. Christopher W. Hughes, *Japan's Re-emergence as a "Normal" Military Power* (Oxford: Oxford University Press for the International Institute for Strategic Studies, 2004), pp. 90–92. On the difficulty of restraining the export of civilian but dual-use technologies, see Reinhard Drifte, *Arms Production in Japan: The Military Applications of Civilian Technology* (Boulder, CO: Westview Press, 1986), esp. pp. 73–83. Drifte also points to the economic incentives for private industry to engage in military-related R&D in the expectation of receiving government procurement should it prove successful. Additionally, he notes that sooner or later dual-use technologies end up finding an application for weapons systems (esp. pp. 35, 40–41).

7. For comprehensive information on outer space treaties and agreements, see United Nations, *United Nations Treaties and Principles on Outer Space*, ST/SPACE/11 (New York: United Nations, 2002); and for the status of those treaties and agreements across countries, see also United Nations, *United Nations Treaties and Principles on Outer Space and Other Related General Assembly Resolutions—Addendum—Status of International Agreements Relating to Activities in Outer Space as of 1 January 2007*, ST/SPACE/11/Rev.1/Add.1/Rev.1 (Austria: United Nations, March 2007). For general information, see the official Web site of the UN Office for Outer Space Affairs, available online at www.unoosa.org (accessed 1 October 2007).

8. For an overview of some of the earliest interpretive problems discussed here, see Paul G. Dembling and Daniel M. Arons, "The Evolution of the Outer Space Treaty," reprinted in *Space Law*, edited by Francis Lyall and Paul B. Larsen (Surrey, UK: Ashgate Publishing, 2007), pp. 164–167. See also Christopher M. Petras, "Military Use of the International Space Station and the Concept of 'Peaceful Purpose,'" *Air Force Law Review*, Spring 2002, available online at www.findarticles.com (accessed 6 October 2007), esp. ¶IV; Joan Johnson-Freese, *Space as a Strategic Asset* (New York:

Columbia University Press, 2007), pp. 107–109; and see National Space Strategy Planning Group (NSSPG), *Report by the National Space Strategy Planning Group—Toward Establishment of New Space Development and Utilization System* (Tokyo: NSSPG, August 2005), esp. pp. 10–12.

9. Interestingly, under the Reagan administration, the interpretation of "peaceful purposes" was defined as "defensive" for the purpose of facilitating the then Strategic Defense Initiative (SDI).

10. Interview, Takashi Tachibana, JAXA, Tokyo, 1 October 2003, available online at www.jaxa.jp (accessed 18 August 2005). Tachibana, one of Japan's most prominent investigative journalists and science and technology commentators, pointed to the market-to-military trend with his remarks that while Japan gave up the militarization of space technology as a matter of national policy a long time ago, things had nevertheless changed in the aftermath of the launch of the spy satellites. See also Andrew L. Oros, "Explaining Japan's Tortured Course to Surveillance Satellites," *Review of Policy Research* 24(1), 2007, p. 35.

11. For the following section, see Space Activities Commission (SAC), *Fundamental Policy of Japan's Space Activities (Provisional Translation)*, revised 24 January 1996, available online at www.mext.go.jp (accessed 22 December 2006), pp. 1–17 (online version). The SAC released regular updates to space activities every five years or so. Following the administrative reforms discussed later in the book, the SAC ceded power to frame overall national technology goals for space development to the Council on Science and Technology Policy (CSTP).

12. Kazuto Suzuki, "Administrative Reforms and the Policy Logics of Japanese Space Policy," *Space Policy* 21(1), 2005, p. 13.

13. Council on Science and Technology Policy (CSTP), *Bunya Betsu Suishin Senryaku (An)* [Promotion Strategy by Area (Proposal)] (Tokyo: CSTP, 21 September 2001), pp. 84–86.

14. Council on Science and Technology Policy (CSTP), *Kongo no Uchū Kaihatsu Riyō ni Kan suru Torikumi no Kihon ni Tsuite* [About the Basics of Future Space Development and Utilization] (Tokyo: CSTP, 19 June 2002), esp. pp. 6–10.

15. Council on Science and Technology Policy (CSTP), *The Basic Strategy for Space Development and Utilization* (Tokyo: CSTP, 9 September 2004), pp. 1–20.

16. Interview, Takeo Kawamura, former minister for Education, Culture, Sports, Science, and Technology (MEXT), Tokyo, 28 March 2006. See also Paul Kallender, "Lurching into Military Space, Japan Rushes to Develop Spy Satellites," *American Chamber of Commerce Journal*, March 1999, p. 29; and Paul Kallender-Umezu, "Japan Ruling Party to Seek Military Role in Space," *Space News*, 6 April 2006, available online at www.space.com (accessed 21 July 2008).

17. For the discussion below, see National Space Strategy Planning Group (NSSPG), *Report by the National Space Strategy Planning Group*, esp. pp. 5–13, 25–30, 61–

66. For an account see also Kazuto Suzuki, "Transforming Japan's Space Policy-Making," *Space Policy* 23, 2007, pp. 76–80.

18. Interview, Takeo Kawamura, chief cabinet secretary, in Paul Kallender-Umezu, "Profile: Revamping Japan's Space Management," *Space News*, 12 November 2008.

19. In the report, space development for national security was discussed broadly in four categories: energy, diplomacy, social order, and defense (pp. 12–13, 26).

20. Jiyū Minshūto Seimu Chōsakai, Uchū Kaihatsu Tokubetsu Iinkai [Policy Affairs Research Council of the Liberal Democratic Party, Special Committee on Space Development], *Aratana Uchū Kaihatsu Riyō Seido no Kōchiku ni Mukete: Heiwa Kokka Nihon Toshete no Uchū Seisaku (An)* [Toward the Establishment of a New Space Development and Utilization System: Japanese Space Policy as a Peaceful Nation (Proposal)], September 2006.

21. Interview, Kazuto Suzuki, Tokyo, 27 July 2007.

22. Interview, Ichiro Taniguchi, "Nihon no Uchū Kaihatsu Riyō to Bijinesu" [Business and Japan's Space Development and Industrialization], 11 July 2007, available online at www.jaxa.jp (accessed 30 September 2008).

23. Nippon Keidanren, *Uchū Kaihatsu Riyō no Sōki Chakujitsu na Suishin o Nozomu* [Towards the Prompt Resumption and Steady Progress of Space Development and Utilization], 22 June 2004, esp. ¶4.1, available online at www.keidanren.or.jp (accessed 22 December 2006).

24. Richard J. Samuels, *"Rich Nation, Strong Army": National Security and the Technological Transformation of Japan* (Ithaca, NY: Cornell University Press, 1996), pp. 77–78.

25. Unless otherwise indicated, the following section draws on Craig Covault, "Japan's Growing Space Effort: Japan Challenging Western Leadership in Space," *Aviation Week & Space Technology* (hereafter *AWST*), 14 July 1986, p. 18; Roy Garner, "How Japan Is Finding Space for Private Enterprise," *Financial Times*, 7 April 1988; "Advanced Technology Moves Japan Toward Launcher Market," *AWST*, 9 March 1987, p. 132; and Paul Kallender, "Japanese Firms Poised to Enter World Markets," *Space News*, 4 August 1997; Eiichiro Sekigawa, "Japan's H-2A Launcher Prompts Manifest Changes," *AWST*, 14 June 1999, p. 21; and Paul Kallender, "The Decline of Japan's Space Program," SPACE.com, 27 June 2000, available online at www.space.com (accessed 5 May 2006).

26. See, for example, the comments by the former director of the STA in "Newsmaker Forum: Toichi Sakata," *Space News*, 30 July 1995. See also the Takashi Matsui, president of the National Space Development Agency of Japan (NASDA), statement before the Committee on Science, U.S. House of Representatives, 19 October 1995; and Dr. Chiaki Mukai, MD, PHD, Astronaut, NASDA, statement before the Committee on Science, U.S. House of Representatives, 19 October 1995, both in U.S.–Japanese Cooperation in Human Spaceflight—Hearing before the Committee on Science, U.S.

House of Representatives, 104th Congress, First Session, No. 22, 19 October 1995 (Washington, DC: U.S. Government Printing Office, 1996), pp. 19–38.

27. See comments by the former president of NASDA in "Newsmaker Forum: Takashi Matsui," *Space News*, 8 January 1996.

28. Nippon Keidanren, "Kokumin Seikatsu no Shitsu Kōjyō ni Shi Suru Uchū Kaihatsu no Suishin o Yōbō" [A Call for Space Budget Increase—Improving Our Quality of Life on Earth], 21 June 1995, available online at www.keidanren.or.jp (accessed 30 June 2006).

29. On the budget and its effect on projects, see Eiichiro Sekigawa, "Japan Boosts Budget for Space by 7.2%," *AWST*, 7 March 1994, p. 23; Eiichiro Sekigawa, "Japan Approves $2 Billion in Space Agency Spending," *AWST*, 18 March 1996 p. 58; Eiichiro Sekigawa, "Japan Seeks 6.5% Boost in Space Budget," *AWST*, 7 October 1996, p. 73; Eiichiro Sekigawa, "Japan Forced to Cut Hope, Other Programs," *AWST*, 4 August 1997, p. 28; Paul Kallender, "Japan Suspends Projects Due to Budget Proposal," *Space News*, 3 February 1997; Paul Kallender, "Japan Cuts Spending: Hope Killed," *Space News*, 21–27 July 1997; Paul Kallender, "Solar-B, Shuttle Denied Full Development Funding," *Space News*, 12 January 1998. Also see the condition and direction of Japan's space budget by Gilbert Kirkham, *Japan's Space Budget Request for JFY98: Little Change for Now*, Report Memorandum #98-04 (Tokyo: National Science Foundation, 19 February 1998), available online at www.nsftokyo.org/rm98-04.html (accessed 5 August 2005).

30. Peter B. de Selding, "Japanese Satellite Industry Still Hampered by 1990 Trade Agreement," *Space News*, 21 December 2004.

31. There is a vast literature on the catastrophic decline of the commercial launch and satellite market since the late 1990s as recognized by top space-related officials. One good summary was provided by Keiji Tachikawa, "JAXA Vision—JAXA 2025, (presentation to the Foreign Correspondents' Club of Japan, Tokyo, 28 April 2005), esp. slide 6.1.

32. Interviews, Masakazu Iguchi, chairman of SAC, Tokyo, 2003.

33. Steven Berner, "Japan's Space Program: A Fork in the Road?" RAND National Security Research Division, Santa Monica, CA, 2005, available online at www.rand.org (accessed 22 December 2006), p. 11.

34. Paul Kallender, "Japan Takes on Europe, U.S., Public Cash Infusion for Satellite Component Makers," *Space News*, 23–29 March 1998.

35. For the brief discussion below, see Chapter 6 in this book, as well as "SDIO Last Week," *AWST*, 20 June 1988; "Japanese Armaments: Meet the New Arms Exporters," *Economist*, 6 August 1988; "Mitsubishi Heavy Submits SDI-Related Research Bid," *Japan Economic Journal*, 13 August 1988; "USA, South Korea and Japan Discuss Cooperation to Counter DPRK Missile Threat," *BBC Summary of World Broadcasts*, 18 September 1993; and "Japan and U.S. Mull Joint Development of Anti-Missile Defense System," *Nikkei Weekly*, 27 September 1993.

36. Michael J. Green, *Arming Japan: Defense Production, Alliance Politics, and the Postwar Search for Autonomy* (New York: Columbia University Press, 1995), pp. 146–148.

37. Paul Kallender, "Stimulus Package Bolsters Japan's Space Programs," *Space News*, 25 January 1999; and Paul Kallender, "Firms Vie to Win Japan's Biggest Satellite Deal," *Space News*, 8 February 1999; Eiichiro Sekigawa, "Japan Considers Bigger Defense," *AWST*, 6 September 1999, p. 25.

38. See Michael J. Green, "The Japanese Defense Industry's Views of U.S.–Japan Defense Technology Collaboration: Findings of the MIT Japan Program Survey," MITJP 94-01 (Cambridge, MA: MIT Japan Program, Center for International Studies, January 1994), pp. 3, 17–20.

39. Interview, Sakamoto Norihiro, 17 May 2004.

40. See Paul Kallender-Umezu, "Japanese Industry Group Prods Government for New Programs," *Space News*, 21 June 2004.

41. The following section draws on an interview with Takashi Inoue, manager of Space, Energy, and Technology Policy Group, Nippon Keidanren, Tokyo, 19 May 2004.

42. LDP backers include, for example, Fukushiro Nukaga, who was appointed director general of the Japan Defense Agency in July 1998. He subsequently headed the LDP's Policy Affairs Research Council (PARC) and was thus in a position to guide the LDP's policy on a number of fronts, including defense and constitutional reform.

43. Interviews, Kobayashi Minoru, Director, Space Systems Department Aerospace Headquarters, MHI; member, Planning Subcommittee of the Space Activities Promotion Council, Tokyo, 17 June 2004 and 6 July 2004. See also Paul Kallender-Umezu, "Industry Wants to Allow Japanese Military to Use Space Technology," *Space News*, 21 June 2004.

44. Nippon Keidanren, "Looking to Japan's Future: Keidanren's Proposal on Constitutional Policy Issues," 18 January 2005, available online at www.keidanren.or.jp (accessed 24 September 2007), ¶3.

45. See "Ishiba Sees Arms Export Ban lifted," *Japan Times*, 15 January 2004; "Koizumi Eyes Relaxing Arms-Export Ban Just for Missile Shield," *Kyodo News International*, 14 January 2004; David Isenberg, "Japan Eyes Eased Ban on Military Exports," *Asia Times*, 4 August 2004, available online at www.atimes.com (accessed 22 December 2006); and Mark Wuebbels, "Recent Developments in the Region: Japan Revises the Three Arms Export Principles," *Asian Export Control Observer* 5, December 2004–January 2005, pp. 10–12, available online at cns.miss.edu (accessed 21 December 2006).

46. See Nippon Keidanren, *Kongo no Uchū Kaihatsu Riyō ni Kan Suru Yōbō—Gaiyō* [Requests for the Future Development and Utilization of Space—Outline], 20 May 2003, available online at www.keidanren.or.jp (accessed 22 December 2006).

47. The discussion on Keidanren's first proposal relies heavily on two interviews with policy expert Setsuko Aoki, Department of Policy Management, Keio University,

Tokyo, 7 July 2004 (e-mail) and 20 July 2004. See also Nippon Keidanren, *Uchū Kaihatsu Riyō no Sōki Chakujitsu na Suishin o Nozomu.*

48. Interview, Takashi Inoue, manager of Space, Energy, and Technology Policy Group, Nippon Keidanren, Tokyo, 15 June 2004.

49. Kallender-Umezu, "Industry Wants to Allow Japanese Military to Use Space Technology."

50. For an overview of the FSX row, see Green, *Arming Japan*, pp. 86–124.

51. U.S.–Japan Industry Forum for Security Cooperation, "IFSEC Joint Report: Revised U.S.–Japan Statement of Mutual Interests," 2003, available online at www.keidanren.or.jp (accessed 2 July 2005), pp. 1–6 (online version).

52. IFSEC recommends the conclusion of government Memorandums of Understanding (MOUs) to facilitate industry discussion in specified mission or product areas being explored in Washington as part of the Defense Trade Security Initiative (DTSI).

53. The following draws on an e-mail interview, Minoru Kobayashi, director of MHI Space Systems Department, member of Keidanren Space Activities Promotion Council Planning Subcommittee, 6 July 2004.

54. Interview, Satoshi Tsuzukibashi, deputy director of Space and Technology Policy at the Environment, Science and Technology Bureau, and also deputy director of the Office of Defense Production Committee, Nippon Keidanren, Tokyo, 18 April 2006.

55. Nao Shimoyachi, "Japan to Lift Arms-Export Ban for U.S. Missile Shield Project," *Japan Times*, 8 December 2004.

56. *Uchū Kaihatsu Riyō Suishin ni Muketa Dai 3 Ki Kagaku Gijutsu Kihon Keikaku ni tai suru Yōbō* [Request Regarding the Promotion of Space Development and Utilization in the Third Basic Science and Technology Plan], 2 March 2005, available online at www.keidanren.or.jp (accessed 21 December 2006).

57. For the timeline of the lobbying activity, see interview, Ichiro Taniguchi, "Nihon no Uchū Kaihatsu Riyō to Bijinesu" [Business and Japan's Space Development and Industrialization], 11 July 2007, available online at www.jaxa.jp (accessed 30 September 2008).

Chapter 3

1. Some of the basic information in this chapter is from Japan Science and Technology Agency (JST), Directory Database of Research and Development Activities (ReaD), available online at read.jst.go.jp (accessed 1 June 2009), which outlines principal institutions and their research and development efforts; Saadia M. Pekkanen, *Picking Winners? From Technology Catch-up to the Space Race in Japan* (Stanford, CA: Stanford University Press, 2003), pp. 171–173; and Steven Berner, "Japan's Space Program: A Fork in the Road?" RAND National Security Research Division, Santa

Monica, CA, 2005, available online at www.rand.org (accessed 22 December 2006), esp. pp. 3–4.

2. For an overview, see National Space Strategy Planning Group (NSSPG), *Report by the National Space Strategy Planning Group—Toward Establishment of New Space Development and Utilization System* (Tokyo: NSSPG, August 2005), pp. 41–46.

3. See "Uchyū Kihon Hōan" [Basic Space Bill], 169th Diet Session, Bill No. 17, esp. ¶25–26, available online at www.shugiin.go.jp (accessed 21 May 2008).

4. See, for example, interview, Takeo Kawamura, chief cabinet secretary, in Paul Kallender-Umezu, "Profile: Revamping Japan's Space Management," *Space News*, 12 November 2008; and also Strategic Headquarters for Space Policy (SHSP), "Wagkuni no Uchū Kaihatsu Riyō Taisei no Arikata ni Suite" [Concerning the State of Japan's Space Development and Utilization System], n.d., Materials 1-1, available online at www.kantei.or.jp (accessed 2 June 2009).

5. Keiko Chino and Koji Masuda, "Govt Space Policy to Promote National Interest," *Daily Yomiuri*, 1 May 2009.

6. Background information is from the Cabinet Office, Government of Japan, which outlines the Council for Science and Technology Policy (CSTP) at www8.cao.go.jp (accessed 28 December 2006); and the Ministry of Education, Culture, Sports, Science and Technology (MEXT) at www.mext.go.jp (accessed 18 August 2005), which focuses on the SAC and other space-related activities.

7. Background information is from MEXT at www.mext.go.jp (accessed 18 December 2006); and also the Japan Aerospace Exploration Agency (JAXA) at www.jaxa.jp, which traces the history of ISAS, NAL, and NASDA (accessed 18 December 2006). See also Eiichiro Sekigawa, "Koizumi Consolidates Japan's Space Activities," *Aviation Week & Space Technology* (hereafter *AWST*), 3 September 2001, p. 41.

8. Interview, Shinichi Isa, deputy director, MEXT Space Utilization Division, Tokyo, 2 September 2005. Statistics are from the official MEXT Web site at www.mext.go.jp (accessed 11 August 2005). METI follows at 17 percent of the total government science and technology-related expenditures.

9. Both NAL and NASDA were under the STA's jurisdiction before becoming part of MEXT.

10. The Kakuda Space Propulsion Laboratory merged with the Kakuda Propulsion Center (of NASDA) to form the Kakuda Space Center (KSPC), whose main mission today is the development and testing of liquid-fuel rocket engines.

11. See Eiichiro Sekigawa, "Japan Boosts Budget For Space by 7.2%," *AWST*, 7 March 1994, p. 23; and Eiichiro Sekigawa and Michael Mecham, "New Start, Big Start," *AWST*, 29 September 2003, p. 22.

12. See Paul Kallender, "NASDA Sets Reform Goals," *Space News*, 5 July 1999, p. 4; and Paul Kallender, "Profile: Keiji Tachikawa—'Restoring Faith in Space,'" *Space News*, 27 June 2005.

13. Paul Kallender, "Japan Takes First Step Toward Merger of Space Agencies," *Space News*, 12 November 2002.

14. Interview, Isao Uchida, president of NASDA, printed in Paul Kallender, "Isao Uchida/President, National Space Development Agency of Japan," *Space News*, 24 February 1997. Prior to the merger, the two organizations did team up and conduct research and development together with respect to launch vehicles, such as the *J* rocket program and the *SELENE* lunar exploration program. See Michael Mecham, "Japan's Launch Capacity to Grow with J-I, M-5," *ASWT*, 20 March 1995, pp. 57–58.

15. Background information about JAXA is from www.jaxa.jp (accessed 26 August 2008). See Jim Banke, "China Launches Its First Piloted Spaceflight," *Space News*, 15 October 2003.

16. "Yamanouchi to Head Japan Aerospace Exploration Agency," *Space News*, 26 May 2003, p. 4; Paul Kallender, "Japan Seeks New Direction for Future Space Program," *Space News*, 8 April 2002; and Eiichiro Sekigawa, "Space Streamlining, Japan's New Space Agency—JAXA Will Unify Technology, Exploration and Aeronautics," *AWST*, 9 June 2003, p. 34.

17. Interview, Toshio Matsumoto, deputy director of ISAS, Tokyo, 9 October 2002. See also Paul Kallender, "Japan's ISAS Plans Ambitious New Series of Missions," *Space News*, 10 March 2003.

18. Jurisdictional splits and the potential for in-fighting over who was going to get what share of the *IGS* program caused former foreign minister Taro Nakayama, a prime mover in ramming through Japan's decision to develop *IGS*, to publicly complain the ministries and agencies were pursuing their own interests and not cooperating at all. See "Japanese Spy Satellite Proposal Faces Political, Budget Hurdles," *Daily Yomiuri*, 17 September 1998.

19. Law Concerning Japan Aerospace Exploration Agency, Number 161 of 13 December 2002. See, specifically, Articles 11 (Consent of the Space Activities Commission to the Appointment of Executives), and Article 26 (Competent Ministers). The unofficial provisional translation in English is available from JAXA at www.jaxa.jp (accessed 30 June 2008).

20. See Basic Space Law, Supplementary Provisions, Article 3 (in Appendix II).

21. For background on CRL, see International Technology Research Institute, World Technology Evaluation Center (WTEC), *WTEC Panel Report on Global Satellite Communications Technology and Systems*, Baltimore, MD, December 1998, especially appendix C, pp. 192–195, available online at www.wtec.org (accessed 1 October 2006). The report was compiled after a series of site visits by experts to leading Japanese satellite and telecoms labs, developers, and companies, such as CRL, ISAS, MITI, Melco, the MPT, NASDA, NEC, and Toshiba. See also briefly Pekkanen, *Picking Winners?* p. 172.

22. For a discussion of the failures, see Eichiro Sekigawa and James R. Asker, "Japanese Satellite in Errant Orbit," *AWST*, 5 September 1994; Craig Covault and Eiichio Sekigawa, "Japanese H-2 Failure Ruins Satcom Research Mission," *AWST*, 2 March 1998; and Paul Kallender, "Japan's Satellite Success Rests on Launch of ETS-7," *Space News*, 17 November 1997.

23. For an overview about the history and organization of NICT, see online at www.nict.go.jp (7 July 2005). For information on the KSRC, see the introduction at the official Web site at http://ksrc-e.nict.go.jp (accessed 30 June 2008).

24. Paul Kallender, "Japanese Lab Forms Unit to Develop Microsatellites," *Space News*, 14 October 2002

25. See Ministry of Internal Affairs and Communications (MIC), Space Communication Policy Division, "Outline of the Report of Study Group on Space Communications in the Ubiquitous Network Era (Excerpts)—Toward the Realization of Ubiquitous Space-Net Program (USN Program)," Tokyo, MIC, August 2005 (accessed 10 May 2006), slides 1–6.

26. E-mail communication, former NICT researcher, 29 October 2008. The researcher is now working on minisatellite technologies in a university department.

27. Takashi Iida, "National Security: How We Should Push Forward R&D of Satellite Communications Technology as a Nucleus of Network Centric Defense System," *Space Japan Review* 12(1–3), No. 59–60, Dec/Jan–Feb/Mar 2008/2009, pp. 1–13, available online at satcom.nict.go.jp (accessed 12 July 2009).

28. This brief historical overview is from Pekkanen, *Picking Winners?* pp. 172–173; and Norihiko Saeki, "COTS Policy & 'Space on Demand' in Japan," METI, Aerospace and Defense Industry Division, Manufacturing Industries Bureau, Tokyo, 29 October 2007, slide 2.

29. For background and organizational information see also METI, "Organizational Chart," available online at www.meti.go.jp (accessed 1 October 2008).

30. For background and organizational information see USEF, "About USEF," available online at www.usef.or.jp (accessed 1 July 2009). See additionally, Paul Kallender, "Japanese Firms Urged to Cut Satellite Costs—Institute Proposes Mission to Test New Parts," *Space News*, 20 April 1998; Paul Kallender, "Japan Takes on Europe, U.S., Public Cash Infusion for Satellite Component Makers," *Space News*, 23–29 March 1998; and "Under Fire, President Quits," *New York Times*, 25 March 1998; "Losses and Scandal Claim Mitsubishi Scalp," *BBC News Online—World: Asia-Pacific*, 25 March 1998; and interview, Ichiro Taniguichi, "Nihon no Uchū Kaihatsu Riyō to Bijinesu" [Business and Japan's Space Development and Industrialization], 11 July 2007, available online at www.jaxa.jp (accessed 30 September 2008).

31. See descriptions by USEF, "Projects," available online at www.usef.or.jp (accessed 1 July 2009); and also "Satellite Returns Safely to Earth Successfully," *Daily Yomiuri*, 31 May 2003.

32. For background and organizational information, see NEDO, "About NEDO," available online at www.nedo.go.jp (accessed 27 July 2009). For project descriptions, see NEDO, "Development of Fundamental Technologies for Next-Generation Satellites," available online at www.nedo.go.jp (accessed 27 July 2009); NEDO, *Outline of NEDO: New Energy and Industrial Technology Development Organization 2009–2009* (Tokyo: NEDO, 2008), pp. 36–39; and NEDO, "Kōdo Seizō Gijutsu to Kakushinteki Sekkei no Yūgō ni yoru Hanyō Kogata Eisei no Kenkyū Kaihatsu—Heisei 15 Nendō Saitaku" [The Fusion of Advanced Manufacturing Technology and Innovative Plans for Research and Development of Multifunctional Small Satellites—Adopted 2003], Tokyo, May 2005.

33. See Saeki, "COTS Policy & 'Space on Demand' in Japan," esp. slides 5–9, 14–20.

34. Ministry of Economy, Trade, and Industry (METI), *Uchū Sangyōka Wakingu Gurupu—Chūkan Hōkoku* [Space Industrialization Working Group—Midterm Report] (Tokyo: METI, August 2004), pp. 3–4.

35. See Japan Resources Observation System and Space Utilization Organization (JAROS), "About JAROS" (as of April 2007), pp. 1–5, available online at www.jaros.or.jp (accessed 29 July 2009).

36. On MLIT, see Paul Kallender-Umezu, "Japan Still Far from Setting H-2A Return to Flight Date," *Space News*, 7 June 2004; and Paul Kallender-Umezu, "Japanese Quasi-Zenith Satellite System May Face Delays," *Space News*, 24 August 2004.

37. Background information is from MOD at www.mod.go.jp (accessed 28 August 2009), including also MOD, "Space Related Defense Policies and Future Topics for Consideration," Tokyo, MOD, November 2008.

38. Paul Kallender-Umezu, "Japanese Quasi-Zenith Satellite System May Face Delays," *Space News*, 24 August 2004; and Paul Kallender, "Impasse Over Japan's QZSS System Persists," *Space News*, 23 December 2004.

39. Official background information in this section on all of Japan's rocket and satellite series (including that broken down by JAXA, ISAS, and NASDA histories, respectively), is from JAXA's official Web site at www.jaxa.jp (accessed 1 October 2007). Some additional information on rocket families, estimated costs, success rates, and so on, is from the official Web site of the *Encyclopedia Astronautica*, available online at www.astronautix.com (accessed 1 October 2007), and also Pekkanen, *Picking Winners?* esp. pp. 173–176.

40. Background information on Nissan is from the following documents: Nissan, *Strategic Reform is the Message at Nissan—Annual Report 1998* (Tokyo: Nissan Motor Company, Ltd., 1998), p. 26, available online at www.nissan-global.com (accessed 1 October 2007); Nissan, *Profile 2007—Nissan: Enriching People's Lives* (Tokyo: Nissan Motor Company, Ltd., 2007), pp. 45–48, available online at www.nissan-global.com (accessed 1 October 2007); Mutsuo Fukushima, "Zero Inspired Today's Innovation," *Japan Times*, 14 January 2004; and Nissan Preparing to Develop First H-2 Solid Rocket Boosters," *AWST*, 4 August 1986, pp. 1–3 (online version).

41. Nissan, "Sale of Nissan's Aerospace Division," press release, available online at www.nissan-global.com (accessed 26 September 2007).

42. Historical and general background on the company is from the official Mitsubishi Heavy Industry (MHI) Web site at www.mhi-ir.jp (accessed 1 October 2007).

43. See MHI, "MHI Officially Launches Mitsubishi Regional Jet Program—Mitsubishi Aircraft Corp. to Conduct MRJ Business Operations," press release No. 1230, 28 March 2008, available online at www.mhi.co.jp (accessed 12 July 2009).

44. Unless otherwise indicated, the following draws on MHI, "Cutting-Edge Technologies for the Coming Space Age," Aerospace Headquarters, along with specific descriptions of the latest main space products, all available online at www.mhi.co.jp (accessed 1 October 2007); Junichi Maezawa and Minoru Kobayashi, "Aircraft and Rockets," *MHI Technical Review* 40(1), February 2003, pp. 1–12; Steven Berner, "Japan's Space Program: A Fork in the Road?" esp. pp. 6–11; "Planned Heavy Lift H-2A Accorded Its Own Name," *Space News*, 5 September 2005; and Frank Morring, Jr., "*H-IIA* Changes," *AWST*, 14 March 2005, p. 17.

45. Information on Rocket System Corporation (RSC) is from the official Web site at www.h2a.jp (accessed 1 October 2007); and see also Frank Morring, Jr., "Japan's Rocket System Corp. Will Be Disbanded," *AWST*, 28 April 2003, p. 17

46. "Planned Heavy Lift H-2A Accorded Its Own Name," *Space News*, 6 September 2005.

47. Interview, Satoshi Tsuzukibashi, Office of Defense Production Committee, Nippon Keidanren, Tokyo, 18 April 2006.

48. See, for example, "Mitsubishi Heavy Industries and Boeing Complete MB-XX Cryogenic Upper Stage Rocket Engine Thrust Chamber Assembly Test Program," August 12, 2002, available online at www.boeing.com (accessed September 1, 2005); Lon Rains, "Arianespace, Boeing, Mitsubishi Weigh Alliance," *Space News*, 24 June 2003; "Arianespace, Boeing and Mitsubishi Alliance," Agence France Presse (AFP), 16 June 2003; and "Arianespace, Boeing Launch Services and Mitsubishi Heavy Industries Announce a New Launch Services Alliance," Arianespace press release, 30 July 2003, available online at www.arianespace.com (accessed 1 October 2005).

49. Unless otherwise indicated, background historical details and general information on space development and products are from the official IHI Corporation Web site at www.ihi.co.jp; IHI Aerospace at www.ihi.com.jp/ia; and Galaxy Express Corporation at www.galaxy-express.co.jp (accessed 1 October 2007). See also Paul Kallender, "Japanese Company Considers New Small Launcher," *Space News*, 25 March 2002, available online at www.space.com (accessed 11 October 2007).

50. "Nissan Motor to Sell Aerospace Unit to IHI," *Japan Times*, 14 February 2000; and "Nissan to Sell Aerospace Division to IHI," *Japan Times*, 11 April 2000.

51. Galaxy Express Corporation, "*GX* Launch Vehicle: Easy Access to Space," *GX*-VGA-07-0003, n.d., available online at www.galaxy-express.co.jp (accessed 1 October

2007); Galaxy Express Corporation, "Outline of *GX* Launch Services," *GX*-VGA-07-0005, January 2007 (Revision A), available online at www.galaxy-express.co.jp (accessed 1 October 2007); and Paul Kallender, "Japan Revives Work on Galaxy Express Rocket Program," *Space News*, 24 March 2003, available online at www.space.com (accessed 11 October 2007).

52. See, for example, "NASDA, IHI Allay Concerns About Launch of *GX* Rocket," *Japan Times*, 1 September 2002; and "Revolutionary Engine Makes New *GX* Rocket Too Heavy," *Japan Times*, 6 July 2003.

53. See USTR, *1996 National Trade Estimate—Japan*, available online at www.ustr.gov (accessed 1 October 2007).

54. The following overview draws on Paul Kallender, "NASDA Bars NEC from Competing for New Awards," *Space News*, 16 November 1998; Paul Kallender, "*ETS-VII/KIKU-7* Abandons COMETS; Buys NASA Telemetry Services," *SpaceDaily*, 8 April 1998 (available online at www.spacer.com, accessed 1 January 2007). See also "NEC Suspected of Cheating NASDA," *Daily Yomiuri*, 7 November 1998; "NEC to Calculate Refund Owed to Defense Agency, NASDA," *Japan Times*, 19 November 1998; "Melco Wins *ETS-VIII/KIKU-8* Award," *AWST*, 17 February 1997, p. 56; "NASDA Picks Melco over Toshiba for Next Engineering Test Sat," *Aerospace Daily & Defense Report*, 12 February 1997, p. 226; Eiichiro Sekigawa and Michael Mecham, "NASDA Plans November Launch for *ETS-VII/KIKU-7*, TRMM," *AWST*, 14 April 1997, p. 61; Paul Proctor, "Arm Wrestling," *AWST*, 30 March 1998, p. 13; Eiichiro Sekigawa, "Mitsubishi Recce Plan Gains Ground in Diet," *AWST*, 9 November 1998, p. 34; and "Japan Signs Recce Accord," *AWST*, 11 October 1999, p. 34.

55. See generally Marco Antonio Caceres, "Investor Disinterest Stymies Commercial Satellite Industry," *AWST*, 15 January 2001, p. 161.

56. Historical details and general information on space development and products are from the official Melco Web site at global.mitsubishielectric.com (accessed 1 October 2007); and SCC Web site at www.superbird.co.jp (accessed 1 October 2007). Rankings of the world's top satellite operators are from "Space New Top 50: 2004," *Space News*, 2 August 2004, available online at www.space.com (accessed 1 October 2007).

57. See generally "Mitsubishi Electric Diversifying Efforts for Space Advances," *AWST*, 25 August 1986, p. 80; "Mitsubishi Entering Global Satellite Market," *Daily Yomiuri*, 17 May 1997; "Satellites: Exploiting Space, Connecting Asia," *International Herald Tribune*, 11 March 2002; Mitsubishi Public Affairs Committee, "Flying High into the Commercial Space: Opening a New Door to Commercial Space," *Mitsubishi Monitor* 20(1), February–March 2006 (online version), available online at www.mitsubishi.com (accessed 1 October 2007); Michael Mecham, "Japanese Cautious About Satellite Plans," *AWST*, 23 August 1999, p. 45; and "Profile: Hiroshi Kimura, Group Executive Vice President, Electronic Systems, Mitsubishi Electric Co.," *Space News*, 25 February 2002.

58. Intelsat is one of the world's largest provider of fixed satellite services. It was originally formed in 1964 by eleven nations, including Japan, as an intergovernmental organization to develop and operate global satellite communication services. It was privatized in 2001. Like Intelsat, Eutelsat was originally set up in 1977 as an intergovernmental organization to provide a satellite-based telecommunication infrastructure in Europe. It was also privatized in 2001, and it subsequently regrouped in 2005 as a new entity known as Eutelsat Communications with a primary focus on the provision of fixed satellite services. The Australian government sold Aussat to Optus Communications at the end of 1991, which started operations in 1992 as the country's private telecommunications carrier offering a broad range of communication services. In 2001, Optus became a wholly owned subsidiary of Singapore Telecommunication (SingTel), with a focus on the construction of fixed, mobile, and satellite networks, especially in the Asian-Pacific region. For additional details, see the official Web site of Intelsat at www.intelsat.com, Eutelsat at www.eutelsat.com, and SingTel Optus at www.optus.com.au (all accessed 1 October 2007).

59. Melco was only a subcontractor for *ALOS/Daichi*, and for *JEM/Kibō*.

60. While Melco had responsibility for the overall payload, it teamed with Space Systems/Loral (SS/L), a subsidiary of Loral Space & Communications, which was responsible for building and testing the *Optus-C1* under contract from Melco. See Loral Space & Communications, "Loral and Mitsubishi Electric Complete In-Orbit Testing of Optus C1: Multi-band Military and Commercial Satellite Begins Service," press release, 21 August 2003, available online at investor.loral.com (accessed 1 October 2007). See also Michael Mecham, "Mitsubishi Wins Key Optus/Military Contract," *AWST*, 19 October 1998, p. 78

61. Melco, "Mitsubishi Electric Receives Order for Superbird 7 Communications Satellite from Space Communications Corporation," press release No. 2373, 1 November 2005, available online at global.mitsubishielectric.com (accessed 1 October 2007); and Paul Kallender-Umezu, "Mitsubishi Electric to Build Superbird-7 for SCC—Deal is Japanese Maker's First Domestic Commercial Satellite Contract," *Space News*, 7 November 2005.

62. SKY Perfect JSAT Corporation, "Making Space Communications Corporation (SCC) a Wholly Owned Subsidiary," Sky Perfect JSAT Corporation (9412), 12 March 2008, esp. slides 4–13, available online at www.skyperfectjsat.co.jp (accessed 12 July 2009). For additional background and also comments by company officials, see "Interview with CEO—Kiyoshi Isozaki, President & CEO JSAT Corporation," *Space Japan Review*, No. 42, August/September 2005, pp. 5, 8; "Profile: Kiyoshi Isozaki, JSAT Corp., 'Consolidation and Expansion,'" *Space News*, 9 June 2008; and also "Interview with Yutaka Nagai, Director of the Board & Senior Executive Vice President, SKY Perfect JSAT Corporation," *APSCC (Asia-Pacific Satellite Communications Council) Newsletter*, No. 01, January 2009, pp. 24–26.

63. "Japan's Space Communications," *AWST*, 17 April 2006, p. 20.

64. Craig Covault, "Farnborough Air Show Satellite Earth Stations: A $65–83 Billion Market," *AWST*, 5 September 1994, p. 120.

65. Melco has also been involved in supplying radio, optical-infrared, and optical telescopes across Japan. Additionally, one of its other notable contributions has been to the VERA project, which involves the use of the very long baseline interferometry (VLBI) method with the goal of creating a three-dimensional map of the galaxy. Between 2001 and 2002, Melco installed four 20m dual-beam antennas for the VERA project, which is administered by the National Astronomical Observatory of Japan.

66. See Philip J. Klass, "New MSAT to Offer Low-Cost Telephone," *AWST*, 22 May 1995, p. 50.

67. Paul Kallender, "Japan's Adeos-2 Suffers Catastrophic Failure," *Space News*, 11 November 2003; Agence France-Presse (Tokyo), "Japanese Spy Satellite Suffers Critical Power Failure," 27 March 2007, available online at www.spacewar.com (accessed 1 October 2007); and "Japan's First Reconnaissance," *AWST*, 2 April 2007, p. 24.

68. Edward H. Phillips, "Reconnaissance Program," *AWST*, 12 July 1999, p. 13.

69. Melco, "Mitsubishi Electric Completes Japan's Largest Privately-Owned Satellite Assembly and Testing Plant," press release No. 0487, 29 October 1999, available online at global.mitsubishielectric.com (accessed 1 October 2007).

70. Interview, Melco official, Kamakura Works, 24 August 2001. Melco's victory over NEC and Toshiba and its gleaning of the *IGS* contracts necessitated an even more radical overhaul of the Kamakura Works, which the company called KDI-21, or Kamakura Digital Innovation 21. This was aimed mainly at enabling the company to considerably shorten its satellite building time to meet the *IGS* schedule from about five years for a NASDA experimental satellite to less than half that for military satellites. The company claims that the system had helped the company speed up its design to delivery cycle from sixty months in 1990 to twenty-four months by 2001.

71. See generally Melco, "Boeing, Mitsubishi Electric Announce Strategic Alliance," press release, 20 June 2001, available online at global.mitsubishielectric.com (accessed 1 October 2007); "Boeing Has Concluded a Pair of Strategic Agreements with Mitsubishi Electric," *AWST*, 2 July 2001, p. 27; Mitsubishi Public Affairs Committee, "Set to Soar with Boeing: The Fourth-Phase Infrastructure," *Mitsubishi Monitor* 15(4), August–September 2001 (online version), available online at www.mitsubishi.com (accessed 1 October 2007); "Boeing Signs Up Space Partner," *BBC News*, 20 June 2001, available online at news.bbc.co.uk (accessed 1 October 2007).

72. Background details and general information on space development and products for the company are from the official NEC Web site at www.nec.co.jp (accessed 1 October 2007); and NEC Space Systems Division, "NEC Space Activities," internal document (Tokyo: NEC, 1995). Formally, the NEC Group has two core areas of

business—one focusing on integrated IT/networks, which is overseen by the NEC Corporation, and the other on semiconductors, which is overseen by the NEC Electronics Corporation.

73. NEC, "NEC Comments on Refunds to Japan's Defense Agency and the National Space Development Agency of Japan, and Prevention of Future Incidents," press release, 18 November 1998, available online at www.nec.co.jp (accessed 1 October 2007); "NEC Suspected of Cheating NASDA," *The Daily Yomiuri*, 7 November 1998; Paul Kallender, "NEC Weathers Storm to Stay on Top in Japan," *Space News*, 19 July 1999; and Paul Kallender, "NASDA Bars NEC from Competing for New Awards," *Space News*, 16 November 1998.

74. In 1997 top NEC officials expressed confidence that the company was number one in Japanese space business and likely to become a top ten player in the global space business. See, for example, interview with Hiroaki Shimayama, senior vice president, NEC Corporation, reprinted in Newsmaker Forum, *Space News*, 14–20 April 1997; and interview with Kiyoshi Murata, general manager, NEC Space Systems Development Division, reprinted in *ACCJ Journal*, November 1997, pp. 20–27.

75. For an assessment of some of the more recent projects, see, for example, Kazuki Shiibashi, "Journey Begins; Crippled Japanese Asteroid Hunter on Its Way Back to Earth in 2010," *AWST*, 30 April 2007, p. 36; Kazuki Shiibashi, "Laser Finder: JAXA Tests Aimed at Seamless Broadband Space, Ground Emergency Planning Networks," *AWST*, 17 April 2006, p. 29; Kazuki Shiibashi, "Looking Down: JAXA Returns to Earth Observation Missions with ALOS Mapping," *AWST*, 30 January 2006, p. 37; Frank Morring Jr., "No Hope," *AWST*, 15 December 2003, p. 17.

76. See, for example, NEC, "NEC Receives Communications Repeater Order for Orion 2 Satellite," press release, 10 April 1997, available online at www.nec.co.jp (accessed 1 October 2007); NEC, "NEC Wins Order as for Satellite Ground Stations from Paraguay ANTELCO," press release, 16 July 1997, available online at www.nec.co.jp (accessed 1 October 2007); and NEC, "NEC LCD Monitors to Be Used in Today's Launch of NASA Space Shuttle," press release, 29 October 1998, available online at www.nec.co.jp (accessed 1 October 2007).

77. On ICO Global Communications, see generally "Inmarsat Affiliate to Operate Medium-Altitude Satellite System," *AWST*, 23 May 1994, p. 58; Bruce D. Nordwall, "Inmarsat Is Changing," *AWST*, 23 January 1995, p. 56; Joseph C. Anselmo, "Hughes Has Selected Launch Providers," *AWST*, 4 November 1996, p. 92; James R. Asker, "Satellite Markets," *AWST*, 1 December 1997, p. 21; Anthony L. Velocci, Jr., "Market Focus," *AWST*, 2 November 1998, p. 13; Bruce A. Smith, "Tough Times Ahead?" *AWST*, 29 March 1999, p. 23.

78. See NEC, "NEC Wins Ground Systems Order for Global Mobile Satellite Communications Network," press release, 3 March 1997, available online at www.nec.co.jp (accessed 1 October 2007); and also NEC, "ICO's Handset Orders Go to NEC,"

press release, 25 June 1998, available online at www.nec.co.jp (accessed 1 October 2007); and NEC, "ICO System TT&C RFT Hand-Over," press release, 15 October 1998, available online at www.nec.co.jp (accessed 1 October 2007).

79. Paul Kallender, "NEC Aims New Satellite Platform at Teledesic," *Space News*, 9 March 1998.

80. "TRW Has Filed a Lawsuit," *AWST*, 20 May 1996, p. 21; Michael A. Taverna, "ICO Stock Shortfall Raises New Questions," *AWST*, 14 June 1999, p. 90; Joseph C. Anselmo, "ICO Faces Long Climb Out of Bankruptcy," *AWST*, 6 September 1999, p. 29; "The Former ICO Global Communications Emerged from Chapter 11 Bankruptcy Protection," *AWST*, 22 May 2000, p. 20; Bruce A. Smith, "ICO Investments," *AWST*, 17 July 2000, p. 23; Bruce A. Smith, "Merger Off," *AWST*, 12 November 2001, p. 17; and Frank Morring, Jr., "Teledesic Pulls the Plug," *AWST*, 7 October 2002, p. 46.

81. New ICO's satellite assets came under the control of a holding company, ICO-Teledesic Global, which planned to provide wireless Internet satellite communication services. Although rumors emerged that the restructuring would lead to a merger of these two corporations, they remained separate, with New ICO continuing to develop its mobile satellite telephone system and Teledesic pursuing the development of a LEO broadband satellite networked to deliver high-speed Internet access.

82. "Government Drafts Spy-Satellite Plan," *The Nikkei Weekly*, 9 November 1998.

83. Unless otherwise indicated, information on the company's space development and products, are from the official Toshiba Web site at www.toshiba.co.jp (accessed 1 October 2007); Paul Kallender, "Japan Takes Steps to Manufacture Satellites," *Space News*, 7 April 1997; "Contractor Named," *AWST*, 14 October 1985, p. 137; and "Excerpt," *AWST*, 7 July 1986, p. 27.

84. Joseph C. Anselmo, "New Partners," *AWST*, 24 November 1997, p. 17.

85. Toshiba, "Toshiba to Supply Space Systems/Loral with Satellite Solar Array Panels," press release, 17 November 1997, available online at www.toshiba.co.jp (accessed 1 October 2007)

86. Interview, Toshiba official, Tokyo, 2003.

87. For an overview of Toshiba's troubles, see "Loss of Satellite Sparks Complaint by Japan," *AWST*, 21 May 1984; and "Japan Broadcaster Wants Space Agency to Guarantee Satellite," *AWST*, 7 July 1986.

88. See generally Paul Proctor, "Arm Wrestling," *AWST*, 30 March 1998, p. 13; "Melco Wins *ETS-VIII/KIKU-8* Award," *AWST*, 17 February 1997, p. 56; and Joseph C. Anselmo, "Toshiba Corp. Has Lost a Second Contract," *AWST*, 24 March 1997, p. 58.

89. General background information on the company is from Nippon Electric Company (NEC), "NEC and Toshiba Unify Space Business," press release, 13 December 2000, available online at www.nec.co.jp (accessed 1 October 2007); NEC, "Establishment of Joint Venture Company on Space Business," press release, 2 April 2001,

available online at www.nec.co.jp (accessed 1 October 2007); and the official NT Space Web site at www.ntspace.jp (accessed 30 January 2008).

90. "Space Agency Files Hacked," *Yomiuri News*, 14 February 2002.

91. Patricia J. Parmalee, "Identifying Cause of Dash Problems," *AWST*, 25 February 2002, p. 19; and Frank Morring, Jr., "NEC–Toshiba Pays Fine for DASH Experiment Failure," *AWST*, 20 October 2003, p. 17.

92. "Japan's NEC Toshiba to Invest in Galaxy Express," *Space News*, 29 April 2003; Paul Kallender, "Impasse Over Japan's QZSS System Persists," *Space News*, 23 December 2004; Eiichiro Sekigawa and Michael Mecham, "Pick Up Bits: The Last Mission of Japan's ISAS Is Underway and Should Be a Wonder—Taking Samples from an Asteroid," *AWST*, 19 May 2003, p. 40; and Kazuki Shiibashi and Michael Mecham, "Looking Down: JAXA Returns to Earth Observation Missions with ALOS Mapping," *AWST*, 30 January 2006, p. 37.

93. E-mail correspondence, Norihiko Saeki, deputy director, METI's Aerospace and Defense Industry Division, Tokyo, 7 July 2009.

94. Eiichiro Sekigawa and Michael Mecham, "Third Merger Pending for Japan's Aerospace," *AWST*, 13 March 2000, p. 32.

95. See the combined research being conducted by JAXA, ISAS, and IHI Aerospace as evidenced in Yashuhiro Morita (ISAS/JAXA), Takayuki Imoto (JAXA), Hirohito Ohtsuka (IHI Aerospace), and Advanced Solid Rocket Research Team (JAXA), "Research on an Advanced Solid Rocket Launcher in Japan," 2008-g-02, paper presented at the 26th International Symposium on Space Technology and Science, Hamamatsu City, Japan, 2–8 June 2008, available online through Scientific Journal Editing System (SciEd) at www.senkyo.co.jp (accessed 19 July 2009).

Chapter 4

1. See Nao Shimoyachi, "Long-Range Missile Quest off Defense Buildup Plan," *Japan Times*, 10 December 2004; Agence-French Presse (AFP), "Japan Drops Plan on Long-Range Missile After Ally Protests," 8 December 2004, available online at www.spacedaily.com (accessed 17 April 2007); "Tokyo Snuffs Plan to Study Long-Range Missiles After New Komeito Balks," *Asahi Shinbun*, 9 December 2004, available online at www.asahi.com (accessed 17 April 2007); James Brooke, "After Failures, Space Effort in Japan Gets a Lift," *New York Times*, 27 February 2005; and Council on Security and Defense (CSDC), *The Council on Security and Defense Capabilities Report: Japan's Visions for Future Security and Defense Capabilities* (Tokyo: CSDC, October 2004), p. 29.

2. Richard J. Samuels, *"Rich Nation, Strong Army": National Security and the Technological Transformation of Japan* (Ithaca, NY: Cornell University Press, 1996), p. 78.

3. The information on Japan's missiles is from Duncan Lennox, ed., *Jane's Strategic Weapon Systems* 40, January 2004, pp. 115–118. To give a few choice examples: In 1973, MHI won the contract for the development of Japan's first indigenous air-to-surface

missile (ASM)—the ASM-1 (designated Type 80), a short-range (50-kilometer) radar-guided missile that is powered by a solid-propellant motor, and armed with a high explosive warhead. The ASM-1 entered Self-Defense Force (SDF) service in 1983, having been cleared for carriage on the Mitsubishi support aircraft F-1, the F-4J Phantom, and the P-3C Orion aircraft. The longer-range version, the ASM-2 (Type 93), that followed in 1993, has a range of 100 kilometers, is equipped with an advanced Fujitsu infra-red terminal guidance system, and is cleared for carriage on F-1, F2, P-3C Orion, and F-15J Eagle aircraft. Importantly, the ASM-2 utilizes guidance and propulsion technology from another set of missiles—surface-to-surface missiles (SSM)—also produced by MHI. The SSM-1A (Type 88), a ground-launched anti-ship missile, capitalized on the ASM-1 design but was fitted with a turbojet engine for a longer range of 150 kilometers. It entered into production and SDF service in 1988. The SSM-1B (Type 90) is the ship-launched version produced in 1990 and is fitted to destroyers of various classes.

4. "The Back Page: Rocket Genealogy," *Crosslink* 6(2), Spring 2005, available online at www.aero.org (accessed 23 May 2007).

5. See Federation of American Scientists (FAS), "Special Weapons Primer: Ballistic Missiles," Section I (Ballistic Missile Basics), available online at www.fas.org (accessed 28 April 2008).

6. U.S. Congress, Office of Technology Assessment (OTA), *Technologies Underlying Weapons of Mass Destruction, OTA-BP-ISC-115* (hereafter *Technologies Underlying WMD*), (Washington, DC: U.S. Government Printing Office, December 1993), esp. 11–13, 197–198. See also Paul Mann, "Asia Opts for Modernization," *Aviation Week & Space Technology* (hereafter, *AWST*), 21 January 2002, p. 62. Rocket technology is one means of possessing "high-end" delivery systems (such as piloted combat aircraft and, of relevant interest here, ballistic missiles and cruise missiles) for a range of warheads (such as chemical, biological, and nuclear). In comparison especially to simpler systems (such as automobiles, civil boats and aircraft, and even humans), if all goes as planned, high-end delivery systems allow even a weak country to do potentially greater damage to a larger number of targets over longer ranges with far more control, reliability, and precision. Ballistic and cruise missiles might well also allow precisely such a country to gain psychological advantages over even far stronger countries because such missiles have conventionally been harder to detect and defend against.

7. We follow the conventional characterization of ranges for land-based missiles capable of delivering warheads, that is, an ICBM has a range greater than 5,500 kilometers; a Medium-Range Ballistic Missile (MRBM) has a range between 1,000 and 3,000 kilometers; a Short-Range Ballistic Missile (SRBM) has a range of less than 1,000 kilometers; and an Intermediate-Range Ballistic Missile (IRBM) has a range between 3,000 to 5,500 kilometers. In light of this, it is also helpful to keep some distances in mind: Tokyo to Beijing about 2,100 kilometers; Tokyo to Pyongyang about 1,280 kilometers; Tokyo to Beijing about 2,100 kilometers; Tokyo to Shanghai about

1,760 kilometers; Tokyo to Seoul about 1,150 kilometers; Tokyo to Moscow about 7,470 kilometers; and Tokyo to San Francisco about 8,260 kilometers.

8. See, for example, "Foreign Missile Developments and the Ballistic Missile Threat," Statement for the Record to the Senate Foreign Relations Committee on Foreign Missile Developments and the Ballistic Missile Threat to the United States Through 2015 by Robert W. Walpole, national intelligence officer for strategic and nuclear weapons, 16 September 1999, available online at www.cia.gov (accessed 30 April 2008); *Report of the Commission to Assess the Ballistic Missile Threat to the United States* (hereafter *Rumsfeld Commission Report*), 15 July 1998, available online at www.fas.org (accessed 30 April 2008), esp. ¶II.D.4; U.S. Congress, OTA, *Technologies Underlying WMD*, esp. p. 227; Joseph C. Anselmo, "China Controversy Has Complex Roots," *AWST*, 1 June 1998, esp. p. 24; National Defense Industry Association (NDIA), *Feasibility of Third World Advanced Ballistic and Cruise Missile Threat—Volume 1: Long Range Ballistic Missile Threat*, June 1999, available online at www.fas.org (accessed 30 April 2008), pp. 86, 88; System Assessment Group of NDIA Strike, Land Attack and Air Defense Committee, *Feasibility of Third World Advanced Ballistic & Cruise Missile Threat—Volume 1: Long Range Ballistic Missile Threat*, October 1998, slide 89; Lennox, *Jane's Strategic Weapon Systems*, p. 7; and Joan Johnson-Freese, *Space as a Strategic Asset* (New York: Columbia University Press, 2007), esp. p. 33.

9. See generally Eric Heginbotham and Richard J. Samuels, "Mercantile Realism and Japanese Foreign Policy," *International Security* 22(4), 1998, pp. 171–203.

10. System Planning Corporation, "Non-Proliferation Issues: Japan," Appendix III, Unclassified Working Papers, *Rumsfeld Commission Report*. This theme resonates nicely all around in Samuels, *"Rich Nation, Strong Army."*

11. U.S. Congress, OTA, *Technologies Underlying WMD*, pp.227–228.

12. National Intelligence Council (NIC), *Foreign Missile Developments and the Ballistic Missile Threat Through 2015* (Unclassified Summary of a National Intelligence Estimate), December 2001, available online at www.fas.org (accessed 30 April 2008), esp. p. 1.

13. On Japan's ballistic missile capability, see especially Selig S. Harrison, "Missile Capabilities in Northeast Asia: Japan, South Korea and North Korea"; David C. Isby, "Barriers to Proliferation and Pathways to Transfer: Building Ballistic Missile Capabilities Under MTCR"; and System Planning Corporation, "Non-Proliferation Issues: Japan"; all in appendix III, Unclassified Working Papers, *Rumsfeld Commission Report*. See also missile options for Japan by Selig S. Harrison, "Japan and Nuclear Weapons," in *Japan's Nuclear Future: The Plutonium Debate and East Asian Security*, edited by Selig S. Harrison (Washington, DC: Carnegie Endowment for International Peace, 1996), pp. 21–24; and Michael D. Swaine with Loren H. Runyon, "Ballistic Missiles and Missile Defense in Asia," *NBR Analysis* 13(3), 2002, esp. p. 9.

14. We model this approach based on an assessment of the minimal technical components necessary for the construction of ballistic missiles. See Central Intelligence Agency, "Prospects for the Worldwide Development of Ballistic Missile Threats to the Continental United States," 17 November 1993, available online at www.fas.org (accessed 30 April 2008).

15. The brief background here relies largely on the published works (abstracts and/or full articles as indicated) of a critical player in Japan's space program who went on to become JAXA's director, namely Yasunori Matogawa. For historical overviews see, Yasunori Matogawa, "Another Destiny of Rocketry in Japan—Festival Rockets in Japanese Shrines," Abstract 342, 1995, in Hervé Moulin, ed., *Rocketry & Astronautics: IAC History Symposia 1967–2000—Abstracts & Index* (Paris: International Academy of Astronautics, 2004), pp. 224–225; and Yasunori Matogawa, "Another Destiny of Rocketry in Japan—Festival Rockets in Japanese Shrines," in *History of Rocketry and Astronautics—AAS History Series*, vol. 23, edited by Donald C. Elder and Chrisophe Rothmund, 2001, pp. 315–326; and Frank Winter and Kubozono Akira, "Festival Rockets in Thailand, Laos, Japan and China: A Case Study of Early Technology Transfer? Part I," Abstract 405, 1998, in Hervé Moulin, ed., *Rocketry & Astronautics: IAC History Symposia 1967–2000—Abstracts & Index* (Paris: International Academy of Astronautics, 2004), pp. 247–248.

16. For general overviews see, Yasunori Matogawa, "'*Pencil*' Rocket and Hideo Itokawa—Pioneering Work of Japanese Rocketry," Abstract 335, 1995, in Hervé Moulin, ed., *Rocketry & Astronautics: IAC History Symposia 1967–2000—Abstracts & Index* (Paris: International Academy of Astronautics, 2004), pp. 221–222; and Yasunori Matogawa, "Lessons from Half a Century Experience of Japanese Solid Rocketry Since Pencil Rocket," *Acta Astronautica* 61, 2007, pp. 1107–1115 ; Yasunori Matogawa, "A Survey of Rocketry for Space Science in Japan," Abstract 290, 1993, in Hervé Moulin, ed., *Rocketry & Astronautics: IAC History Symposia 1967–2000—Abstracts & Index* (Paris: International Academy of Astronautics, 2004), pp. 205–206; and Yasunori Matogawa, "A Survey of Rocketry for Space Science in Japan," *History of Rocketry and Astronautics—AAS History Series*, vol. 22, edited by Philippe Jung, 1998, pp. 203–224.

17. For the remainder of this section the military-related background relies on Yasunori Matogawa, "Japanese Solid Rockets in World War II," Abstract 347, 1996, in Hervé Moulin, ed., *Rocketry & Astronautics: IAC History Symposia 1967–2000—Abstracts & Index* (Paris: International Academy of Astronautics, 2004), pp. 226–227; and Yasunori Matogawa, "Japanese Solid Rockets in World War II," in *History of Rocketry and Astronautics—AAS History Series*, vol. 25, edited by Hervé Moulin and Donald C. Elder, 2003, pp. 123–136; Yasunori Matogawa, "Japanese Liquid Rockets in World War II," Abstract 375, 1997, in Hervé Moulin, ed., *Rocketry & Astronautics: IAC History Symposia 1967–2000—Abstracts & Index* (Paris: International Academy of Astronautics, 2004), p. 236; and Yasunori Matogawa, "Japanese Liquid Rockets in

World War II," in *History of Rocketry and Astronautics—AAS History Series*, vol. 26, edited by Donald C. Elder and George S. James, 2005, pp. 111–125; and Toshio Masutani, Hisao Saijo, Hidemaro Wachi, "Military Rockets in 1930's and 1940's in Japan," IAC-050E4.4.03, IAC 2005, October 17–21, Fukuoka, Japan.

18. Yasunori Matogawa, "Ohka: Japanese Rocket-Propelled Attack Glider in World War II," Abstract 406, 1998, in Hervé Moulin, ed., *Rocketry & Astronautics: IAC History Symposia 1967–2000—Abstracts & Index* (Paris: International Academy of Astronautics, 2004), p. 248; and Yasunori Matogawa, "Ohka: Japanese Rocket-Propelled Attack Glider in World War II," in *History of Rocketry and Astronautics—AAS History Series*, vol. 27, edited by Kerrie Dougherty and Donald C. Elder, 2007, pp. 273–294.

19. Yasunori Matogawa, "Shusui: Japanese Rocket Fighter in World War II," Abstract 431, 1999, in Hervé Moulin, ed., *Rocketry & Astronautics: IAC History Symposia 1967–2000—Abstracts & Index* (Paris: International Academy of Astronautics, 2004), pp. 259–260; and Yasunori Matogawa, "Shusui: Japanese Rocket Fighter in World War II," in *History of Rocketry and Astronautics—AAS History Series*, vol. 28, edited by Frank H. Winter, 2007, pp. 177–187.

20. Yasunori Matogawa, [No Title], Y. M. Column Archive, 30 May 2001, available online at www.planetary.or.jp (accessed 15 May 2008).

21. Brian Harvey, *The Japanese and Indian Space Programmes: Two Roads into Space* (Chichester, UK: Springer-Praxis, 2000), pp. 1–64; Steven Berner, "Japan's Space Program: A Fork in the Road?" RAND National Security Research Division, Santa Monica, CA, 2005, pp. 1–37, available online at www.rand.org (accessed 22 December 2006); and Frank H. Winter, *Frontiers of Space* (Cambridge, MA: Harvard University Press, 1990), pp. 102–103.

22. From this point on, unless otherwise indicated, the facts, figures, and historical details on Japanese SLVs are from the official version published by Japan Aerospace Exploration Agency (JAXA), *Nihon no Uchyū Kaihatsu no Rekishi: Uchyūken no Monogatari* [The History of Japan's Space Development: The ISAS Story, hereafter *Uchyūken no Monogatari*], Chapters 1–10, available only online at www.isas.jaxa.jp (accessed 16 April 2008) (cited here by chapter numbers); and from the mission and activities timelines of the Institute of Space and Astronautical Science (ISAS), National Aerospace Laboratory of Japan (NAL), National Space Development Agency of Japan (NASDA), and JAXA for past, current, and projected transport systems/rockets also available online through JAXA at www.jaxa.jp (accessed 24 June 2008).

23. From this point on, unless otherwise indicated, additional SLV technical data on structures, design, performance, and costs on SLVs is taken from Steven J. Isakowitz, Joseph P. Hopkins, Jr., and Joshua B. Hopkins, *International Reference Guide to Space Launch Systems, Third Edition* (Reston, VA: American Institute of Aeronautics and Astronautics [AIAA], 1999), pp. 139–173, 245–255; and also Steven J. Isakowitz,

Joshua B. Hopkins, and Joseph P. Hopkins, Jr., *International Reference Guide to Space Launch Systems, Fourth Edition* (Reston, VA: American Institute of Aeronautics and Astronautics [AIAA], 2004), pp. 167–186, 263–274; Jonathan McDowell, *Jonathan's Space Home Page*, available online at www.planet4589.org; and *Encyclopedia Astronautica*, available online at www.astronautix.com (accessed 24 June 2008). Information on missiles is from "missile control system," available online at www.aerospace web.org (accessed 24 June 2008). In the case of any discrepancies among the cited sources above on dates, details, or any other information, we follow the official (JAXA) and secondary Japan sources identified.

24. The discussion on some of the key advantages and disadvantages of solid-propellant and liquid-propellant rockets draws exclusively on George P. Sutton and Oscar Biblarz, *Rocket Propulsion Elements,* 7th edition (New York: John Wiley & Sons, 2001), esp. pp. 25, 28, 628–630, and esp. tables 17-1 to 17-4, which provide a comprehensive listing of factors.

25. All this is not to say that there are no problems with solid-propellant rockets. They do present explosion and fire hazards, are vulnerable to impact, cannot be tested prior to use, need safety provisions to prevent inadvertent ignition, cannot also change predetermined thrust or duration once ignited (only some motors can be stopped at random but then become disabled for good), and require extensive environmental permits and safety feature for transport on public conveyance.

26. For the background, see Mutsuo Fukushima, "Zero Inspired Today's Innovations, Warplane's Engineers Later Excelled in Auto, Rocket Sectors," *Japan Times,* 14 January 2004; and also Samuels, "*Rich Nation, Strong Army,*" pp. 102, 110–111, 116–118. Nakajima itself grew out of Nippon Hikōki, which was established in 1917.

27. For the background on Itokawa, see JAXA, *Uchyūken no Monogatari,* chapter 1; Matogawa, "'*Pencil*' Rocket and Hideo Itokawa," pp. 121–132; Yasunori Matogawa, "The Birth of *Pencil* Rocket," Y. M. Column Archive, 16 March 2005, available online at www.planetary.or.jp (accessed 15 May 2008); Yasunori Matogawa, "Lessons from Half a Century Experience of Japanese Solid Rocketry Since Pencil Rocket," *Acta Astronautica* 61, 2007, pp. 1107–1115; and additional anecdotal information from JAXA, "Prof. Itokawa, 'The Father of Japanese Rocketry,'" available online at www.isas.jaxa.jp (accessed 8 May 2008).

28. See, for example, the remarks by a NASDA official on Japan maintaining control of its core rocket technology, including engines and key electronic components in line with official objectives of having independent space technologies in Eriko Arita, "Space Only the First Frontier for *H-IIA,*" *Japan Times,* 10 September 2002.

29. Along with heavy use of JAXA, *Uchyūken no Monogatari,* chapters 1–6, 8–9, additional background information on the early rocket series in this section is from Yasunori Matogawa, "A Survey of Rocketry for Space Science in Japan," *History of Rocketry and Astronautics—AAS History Series,* vol. 22, edited by Philippe Jung, 1998,

pp. 203–224; Ryojiro Akiba, Hiroki Matsuo, and Tomifumi Godai, "A History of Japanese Space Launch Vehicles—From the *Pencil* Roxcket [sic] to the M-V and H-IIA" (paper presented at the Sixth International Symposium, Propulsion for Space Transportation for the XXIst Century [AAAF 16_80_P], Versailles, France, 14–16 May 2002), pp. 1–16; and interview, Yasunori Matogawa, available online in "Pencil Rocket Story—The Dawn of Japanese Space Development," 21 May 2005, at www.jaxa.go.jp (accessed 24 June 2008), pp. 1–15.

30. On the *Pencil*, see JAXA, *Uchyūken no Monogatari*, chapter 1.

31. On the *Baby*, see JAXA, *Uchyūken no Monogatari*, chapter 1.

32. On the *Kappa*, see JAXA, *Uchyūken no Monogatari*, chapter 1.

33. See the information on Japan's sounding rockets and their identified functions by JAXA, specifically, at www.isas.ac.jp (accessed 22 June 2008).

34. On the *Lambda*, see JAXA, *Uchyūken no Monogatari*, chapter 3.

35. On the *Mu* (*M*), see JAXA, *Uchyūken no Monogatari*, chapters 4–6, 8–9.

36. As the SLV history by ISAS shows, the increasing payload capability of each vehicle has become evident in the growing weight of each satellite. For example, the *Shinsei* ionosphere observation satellite launched in 1971 weighed 66 kilograms, and the *Tansei-3* launched in February 1977 weighed 129 kilograms.

37. See Akiba, Matsuo, and Godai, "History of Japanese Space Launch Vehicles," esp. pp. 3–4.

38. See R. Lloyd and G. P. Thorp, "A Review of Thrust Vector Control Systems for Tactical Missiles" (paper presented at the AIAA and Society of Automotive Engineers, Fourteenth Joint Propulsion Conference, Las Vegas, Nevada, 25–27 July 1978), p. 1, available online at www.aiaa.org (accessed 20 July 2009); and C. Tirres and F. D. Cantrell, "A Proposed Large Solid Rocket Test Cell" (paper presented at the AIAA, Thirteenth Aerodynamic Testing Conference, San Diego, CA, 5–7 March 1984), p. 172, available online at www.aiaa.org (accessed 20 July 2009).

39. For the U.S. missiles, see Federation of American Scientists (FAS), "Trident II D-5 Fleet Ballistic Missile," available online at www.fas.org (accessed 24 June 2009); and Stewart B. Larsen, "History of Minuteman System Update" (paper presented at the AIAA, Thirty-third Joint Propulsion Conference and Exhibit, Seattle, WA, 6–9 July 1997), pp. 1–14, available online at www.aiaa.org (accessed 20 July 2009).

40. Nissan had been instrumental in manufacturing the series of new technologies for further improving steering and maneuverability. See especially "Nissan Preparing to Develop First H-2 Solid Rocket Boosters," *AWST*, 4 August 1986, p. 80.

41. System Planning Corporation, "Non-Proliferation Issues: Japan," appendix III, Unclassified Working Papers, *Rumsfeld Commission Report*.

42. "Japan Presses on with M-5 Project Despite Motor Casing Problems," *AWST*, 10 August 1992, p. 48.

43. Out of sequence, the *M-V-8* was launched earlier in February 2006, followed by the last *M-V-7* in September 2006.

44. Paul Kallender, "Japan's M-5 Refit on Track," *Space News*, 29 September–5 October 1997.

45. Selig S. Harrison, "Missile Capabilities in Northeast Asia: Japan, South Korea and North Korea," in appendix III, Unclassified Working Papers, *Rumsfeld Commission Report.*

46. On cost concerns, see "Last M-V Rocket Delivers Satellite to Observe the Sun," *Japan Times*, 24 September 2006; and "M5 Roketto Kōkei ni Shinchō Shisei" [Cautious Attitude to M5 Succession], *Nikkei Sangyō Shinbun*, 31 July 2006.

47. The following section draws on Paul Kallender, "Japanese Company Considers New Small Launcher," *Space News*, 25 March 2002, which uses interviews with Yasunori Matogawa (then director of the Kagoshima Space Center), Junichiro Kawaguchi (a leading member of ISAS's rocket group), and Kazunori Kawasaki (general manager of Tokyo-based Ishikawajima-Harima Aerospace's [IA] Space Systems Department). For the continued focus on solid-rocket research, see also Morita Yasuhiro, "Kotai Roketto no Kenkyū—Sekai Ichi kara Sekai Ichi e no Chōsen" [Solid Rocket Research: From the World's Number 1 to the Challenge of Staying the World's Number 1], October 2007, available online at www.jaxa.go.jp (accessed 7 June 2008).

48. See MHI, "Our History," available online at h2a.mhi.co.jp (accessed 24 June 2008). Some portions in the remainder of this section also draw on this document. The most notable early rockets were in the *LS-A* series, a two-stage rocket, with one liquid and one solid stage that was developed by MHI. According to the company, it was designed explicitly to make ballistic flight observations possible. The *LS-A* was a sub-orbital version of the *Lambda* and was launched four times between August 1963 and November 1965. Japan then went to make steady but plodding progress in developing the *LS-B* and later the *LS-C* rockets. The *LS-C* rockets, eight of which were launched between 1968 to 1974, were designed and tested for improving satellite launches.

49. See United States of America and Japan, "Exchange of Notes Constituting an Agreement Concerning Cooperation in Space Activities for Peaceful Purposes (with attachment)," Tokyo, 31 July 1969. Registered by the United States of America, 4 March 1970. As printed in United Nations, *United Nations Treaty System 1970*, no. 10342, pp. 86–87. For the reentry technology quote, see section B of the attached agreement.

50. For the *N*, see JAXA, "Space Transportation Systems: Past Projects," available online at www.jaxa.jp (accessed 16 April 2008).

51. For the *H-I*, see JAXA, "Space Transportation Systems: Past Projects," available online at www.jaxa.jp (accessed 16 April 2008).

52. On the progressive indigenization of key rocket technologies especially with respect to the *H-I*, Berner, "Japan's Space Program: A Fork in the Road?" pp. 5–7.

53. For the *H-II*, see JAXA, "Space Transportation Systems: Past Projects," available online at www.jaxa.jp (accessed 16 April 2008).

54. Michael Mecham, "H-2 Launch Moves Japan Toward Goal," *AWST*, 14 February 1994, p. 30; and Berner, "Japan's Space Program: A Fork in the Road?" pp. 8–11.

55. Michael Mecham, "Japan Space Programs Keyed to H-2 Success," *AWST*, 31 January 1994, p. 50.

56. "Rabbits' Ears and Doves' Dreams: Stern Eyes Cast on the H-2—Military Diversion Feared by the International Community Because No Technical Boundary Can Be Drawn Between Military and Peaceful Uses," *Mainichi Shinbun*, 24 August 1994, available online at www.globalsecurity.org (accessed 24 June 2008); System Planning Corporation, "Non-Proliferation Issues: Japan," appendix III, Unclassified Working Papers, *Rumsfeld Commission Report*; and "Ballistic, Cruise Missile, and Missile Defense Systems: Trade and Significant Developments, June 1994–September 1995 (section on Japan), *Nonproliferation Review* 2(2), Winter 1995, p. 137.

57. For a general overview of the problems, see Eiichiro Sekigawa and Michael Mecham, "High H-2 Cost Worries NASDA," *AWST*, 19 July 1993, p. 66; Eiichiro Sekigawa and James R. Asker, "Japanese Satellite in Errant Orbit," *AWST*, 5 September 1994, p. 47; Michael Mecham, "ADEOS-1 Loss Sets Back Global Environmental Studies," *AWST*, 7 July 1997, p. 31; Craig Covault and Eiichiro Sekigawa, "Japanese H-2 Failure Ruins Satcom Research Mission," *AWST*, 2 March 1998, p. 34; Paul Kallender, "Japan Races Against Clock to Put Comets in Safe Orbit," *Space News*, 2–8 March 1998; "*H-II* Failure a Big Step Back for Space Program," *Japan Times*, 17 November 1999; and Michael Mecham and Eiichiro Sekigawa, "NASDA Kills H-2 in Favor of H-2A," *AWST*, 13 December 1999, p. 38.

58. For the *H-IIA*, see JAXA, "Space Transportation Systems: In Operation," available online at www.jaxa.jp (accessed 16 April 2008); and also MHI on the *H-IIA* launch services, available online at h2a.mhi.co.jp (accessed 5 July 2008).

59. "Industry Observer LE-7 Tests to Resume," *AWST*, 30 September 1991; and on launch vehicle mishaps, see especially David M. Harland and Ralph D. Lorenz, *Space Systems Failures: Disasters and Rescues of Satellites, Rockets and Space Probes* (Chichester, UK: Springer-Praxis, 2005), pp. 3–174.

60. Additional information on the *H-IIA* draws extensively on one of the authors' own sources: Paul Kallender, "Japanese Officials Planning H2 Rocket Upgrade," *Space News*, 24 June 1996; Paul Kallender, "H2A Passes Key Developmental Tests," *Space News*, 9 June 1997; Paul Kallender, "Cracks Appear During H2A Engine Firing," *Space News*, 17 November 1997; Paul Kallender, "H2A Engine Problem Threatens H2 Launch Date," *Space News*, 16 February 1998; Paul Kallender, "Hiroshi Imamura—Executive Vice President Rocket System Corp.," *Space News*, 2 February 1998; Paul Kallender, "With H2A on Track, Marketing Drive Begins," *Space News*, 28 September 1998; Paul Kallender, "H2A Success Boosts Confidence in Rocket," *Space News*, 23 September 2002;

Paul Kallender, "Two H-2A Successes Pave Way for Planning Japan's Future in Space, *Space News*, 23 December 2002; Paul Kallender, "MHI to Assume Control of H-2A Rocket Over Three-Year Period," *Space News*, 9 September 2002; Paul Kallender, "Japan, Mitsubishi Sign H2A Privatization Deal," *Space News*, 26 February 2003; Paul Kallender, "H-2A Failure Deals Blow on Several Fronts to Japanese Space Program, *Space News*, 23 December 2003; Paul Kallender, "H-2A Launch Failure Delays Launch of Three Satellites," *Space News*, 20 January 2004; Paul Kallender, "Solid-Rocket Motor Design Flaw Cited in November H-2A Failure," *Space News*, 22 March 2004; Paul Kallender, "Japan May Accelerate Follow-on Solid Rocket Booster for H-2A," *Space News*, 31 March 2004; Paul Kallender, "Restoring Faith in Space—Profile: Keiji Tachikawa, President of Japan Aerospace Exploration Agency," *Space News*, 27 June 2005; Paul Kallender, "Japan Still Far from Setting H-2A Return to Flight Date," *Space News*, 7 June 2004; and Paul Kallender, "Japan Readies Modified H-2A for Return to Flight," *Space News*, 21 February 2005. All *Space News* articles are also available online at www.space.com (accessed 24 June 2008).

61. Eichiro Sekigawa and Michael Mecham, "Japan Sets New Space Missions," *AWST*, 30 October 1995, p. 62.

62. Paul Kallender and Patrick Seitz, "Hughes Nears H2A Deal, *Space News*, 8 July 1996; and Patrick Seitz, "Japan's RSC Eyeing Launch Deal with Loral," *Space News*, 9 September 1996. Warren Ferster, "Booster Exports Approved for Japanese H2A Rocket," *Space News*, 2 December 1996; "U.S. Strap-ons Picked for H-2A," *AWST*, 17 March 1997; and Warren Ferster and Paul Kallender, "Thiokol Protests Nissan H2A Role," *Space News*, 27 January 1997.

63. Sumiko Oshima and Michael Mecham, "Japan Studies H-IIA for Heavy-Lift Market," *AWST*, 22 April 2002, p. 30; Michael Mecham and Frank Morring, Jr., "Simplified Lift: JAXA Looks to H-IIB Design as Backup for HTV Launches," *AWST*, 28 November 2005, p. 66.

64. For the *H-IIB*, see JAXA, "Space Transportation Systems: In Operation," available online at www.jaxa.jp (accessed 11 September 2009).

65. Eichiro Sekigawa and Michael Mecham, "NASDA Completes H-2A Design, Begins Assembly," *AWST*, 3 August 1998, p. 28; and Sumiko Oshima, "Japan's Launcher Back at Tanegashima," *AWST*, 20 April 2001, p. 30.

66. Interview, Atsutaro Watanabe, deputy director, NASDA's Office of Space Transportation Systems, Launch Systems Department, June 1996.

67. Sumiko Oshima and Eichiro Sekigawa, "Japan Designates H-IIA for Science Satellites," *AWST*, 28 January 2002, p. 41; Eiichiro Sekigawa and Michael Mecham, "Mitsubishi Assumes Control of Japan's H-IIA Launchers," *AWST*, 2 December 2002, p. 51.

68. Interview, Yoshihide Kanie, vice president and general manager at RSC, 8 March 2002. At the time even Kanie said the chances for the *H-IIA* securing commercial launch contracts were extremely limited and that RSC was focused on the Japanese

government market. At that point, RSC faced a difficult situation with a massive withdrawal of shareholders. See also Jason Bates, "Hughes Terminates H-2A Deal," *Space News*, 5 June 2000.

69. "Japan's Space Communications," *AWST*, 17 April 2006; and Arianespace, "Arianespace to Launch Japanese Satellite SUPERBIRD-7," press release, 10 April 2006, available online through www.arianespace.com.

70. Mitsubishi Heavy Industries (MHI), "MHI Receives Order for Launch Services of Korea Multipurpose Satellite-3," press release No. 1270, 12 January 2009, available online at www.mhi.co.jp (accessed 27 June 2009).

71. For the brief discussion on MHI below, see Toru Sugawara, "Aerospace Firms Target World Market," *Nikkei Weekly*, 26 January 2009; MHI, "MHI and Arianespace to Jointly Offer Satellite Launch Services," News, April 2007, available online at hwa.mhi.co.jp (accessed 27 June 2009); Agence France Presse, "Japan's Mitsubishi Heavy Aims to Cut Rocket Launch Costs: Company," Tokyo, 7 January 2008; Peter B. de Selding, "Sea Launch Bankruptcy Stokes Fears of Rising Launch Prices," *Space News*, 29 June 2009.

72. See H-IIA Launch Services, available online at h2a.mhi.co.jp (accessed 29 August 2009).

73. Marco Antonio Caceres, "Expendables Face Tough Market," *AWST*, 15 January 2001, p. 145; John Edwards, "ELV Market on the Up," *AWST*, 17 January 2005, p. 135; Peter B. de Selding, "Marketing the H-2A on a Global Stage," *Space News*, 10 May 2005; and Frank Morring, Jr., H-IIA Changes," *AWST*, 14 March 2005, p. 17.

74. The SRB-A design and manufacturing flaw that turned out to be the real culprit in the blowup was considered rather mysterious and "beyond prediction" by either JAXA or its contractor IHI, because until that point seven SRB-As tests and eleven SRB-As in actual operation (two each on the first five *H-IIA*s as well as one on the destroyed rocket) had proved successful. In the review process, there was also a more concerted movement toward a follow-on booster, the SRB-A2, which was considered more appropriate for heavy-lift variants of the launcher and work on which had begun in 2002.

75. On both *DASH* and USERS, see Stephen Clark, "Japanese H-2A Rocket Launches Two Satellites," *Spaceflight Now*, 10 September 2002, available online at spaceflight.com (accessed 7 July 2008); Sumiko Oshima, "Second H-IIA Flight Uses New Boosters," *AWST*, 4 February 2002, p. 41; Patricia J. Parmalee, "Identifying Cause of DASH Problems," *AWST*, 25 February 2002, p. 19; Patricia J. Parmalee, "H-IIA Program," *AWST*, 25 February 2002, p. 19; Frank Morring, Jr., "NEC-Toshiba Pays Fine for DASH Experiment," *AWST*, 20 October 2003, p. 17; "Japan's Space Program—Part 2: First Japanese Reusable Spacecraft to Carry Science, Technology Payload," *AWST*, 20 August 1990, p. 70; "The Unmanned Space," *AWST*, 9 June 2003, p. 19; and Craig Covault, "Japanese Technology Missions Readied," *AWST*, 7 December 1998. Additional

information is from the official Web site of USEF, available online at www.usef.or.jp (accessed 16 June 2008).

76. For the *J-I*, see JAXA, "Space Transportation Systems: Past Projects," available online at www.jaxa.jp (accessed 16 April 2008); as well as one of the authors' own sources: Paul Kallender, "Japan's J-1 Launcher Might Get More Work," *Space News*, 13 January 1997; Paul Kallender, "Japan's NASDA Shoots for Advanced J-1 Launcher," *Space News*, 2 June 1997; Paul Kallender, "J-1 Updates May Involve Non-Japanese Suppliers," *Space News*, 26 January 1998; Paul Kallender, "Critical Report Hammers J-1 Program Costs," *Space News*, 18 May 1998; with all *Space News* articles available online at www.space.com (accessed 24 June 2008).

77. For general trends and concerns below, see also Michael Mecham, "Japan Eyes Low-Cost J-1," *AWST*, 2 June 1997, p. 67; Patricia M. Parmalee, "NASDA J-1 Launch Set," *AWST*, 23 April 2001, p. 19; Bruce A. Smith, "J-1 Plan Selected," *AWST*, 10 May 1999, p. 21; "Japan Science," *AWST*, 4 March 1996, p. 17; and Paul Proctor, "Japan Developing New Solid Boosters—Japan to Offer J-1 Solid Rocket for Launch of Small Payload," *AWST*, 10 August 1992, p. 47.

78. For an overview of high costs across the *H-II*, *J-I*, and *M-V* that made them commercially uncompetitive worldwide see Marco Antonio Caceres, "Launch Vehicles: Mostly Thriving," *AWST*, 12 January 1998, p. 125.

79. Selig S. Harrison, "Missile Capabilities in Northeast Asia: Japan, South Korea and North Korea," in appendix III, Unclassified Working Papers, *Rumsfeld Commission Report*.

80. "HYFLEX Assembly Nearly Finished," *AWST*, 12 June 1995, p. 47; Michael Mecham, "Last-Minute Sinking of HYFLEX Mars J-1 Debut," *AWST*, 19 February 1996, p. 27.

81. Eiichiro Sekigawa, "Japanese Watchdog Agency Questions Need for J-1," *AWST*, 11 May 1998, p. 41.

82. See "Foreign Missile Developments and the Ballistic Missile Threat," Statement for the Record to the Senate Foreign Relations Committee on Foreign Missile Developments and the Ballistic Missile Threat to the United States Through 2015 by Robert W. Walpole, national intelligence officer for strategic and nuclear weapons, 16 September 1999, available online at www.cia.gov (accessed 30 April 2008) [emphasis ours]. See also Johnson-Freese, *Space as a Strategic Asset*, pp. 44–45, who similarly points out that missiles do not have to be highly reliable to have a deterrent and psychological impact as compared especially to the stringent requirements of SLVs launched with commercial payloads.

83. James Brooke, "After Failures, Space Effort in Japan Gets a Lift," *New York Times*, 27 February 2005.

84. Michael D. Swaine with Loren H. Runyon, "Ballistic Missiles and Missile Defense in Asia," *NBR Analysis* 13(3), 2002, pp. 10, 12.

Chapter 5

1. Unless otherwise indicated in this chapter, the most up-to-date information on Japanese satellite and spacecraft is from the official Web site of the at www.jaxa.jp (accessed 16 July 2008). For some information on the earlier satellites, see Japan Aerospace Exploration Agency (JAXA), *Nihon no Uchyū Kaihatsu no Rekishi: Uchyūken no Monogatari* [The History of Japan's Space Development: The ISAS Story], hereafter *Uchyūken no Monogatari*, Chapters 1–10, available only online at www.jaxa.jp (accessed 16 April 2008). References to this work are cited here only by chapter numbers. For *Ohsumi*, see Chapter 3. See also Frank Morring, Jr., "Japan's Oldest Satellite Reenters," *Aviation Week & Space Technology* (hereafter *AWST*), 25 August 2003, p. 17; and for an earlier overview, see H. Mitsuma, "Historical Aspects of Spacecraft Technology and Its Diffusion in Society in Japan," *History of Rocketry and Astronautics*, 21, 1992, pp. 163–178.

2. Eiichiro Sekigawa, "Recce Recovery," *AWST*, 7 February 2005, p. 38.

3. Spy satellites came out to nearly one-third of space spending which was almost as large as JAXA's entire budget for space exploration, manned space, and operational programs.

4. The brief discussion of the RMA draws on Tomohiro Okamoto, "RMA: Why Is It the Revolution in Military Affairs?" *DRC Annual Report*, 2002, pp. 1–6, available online at www.drc-jpn.org (accessed 18 August 2008); Carl Conetta, "We Can See Clearly Now: The Limits of Foresight in the Pre-World War II Revolution in Military Affairs (RMA)," *Research Monograph No. 12*, Project on Defense Alternatives (Cambridge, MA: Commonwealth Institute, 2 March 2006), esp. appendix; and *Rumsfeld Commission Report*, Attachment 1.B.

5. Jeffrey T. Richelson, *America's Space Sentinels: DSP Satellites and National Security* (Lawrence: University Press of Kansas, 1999), pp. 157–175.

6. William Nolte, "Operation Iraqi Freedom and the Challenge to Intelligence: Keeping Pace with the Revolution in Military Affairs," *Studies in Intelligence*, 48(1), 2004, pp. 1–16 (online version), available online at www.cia.gov (accessed 5 August 2008).

7. Christopher W. Hughes, *Japan's Re-emergence as a 'Normal' Military Power* (Oxford: Oxford University Press for the International Institute for Strategic Studies, 2004), pp. 83–85.

8. Japan Air Defense Fund (JASDF), "Jidō Keikai Kansei Shisutemu (JADGE) no Unyō Kaishi ni Tsuite" [Concerning the Operational Commencent of JADGE], press release, 7 July 2009, available online at www.mod.go.jp (accessed 2 July 2009); and "Japan Upgrades Missile-Detection System," United Press International (UPI), 2 July 2009, available online at www.upi.com (accessed 2 July 2009).

9. System Planning Corporation, "Non-Proliferation Issues: Japan," appendix III, Unclassified Working Papers, *Rumsfeld Commission Report*.

10. See National Space Strategy Planning Group (NSSPG), *Report by the National Space Strategy Planning Group—Toward Establishment of New Space Development and Utilization System* (Tokyo: NSSPG, August 2005), pp. 26–30.

11. For the analysis of the *IGS* see Joan Johnson-Freese and Lance Gatling, "Security Implications of Japan's Information Gathering Satellite (IGS) System," *Intelligence and National Security* 19(3), 2004, pp. 538–552; Andrew L. Oros, "Explaining Japan's Tortured Course to Surveillance Satellites," *Review of Policy Research* 24(1), 2007, pp. 29–48; Andrew L. Oros, *Normalizing Japan: Politics, Identity and the Evolution of Security Practice* (Stanford, CA: Stanford University Press, 2008), esp. pp. 139–146; Hughes, *Japan's Re-emergence as a 'Normal' Military Power*, pp. 85–88; Christopher W. Hughes, "'Supersizing' the DPRK Threat: Japan's Evolving Military Posture and North Korea," *Asian Survey* 49(2), 2009, esp. p. 198; Federation of American Scientists (FAS), "Space Policy Project/World Space Guide—Japan: Information Gathering Satellites—Imagery Intelligence," available online at www.fas.org (accessed 16 April 2009). See also works by one of the authors on point in Paul Kallender, "Lurching into Military Space, Japan Rushes to Develop Spy Satellites," *ACCJ Journal*, March 1999; Paul Kallender, "Japan Considers Spy Satellite in Response to N. Korea," *Space News*, 14 September 1998; Paul Kallender, "Firms Vie to Win Japan's Biggest Satellite Deal," Space News, 8 February 1999; Paul Kallender, "Spy Satellite Launch Marks New Era for Japan in Space," *Space News*, 9 April 2003; and Paul Kallender, "Japan Aims for Operational Military Space Systems by 2006," *Space News*, 2 September 2003. All *Space News* articles are available online at www.space.com (accessed 1 July 2008).

12. For earlier stages of the spy satellite saga that suggest the pivotal role of corporations, see, for example, Tsuyoshi Sunohara, *Tanjyō Kokusan Supai Eisei: Dokuji Jyōhōmō to Nichbei Dōmei* [The Birth of National Spy Satellites: An Independent Intelligence Network and the U.S.-Japan Alliance] (Tokyo: Nihon Keizai Shinbunsha, 2005), esp. pp. 11, 22–23, 25, 44–45. See also "Nissan Jidōsha—Eisei Uchiage Sannyū [Nissan Motors: Joining Satellite Launch], *Nihon Keizai Shinbun*, 6 January 1998; "Tamokuteki Eisei de Dokuji Kōsō—NEC to Mitsubishi Denki—1 Meetoru no Buttai mo Shikibetsu" [Independent Scheme with Multi-Purpose Satellites—NEC, Melco—Distinguishing Objects Also at 1 Meter], *Nihon Keizai Shinbun*, 12 September 1998; and Yamada Shūhei, "Mitsubishi Denki, Eisei Jigyō o Kakudai" [Melco—Expanding Satellite Business], *Nikkei Sangyō Shinbun*, 2 December 1999.

13. Norihide Miyoshi, "Integrated Security HQ Set Up by Defense Agency to Enhance Operations," *Daily Yomiuri*, 25 January 1997; "Japan's New Spy Agency Consolidates Five Units," *Wall Street Journal*, 21 January 1997; Derek de Cunha, "Japan Pushes Hard to Beef Up Its Military Intelligence," *Straits Times (Singapore)*, 4 July 1996; Masahiko Sasajima, "Information Is Power When Responding to Emergencies," *Daily Yomiuri*, 23 April 1998; Nao Shimoyachi, "Spy Satellites Part of Intelligence Quest," *Japan Times*, 27 March 2003; and for an overview of Japan's intelligence trends and

policy structure, see Andrew L. Oros, "Japan's Growing Intelligence Capability," *International Journal of Intelligence and CounterIntelligence* 15, 2002, pp. 1–25.

14. For the trajectory of events, see "Teisatsu Eisei Dōnyū—Shushō, Maemuki Shisei" [Introduction of Reconnaissance Satellites—Forward-Looking Stance of Prime Minister], *Nihon Keizai Shinbun*, 3 September 1998; "'Tamokuteki Eisei' o Kentō" [Investigating "Multi-Purpose" Satellites], *Nihon Keizai Shinbun*, 8 September 1998; Tamokuteki Eisei de Dokuji Kōsō—NEC to Mitsubishi Denki—1 Meetoru no Buttai mo Shikibetsu" [Independent Scheme with Multi-Purpose Satellites—NEC, Melco—Distinguishing Objects Also at 1 Meter], *Nihon Keizai Shinbun*, 12 September 1998; "Government Considers Reconnaissance Satellites," *Daily Yomiuri*, 9 September 1998; Akinori Uchida, "U.S. Backs Japan's Plan to Launch Recon Satellite," *Daily Yomiuri*, 15 September 1998; Eiichiro Sekigawa, "Japan Considers Hybrid Satellite Defense Options," *AWST*, 2 November 1998, p. 29; Michiyo Nakamoto, "Tokyo Aims for Final Frontier," *Financial Times*, 6 November 1998; "Cabinet Gives Go-Ahead to Recon Satellite Plan," *Daily Yomiuri*, 7 November 1998; "Kokunai Hatsu no 'Teisatsu Eisei' Uchiage" [First National "Spy Satellites" Launched], *Asahi Shinbun*, 28 March 2003; "Jyōhō Eisei Uchiage Seikō" [Successful Launch of Information Satellites], *Nihon Keizai Shinbun*, 28 March 2003; "Shin Jyōhō Eisei Kenkyū e" [Toward New Information Satellite Research], *Nihon Keizai Shinbun*, 26 December 2004; Joan Johnson-Freese and Lance Gatling, "Japan Joins the Exclusive Space Spy Club," *YaleGlobal*, 31 March 2003, available at yaleglobal.yale.edu (accessed 7 October 2005); Eiichiro Sekigawa and Michael Mecham, "Japan Preps for Its First Milsat Launch," *AWST*, 27 June 2003, p. 26; and Reiji Yoshida, "Japan Readies Launch of Third Spy Satellite from Kagoshima," *Japan Times*, 9 September 2006.

15. In annual chronological order, see for example Naoaki Usui, "Japan May Include Spy Satellites in Plan," *Space News*, 24–30 January 1994; Naoaki Usui and B. Opall, "Japan Eyes Space for Defense," *Space News*, 1–7 August 1994; Naoaki Usui, "Group in Japan Recommends Spy Satellites," *Space News*, 29 August–4 September 1994; with all *Space News* articles available online at www.space.com (accessed 1 July 2008); "Hata Supports Future Possession of 'Spy Satellites,'" *BBC Summary of World Broadcasts*, 9 June 1994; "Bōeichō, Teisatsu Eisei, Dōnyū no Hōshin—Dokuji no Jyōhō Shūshūryoku Kyōka, 2001 Nen Ikō Uchiage" [JDA Considers Introduction of Reconnaissance Satellites—Emphasis on Independent Information-Gathering Power, Launch From 2001], *Nihon Keizai Shinbun*, 5 May 1996; Eiichiro Sekigawa, "Japan Ponders Building Military Recon Network," *AWST*, 10 June 1996, p. 38; and "USA Reported Opposed to Japan's Plan for Spy Satellites," *BBC Summary of World Broadcasts*, 8 January 1998; Sunohara, *Tanjyō Kokusan Supai Eisei*, p. 22, quotes a figure of about $50,000 in MOFA's 1997–1998 budget allocated for the study of "international Information Gathering Satellites."

16. On managing the relationship with the U.S., see Akinori Uchida, "Washington Asks Tokyo to Buy U.S. Satellite," *Daily Yomiuri*, 15 May 1999; Robert Wall, "Japanese

Recce Program Wary of FS-X Missteps," *AWST*, 23 August 1999, p. 44; and "Japan Signs Recce Accord," *AWST*, 11 October 1999.

17. Interview, Shinya Ona, chairman, Liberal Democratic Party (LDP) Science and Technology Committee, 17 September 1998.

18. See, for example, "Spokesman Says: Japanese Move on Satellite a 'Dangerous Military Action,'" *BBC Summary of World Broadcasts*, 19 November 1998; "NKorea: Radio Criticizes Japan for Planning to Develop Spy Satellites," *BBC Monitoring Asia Pacific*, 9 July 1999; Mary Kwang, "China Military Lashes Out at Japan's Defense Moves," *Straits Times*, 1 September 2000.

19. Information on the specifics of the *IGS* satellites is from Eiichiro Sekigawa, "And So It Begins," *AWST*, 7 April 2003; Craig Covault, "Launch of Japan's Second Pair of Reconnaissance Satellites Delayed," *AWST*, 6 October 2003, p. 33; Frank Morring, Jr., "Reconnaissance Launch," *AWST*, 18 September 2006, p. 15; "Japan's First Reconnaissance," *AWST*, 2 April 2007, p. 24; Frank Morring, Jr., "Watching North Korea," *AWST*, 5 March 2007, p. 15; "Japan's Spy Satellites Inferior to U.S. Commercial Resolution," *Japan Times*, 30 December 2002; Frank Morring, Jr., "Japanese Optical Reconnaissance Satellite Isn't Meeting Spec," *AWST*, 14 July 2003, p. 21; Paul Kallender, "Spy Satellite Launch Marks New Era for Japan in Space," *Space News*, 9 April 2003; and Paul Kallender, "Rocket Problems Postpone Japanese Spy Satellite Launch," *Space News*, 20 October 2003. All *Space News* articles are available online at www.space.com (accessed 1 July 2008).

20. "Tamokuteki Eisei de Dokuji Kōsō—NEC to Mitsubishi Denki—1 Meetoru no Buttai mo Shikibetsu" [Independent Scheme with Multi-Purpose Satellites—NEC, Melco—Distinguishing Objects Also at 1 Meter], *Nihon Keizai Shinbun*, 12 September 1998. One early estimate suggested that the expected life span of the spy satellite ranged from between ten days to two years.

21. "Japan Ignores International Treaty on Registering Spy Satellites," *BBC Monitoring Asia Pacific*, 10 June 2007; and Keisuke Yoshimura, "Japan's Spy Satellites Are an Open Secret," *Japan Times*, 15 June 2007. This refers to the Convention on Registration of Objects Launched into Outer Space, which opened for signature in 1975. Japan signed it in 1983 and, as of 2008, had registered more than 100 spacecraft, including their shape, orbits, and inclinations. See the Annex to the Resolution adopted by the General Assembly, 3235 (XXIX) Convention on the Registration of Objects Launched into Outer Space, available online through the United Nations Office on Outer Space Affairs (UNOOSA) at www.oosa.unvienna.org (accessed 19 August 2009).

22. Frank Morring, Jr., "U.S. Classifies Orbits of Visible Japanese Satellites," *AWST*, 21 April 2003.

23. Tetsuo Hidaka, "Spy Satellites to Watch N. Korea; Extra Surveillance to Reduce Reliance on U.S. Intelligence," *Daily Yomiuri*, 4 March 2003; and Eiichiro Sekigawa, "Japan Ponders Building Military Recon Network," *AWST*, 10 June 1996, p. 38.

24. Strategic Headquarters for Space Policy (SHSP), *Uchū Kihon Keikaku; Nihon No Eichi ga Uchū o Ugokasu* [Basic Space Plan: Wisdom of Japan Moves Space] (Tokyo: SHSP, 2 June 2009), appendix 2.

25. "Japanese Spy Satellite Proposal Faces Political, Budget Hurdles," *Daily Yomiuri*, 17 September 1998.

26. For some of the earliest concerns with the dual-use nature of commercial satellite imagery, see Michael Krepon and Peter D. Zimmerman, *Commercial Observation Satellites and International Security* (London: Palgrave MacMillan, 1990); and Gerald Steinberger, "Dual Use Aspects of Commercial High Resolution Imaging Satellites," Security and Policy Studies No. 37, BESA Center for Strategic Studies, February 1998, available online at www.biu.ac.il (accessed 1 August 2009).

27. For an overview of the key corporate and government actors, see Hirofumi Matsuo, "Firms Compete to be Japan's Eye in the Sky—Mitsubishi, Hitachi Aim to Dominate Satellite Imaging," *The Nikkei Weekly*, 17 June 1996. On Japan and satellite imagery trends, see Ryuichi Otsuka, "High-Resolution Satellite Photos Go on Sale," *Daily Yomiuri*, 5 October 1999; Michael Mecham, "Commercial Imaging to Enter 1-Meter Era," *AWST*, 26 April 1999; Kyle T. Umezu, "EarlyBird Tweaks the Law," *SpaceDaily*, 1997, available online at www.spacedaily.com (accessed 1 July 2008); Joseph C. Anselmo, "Commercial Space's Sharp New Image," *AWST*, 31 January 2000; Vernon Loeb, "Now Everybody Can Snap Up Satellite Secrets," *The Sunday Herald*, 26 September 1999; "Protecting Japan—Eyes in the Sky Vital for Security," *Daily Yomiuri*, 8 June 2004; Tetsuo Hidaka and Koichi Yasuda, "Better Spy Satellite System Needed: Reliance on U.S. Intelligence on Missile Launch Shows Need for Improvement," *Daily Yomiuri*, 31 July 2006; and Paul Kallender, "Consortia to Develop Imagery Markets in Japan," *Space News*, 5 May 1997, available online at www.space.com (accessed 1 July 2008).

28. Oros, *Normalizing Japan*, p. 137.

29. See generally Doug Struck, Akiko Yamamoto, and Sachiko Sakamaki, "Japan to Launch Spy Satellites—Move Is Attempt to Lessen Dependence on U.S. Intelligence," *Washington Post*, 26 March 2003; Tetsuo Hidaka, "Spy Satellites to Watch N. Korea; Extra Surveillance to Reduce Reliance on U.S. Intelligence," *Daily Yomiuri*, 4 March 2003; and Peter B. de Selding and Naoaki Usui, "WEU Plan Strengthens European-Japanese Ties," *Space News*, 9 December 1996.

30. Corporate and historical information on JSI can be found at www.spaceimaging.co.jp (accessed 9 September 2008).

31. Corporate information on GeoEye is from the official Web site at www.geoeye.com (accessed 9 September 2008). See also GeoEye, "GeoEye Signs with Mistsubishi Corporation as GeoEye Regional Affiliate," press release, 2 April 2008; GeoEye, "GeoEye-1 Satellite Launches Into Space From Vandenberg Air Force Base, California," press release, 6 September 2008; and GeoEye, "GeoEye Makes Progress on a

Third-Generation Commercial Earth-Imaging Satellite," press release, 10 June 2008. In addition to that from GeoEye, information on satellite specifications and regulations are from Satellite Imaging Corporation, the official Value Added Reseller of imaging and geospatial data products for GeoEye, "GeoEye-2 Satellite Images and Sensor Specifications," available online at www.satimagingcorp.com (accessed 2 July 2009). See also Stephen Shankland, "Google to Buy GeoEye Satellite Imagery," *CNET News*, 29 August 2008, available online at www.cnet.com; Steve Gelsi, "GeoEye CEO Bullish on Satellite Image Potential," *MarketWatch*, 21 May 2009, available online at www.marketwatch.com; and also "Sales Double at GeoEye," *Washington Business Journal*, 11 August 2009, available online at www.bizjournals.com (both accessed 19 August 2009).

32. Corporate and historical information on DigitalGlobe and HitachiSoft can be found through the official Web site of DigitalGlobe at www.digitalglobe.com (accessed 19 August 2009). Despite its earlier start relative to GeoEye in 1993 with a U.S. government license to obtain high-resolution digital imagery, DigitalGlobe only successfully launched *QuickBird-2* in 2001 and *WorldView-1* in 2007 and is projected to launch *WorldView-2* in 2009. Like GeoEye, DigitalGlobe's three-satellite based system will compete for both government and commercial customers around the world.

33. See here, Berner, "Japan's Space Program," p. 17; Andrew L. Oros, "Explaining Japan's Tortured Course to Surveillance Satellites,"pp. 31–32; Federation of American Scientists (FAS), "Space Policy Project/World Space Guide—Japan: Information Gathering Satellites—Imagery Intelligence" available online at www.fas.org (accessed 16 April 2009); and Sunohara, *Tanjyō Kokusan Supai Eisei*, p. 22.

34. On the capabilities of of private actors and technologies, see Andrew L. Oros, "Explaining Japan's Tortured Course to Surveillance Satellites," esp. pp. 39–40; Steven Berner, "Japan's Space Program: A Fork in the Road?" RAND National Security Research Division, Santa Monica, CA, 2005, available online at www.rand.org (accessed 22 December 2006), pp. 17–20; Federation of American Scientists, "Information Gathering Satellites—Imagery Intelligence," available online at www.fas.org (accessed 27 January 2008); and Paul Kallender, "Japan Faces Major Hurdles in Spy Satellite Development," *Defense News*, 12–18 October 1998, p. 76.

35. See NT Space, "Earth Observation Sensors," at www.nec.com (accessed 28 August 2008); and the description on JAXA's Web site, "Japan Earth Resources Satellite 'Fuyo-1' (JERS-1)," at www.jaxa.jp (accessed 28 August 2008).

36. "Tamokuteki Eisei de Dokuji Kōsō—NEC to Mitsubishi Denki—1 Meetoru no Buttai mo Shikibetsu" [Independent Scheme with Multi-Purpose Satellites—NEC, Melco—Distinguishing Objects Also at 1 Meter], *Nihon Keizai Shinbun*, 12 September 1998.

37. The following draws heavily on a document by Melco, "Tamokuteki Jyōhō Shūshū Shisutemu (kashō) ni kan suru Kentō [Regarding the Investigation of the

Multi-Purpose Information Gathering Satellite System (provisional name)]," J1-1803/FES5503/SSRS98024, October 1998, slides 1–37 (broken down as satellite imagery information uses, slides 3–10; IGS system, slides 11–18; Information Gathering sequence, slides 19–22; development plan, slides 23–29; proposals, slides 30–31; additional attachments on uses). See also Eiichiro Sekigawa, "Mitsubishi Recce Plan Gains Ground in Diet," *AWST*, 9 November 1998, p. 34; and, subsequently, Yamada Shuhei, "Mitsubishi Denki, Eisei Jigyō o Kakudai [Melco Expands Satellite Activities], *Nikkei Sangyō Shinbun*, 2 December 1999.

38. Science and Technology Agency (STA), Ministry of International Trade and Industry (MITI), and Ministry of Posts and Telecommunications (MPT), "Jikō: Jyōhō Shūshū Eisei no Kenkyū" [Subject: Research on the Information Gathering Satellites] (Tokyo: STA/MITI/MPT, 9 November 1998); and Cabinet Office, "Jyōhō Shūshū Eisei no Riyō ni Tsuite" [Concerning the Utilization of the Information Gathering Satellites], Tokyo, 11 November 1998.

39. The size and orbits of the *IGS* system satellites are public knowledge. See, for example, National Space Science Data Center/World Data Center for Satellite Information, SPACEWARN Bulletin, No. 593, 1 April 2003, ¶B, available online at nssdc.gsfc.nasa.gov (accessed 7 April 2008).

40. This section draws on Japan Aerospace Exploration Agency (JAXA), "JAXA no Chikyū Kansoku no Ayumi" [The Path of JAXA's Earth Observation], available online at www.jaxa.jp (accessed 1 July 2008). All other technical details on the EO systems, spacecraft, and institutions discussed here are directly from JAXA at the same official website and also from the Committee on Earth Observation Satellites, "Earth Observation Programs in Japan, National Space Development Agency of Japan, Country Report, CEOS/14/DOC/02, 14 September 2000, pp. 1–4 (online version).

41. Additional background information in the remainder of this section is from Paul Kallender, "Lurching into Military Space, Japan Rushes to Develop Spy Satellites," *ACCJ Journal*, March 1999; Paul Kallender, "Japan Considers Spy Satellite in Response to N. Korea," *Space News*, 14 September 1998; Paul Kallender, "Prime Minister Seeks More Funds to Speed Japan's ALOS effort," *Space News*, 30 November 1998; Paul Kallender, "Satellites Key to Japanese Disaster Management Plan," *Space News*, 17 February 2003; Paul Kallender-Umezu, "Japan Lofts Newest Earth Observing Satellites," *Space News*, 30 January 2006. All *Space News* articles are available online at www.space.com (accessed 1 July 2008). See also Isamu Mishima, "Technology Ready for Observation Satellites," *Daily Yomiuri*, 16 December 1998; and Eiichiro Sekigawa, "Japan Considers Hybrid Satellite Defense Options," *AWST*, 2 November 1998, p. 29;

42. Tariq Malik, "NASA Climate Satellite Crashes in Ocean After Launch Failure," 24 February 2009, available online at www.space.com (accessed 11 February 2010).

43. Paul Kallender-Umezu, "Japan's Ibuki Satellite Carrying on Without OCO Assist," *Space News*, 2 March 2009.

44. See generally Berner, "Japan's Space Program," p. 19–20.

45. On *ALOS/Daichi*, see Kazuo Inaba and Yasushi Horikawa, "Utlization of Data Acquired by 'Daichi' (Advanced Land Observing Satellites) for Maps" (Tokyo: MLIT/JAXA, 16 January 2008), pp. 1–17, available online at www.jaxa.jp (accessed 29 July 2009); and on PALSAR, see especially Japan Resources Observation System and Space Utilization Organization (JAROS), "About JAROS" (as of April 2007), esp. p. 9, available online at www.jaros.or.jp (accessed 29 July 2009).

46. Interview, Hiroshi Fujita, director, Space Utilization Division, STA, Tokyo, 11 September 1998.

47. Interview, Hiroshi Fujita, director, Space Utilization Division, STA, Tokyo, 24 November 1998.

48. Interview, Tsugihiko Katagi, project manager for ALOS/Daichi, NASDA, Tokyo, 24 November 1998.

49. See, for example, Nippon Keidanren, "Jyōhō Shūshū Eisei no Dōnyū no Keii to Kongo no Sukejyuuru nado ni Tsuite" [Regarding the Chronology of Introduction and Future Schedule of the IGS], *Keidanren Kurippu* 96, 25 February 1999, esp. ¶2, available online at www.keidanren.or.jp (accessed 30 July 2008).

50. General background information in this section is from Donald H. Martin, "A History of U.S. Military Satellite Communication Systems," *Crosslink* 3(1), 2002, available online at www.aero.org (accessed 15 May 2008; and Joan Johnson-Freese, *Space as a Strategic Asset* (New York: Columbia University Press, 2007), pp. 34–37.

51. Takashi Iida, "National Security: How We Should Push Forward R&D of Satellite Communications Technology as a Nucleus of Network Centric Defense System," *Space Japan Review* 12 (1–3), nos. 59–60, December–March 2009, pp. 1–13, available online at satcom.nict.go.jp (accessed 12 July 2009).

52. Unless otherwise indicated, all details and background on the actual communication satellite and spacecraft systems, as well as the general trajectory of events, in this section are taken from JAXA, available online at www.jaxa.jp (accessed 1 July 2008); and also Mitsuma, "Historical Aspects of Spacecraft Technology," esp. pp. 174–176.

53. Johnson-Freese and Gatling, "Security Implications," pp. 544–547; Oros, *Normalizing Japan*, p. 137; and Craig Covault, "Ariane Launches Japanese Communications Satellite, "AWST, 7 December 1992. See also more generally Donald H. Martin, *Communication Satellites*, 4th ed. (El Segundo, CA: The Aerospace Press, 2000), pp. 403–404; and Norman Friedman, *The Naval Institute Guide to World Naval Weapon Systems*, 5th ed. (Annapolis, MD: Naval Institute Press, 2006), pp. 5–6.

54. U.S. Government Accountability Office (GAO), "Space Acquisitions: DOD Needs Additional Knowledge as It Embarks on a New Approach for Tranformational

Satellite Communications System," GAT-06-537, May 2006, esp. 1–17, Figure 1; Patricia Maloney Figliola, "U.S. Military Space Programs: An Overview of Appropriations and Current Issues," *CRS Report for Congress*, 7 August 2006, p. 6; Turner Brinton, "U.S. Air Force Scales Back T-Sat," *Space News*, 23 December 2008; Turner Brinton, "Air Force Terminates T-Sat Ground Segment Contract," *Space News*, 9 June 2009; and Brian Berger, "Griffin: Lasers Key to NASA's In-Space Comm Network," *Space News*, 29 September 2008; and Space News Briefs, "U.S. Air Force Seeks Input on Laser Comm Demo Sat," *Space News*, 28 August 2009. All Space News articles are available online at www.spacenews.com (accessed 31 August 2009).

55. See Berry Smutny et al., "5.6 Gbps Optical Intersatellite Communication Link," *Proceedings of the International Society for Optical Engineering (SPIE)*, vol. 7199, 24 February 2009 (online publication), pp. 1–8 (esp. introduction), available online via the Smithsonian Astrophysical Laboratory (SAO)/NASA Astrophysics Data System (ADS) at adsabs.harvard.edu (accessed 28 August 2009).

56. See also Tesat-Spacecom, "German Laser Terminals Successfully Tested in Space," press release, 12 March 2008, available online at www.tesat.de (accessed 28 August 2009); U.S. Missile Defense Agency, "NFIRE Laser Communications Terminal Delivered to MDA," press release, 18 December 2006, available online at www.mda .mil (accessed 28 August 2009); Peter B. de Selding, "U.S.–German Laser Intersatellite Link Performs Well on 2 Spacecraft," *Space News*, 17 March 2008, available online at www.space.com (accessed 1 July 2008); and Peter B. de Selding, "European Firms Launch Dual-Use Sat Laser Projects," 8 October 2007, available online at the official Web site of the Asia-Pacific Multilateral Cooperation in Space Technology and Applications (AP-MCSTA) at www.apsco.int (accessed 30 August 2008).

57. General background information in this section draws on "Industry Observer: Optical Communications Tests," *AWST*, 27 July 1992; Ben Iannotta, "Some See Optical Spectrum as Wave of the Future," *Space News*, 8 July 1996; Peter B. De Selding, "Japanese, ESA Satellites Establish Laser Link," *Space News*, 12 December 2005. All *Space News* articles are available online at www.space.com (accessed 1 July 2008).

58. On *ETS-VI*, see also Japan Aerospace Exploration Agency (JAXA), "Laser Communication Experiment Using ETS-6 Satellite," and "Engineering Test Satellite VI "Kiku No.6 (ETS-VI)," both available online at www.jaxa.jp (accessed 1 July 2008).

59. On the *OICETS*, see also Yuuichi Fujiwara et al., "Optical Inter-Orbit Communications Engineering Test Satellite (OICETS), *Acta Astronautica* 61(11–12) 2007, pp. 163–175; Morio Toyoshima et al., "Reconfirmation of the Optical Performances of the Laser Communications Terminal Onboard the OICETS Satellite," *Acta Astronautica* 55(3–9) 2004, pp. 261–269; JAXA, "Launch Result of the Optical Inter-orbit Communications Engineering Test Satellite (OICETS) and Innovative Technology Demonstration Experiment Satellite (INDEX)," press release, 24 August 2005; JAXA, "Launch Result of the Optical Inter-orbit Communications Engineering Test Satellite

(OICETS) and Innovative Technology Demonstration Experiment Satellite (INDEX)," 24 August 2005; JAXA, "The Status of the Kirari, Optical Inter-orbit Communications Engineering Test Satellite (OICETS)," press release, 25 November 2005; JAXA, "Toward the Era of Optical Communication in Space. Success of the Optical Inter-orbit Communication Experiment between the Optical Inter-orbit Communications Engineering Test Satellite 'Kirari' (OICETS) and the Advanced Relay and Technology Mission (ARTEMIS)," press release, 9 December 2005; JAXA, "Update, First Success of the Optical Inter-orbit Communication over 40,000 km," information release, 13 December 2005; JAXA, "Successful Optical Communication Experiment Between the NICT Optical Ground Station and the Optical Inter-Orbit Communications Engineering Test Satellite 'Kirari (OICETS),'" press release, 7 April 2006; JAXA, "Success with Optical Communication Experiment between the Optical Inter-orbit Communication Engineering Test Satellite 'Kirari' (OICETS) and the Optical Ground Station at German Aerospace Center (DLR)," press release, 9 June 2006; JAXA, "Mission Operation Completed with 'Extra Success,'" information release, 2 November 2006; with all JAXA press and information releases available online at www.jaxa.jp (accessed 1 July 2008).

60. See Japan Aerospace Exploration Agency (JAXA), "Communications and Broadcasting Engineering Test Satellites 'Kakehashi' (COMETS)," available online at www.jaxa.jp (accessed 1 July 2008); Craig Covault and Eiichiro Sekigawa, "Japanese H-2 Failure Ruins Satcom Research Mission," *AWST*, 2 March 1998, p. 34; Joseph C. Anselmo, "Martian Voyage," *AWST*, 15 June 1998, p. 21; and Bruce A. Smith, "Operation Ceased," *AWST*, 15 February 1999, p. 17.

61. See Japan Aerospace Exploration Agency (JAXA), "Data Relay Test Satellite 'KODAMA' (DRTS)," "Overview: 'KODAMA' Data Relay Test Satellite (DRTS) to Dramatically Extend Contact Time and Areas," and "DRTS: Data Relay Test Satellite"; JAXA, "Situation of 'Kodama' (Data Relay Test Satellite: DRTS)," press release, 2 June 2004; and NASDA, "Successful Inter-satellite Coummunications Experiments between DRTS (Kodama) and ADEOS-II (Midori-II)," press release, 21 February 2003, with all documents available online at www.jaxa.jp (accessed 1 July 2008). See also Kenji Numata and Yasuharu Tusenoka, "The Operation Concept of Data Relay Test Satellite Verification Experimental Space Network," NASDA and NEC Corporation (respectively), 1998, pp. 1–8; available online at track.sfo.jaxa.jp (accessed 29 August 2009); Andreas Rudolph and S. Hamer, "Envisat—Kodama (DRTS)—An JAXA/ESA Ka-Band Inter-Operability Demonstration Test," paper presented at the Envisat Symposium, Montreaux, Switzerland, 23–27 April 2007, available online at envisat.esa.int (accessed 30 August 2009); and "Space Agency Shows Satellite to Be Launched in September," *Japan Times*, 28 July 2002.

62. Interview, Yoshiaki Suzuki, executive director, Wireless Communications Department, NICT, Tokyo, 10 August 2005.

63. On *ETS-VIII/Kiku-8*, see Japan Aerospace Exploration Agency (JAXA), "ETS-VIII: Engineering Test Satellite-VIII "Kiku No. 8"; JAXA, "Deployment Result of the Large Deployable Antenna Reflectors of the ETS-VIII Kiku No. 8," press release, 26 December 2006; NICT, "Anomaly in the Communication Mission Equipment of the ETS-VII, or Kiku No. 8," press release, 2 February 2007; JAXA, "Geostationary Orbit Injection of the ETS-VII Kiku No. 8," press release, 9 January 2007; and "Gijutsū Shiken Eisei VIII GAta 'Kiku 8 Go' (ETS-VIII) no Saikin no Seika to Genjyō" [Recent Results and Condition of ETS-VIII Kiku-8], press release, 2 September 2009, pp. 1–15: with all documents available online at www.jaxa.jp (accessed 3 September 2009). See also Naokazu Hamammoto, Shigetoshi Yoshimoto, and Michito Imae, "Overview of the Engineering Test Satellite VIII (ETS-VII) Project," *Journal of the National Institute of Information and Communications Technology* 50(3/4), 2003, pp. 3–11; Craig Covault, "Advanced Satcoms Pace Asian Space Technology," *AWST*, 7 December 1998, p. 50; Paul Kallender, "Japan Develops Mobile System," *Space News*, 12 May 1997; and Paul Kallender, "Japan to Develop Next-Generation Gobal Satellite Communication Constellation," *Japan Press Network*, 13 May 1997. See also conferences such as those sponsored by MPHPT (now MIC), "Third International Forum on Advanced Satellite Communications in the Asia-Pacific Region," Tsukuba City, Japan, 19–20 February 2002 (conference abstract only), available online at www.soumu.go.jp (accessed 28 August 2008).

64. For *WINDS/Kizuna*, see Japan Aerospace Exploration Agency (JAXA), "WINDS: Wideband InterNetworking Engineering Test and Demonstrateion Satellite Kizuna"; and JAXA, "'KIZUNA' Carries out World's Fastest Satellite Data Communication at Speed of 1.2 Gbps," press release, 12 May 2008; JAXA, "WINDS Kizuna Successful Satellite Data Communication at Speed of 622 Mbps by the Active Phased Array Antenna," 16 May 2008. All documents are available online at www.jaxa.jp (accessed 3 September 2009). See also Paul Kallender, "Gigabit Satellite Gets Japan's Comets Mission," *Space News*, 22 June 1998; Frank Morring, Jr., "Winds in Position," *AWST*, 24 March 2008, p. 18; and Frank Morring, Jr., "High-Speed Satellite," *AWST*, 19 May 2008, p. 21. See also conferences such as those sponsored by the Kobe Chamber of Commerce and Industry, "Asian Workshop on Satellite Technology Data Utilization for Disaster Monitoring," Kobe, Japan, 20 January 2005 (conference abstract only), available online at www.aric.or.kr (accessed 28 August 2008).

65. On this front, see Johnson-Freese, *Space as a Strategic Asset*; James Clay Moltz, *The Politics of Space Security: Strategic Restraint and the Pursuit of National Interests* (Stanford, CA: Stanford University Press, 2008); Bruce M. DeBlois, Richard L. Garwin, R. Scott Kemp, and Jeremy C. Marwell, "Space Weapons: Crossing the U.S. Rubicon," *International Security* 29(2), 2004, pp. 50–84; and Michael Krepon (co-founder, The Henry L. Stimson Center), Testimony of Michael Krepon Before the House Committee on Armed Services Subcomittee on Strategic Forces, Space Security,

18 March 2009, available online at www.armedservices.house.gov (accessed 1 August 2009). For an earlier view, see The Aspen Strategy Group, *Anti-Satellite Weapons and U.S. Military Space Policy* (Lanham, MD: The Aspen Strategy Group/University Press of America, 1986).

66. Moltz, *Politics of Space Security*, p. 43.

67. DeBlois et al., "Space Weapons," esp. pp. 55–56. Virtually the same range had also been highlighted under the chairmanship of Donald Rumsfeld, *Report of the Commission to Assess United States National Security Space Management and Organization*, Pursuant to Public Law 106-65, 11 January 2001, esp. p. xiii; and Tom Wilson, "Threats to United States Space Capabilities," prepared for the Commission to Assess United Stated National Security Space Mangement and Organization, January 2001; both available online at www.fas.org (accessed 1 June 2009). According to The Aspen Strategy Group, *Anti-Satellite Weapons and U.S. Military Space Policy*, pp. 11–12, the United States did develop a crude nuclear-armed system as an ASAT weapon in the 1960s. However, it abandoned its development by the mid-1970s in favor of non-nuclear ASAT weaponry.

68. Jieitai Hō, (Shōwa 29, 6 Gatsu 9 Nichi Hōritsu 165 Go [Self-Defense Forces Act (Law No. 165 of 1954)]), [last revision 24 June 2009 by Law No. 55], Article 82.3, esp. ¶3, available online via e-Gov at www.e-gov.go.jp (accessed 1 August 2009) authorizes the destruction of "dandō misairu nado [ballistic missiles etcetera]," which suggests that non-missile objects (debris, satellite, etc.) can also potentially be targeted in situations in which there is an imminent threat to Japanese lives and property. With respect to the destruct order for the North Korean projectile, see the Ministry of Defense (MOD), "Dandō Misairu Nado ni tai suru Hakai Sochi ni Kan Suru Meirei ni Tsuite" [Orders Concerning the Destruct Measures for Ballistic Missiles and Other Objects], press release, 27 March 2009; and operational activity in MOD, "Order for Operation of the Self-Defense Forces Concerning Measures to Destroy Ballistic Missile or Other Objects," current trend and press release, 27 March 2009. See also Alex Martin, "SDF Gets Intercept Order," *Japan Times*, 28 March 2009; Jun Hongo, "What Did Japan's Response to the North's Rocket Prove?" *Japan Times*, 10 April 2009; and "Japan OKs Shoot-Down of Inbound N. Korean Rocket," *DefenseNews*, 27 March 2009, available online at www.defensenews.com (accessed 18 July 2009).

69. See Craig Covault, "Chinese Test Anti-Satellite Weapon," *Aviationweek.com*, 17 January 2007, available online at www.aviationweek.com (accessed 1 July 2007); Craig Covault, "Space Control: Chinese Anti-Satellite Weapon Test Will Intensify Funding and Global Policy Debate on the Military Uses of Space," *AWST*, 22 January 2007, p. 24; Mure Dickie, Stephen Fidler, and Demetri Sevastopulo, "Chinese Space Test Raises U.S. Suspicions," *Financial Times*, 20–21 January, 2007; "China Sets off a New Round of Star Wars," *Financial Times*, 20 January 2007; David A. Fulghum and Amy Butler, "Reassessing Space: U.S. Analysts Sort Through the Fall-

out from China's Satellite Shoot-down," *AWST*, 30 April 2007, p. 27; Ashley J. Tellis, "China's Military Space Strategy," *Survival* 49(3), 2007, esp. pp. 43–44; Vago Muradian, "China Attempted to Blind U.S. Satellites with Laser," *Defense News*, 22 September 2006, available online at www.defensenews.com (accessed 1 November 2007); William B. Scott and Michael J. Coumatos, "Would China Start a War in Space?" *AWST*, 7 January 2008; "Arms Race in Space?" *Japan Times*, 24 January 2007; and "Chugoku, Eisei no Hakai Jikken" [China's Satellite Destruction Test], *Ashai Shinbun*, 20 January 2007.

70. "Disharmony in the Spheres—The Militarisation of Space," *Economist*, 19 January 2008; Thom Shanker, "Missile Strikes a Spy Satellite Falling from Its Orbit," *New York Times*, 21 February 2008; Marc Kaufman and Josh White, "Spy Satellite's Downing Shows a New U.S. Weapon Capability," *Washington Post*, 22 February 2008; "Navy to Shoot Down Errant Spy Satellite with SM-3," *Defense Daily*, 15 February 2008; Graham Warwick, "Pentagon to Shoot Down Crippled Spy Satellite," *Flight International*, 19 February 2008; Marc Kaufman and Josh White, "Navy Missile Hits Satellite, Pentagon Says," *Washington Post*, 21 February 2008; Demetri Sevastopulo, "China Irate After Pentagon Brings Down Spy Satellite," *Financial Times*, 22 February 2008; and Edward Lundquist, "Hitting a Bullet with a Bullet—Missile Defense from the Sea," *Military Space & Missile Forum* 1(3), 2008, pp. 2–6, available online at www.kmimediagroup.com (accessed 31 August 2009.

71. Raytheon, "Raytheon Missile and Radar Played Critical Roles in Satellite Intercept," news release, 25 February 2008, available online at raython.mediaroom.com (accessed 1 August 2008); and also Robert Weisman, "For Raytheon, Missile Mission Turns into Global Showcase—Downing of Satellite Demonstrates Strengths of Company and Country," *Boston Globe*, 6 March 2008.

72. Johnson-Freese, *Space as a Strategic Asset*, pp. 115–116.

73. Ministry of Defense (MOD), "Japan's BMD," February 2009, available online via www.mod.go.jp (accessed 1 August 2009); Interview with Rear Admiral Tomohisa Takei, director general of operations and plans, Maritime Self-Defense Force (MSDF), "The Spear and Shield: Building International Cooperation Against the Ballistic Missile Threat—The U.S.-Japan Alliance and How the Standard Missile-3 Cooperative Development Program Will Improve International Ballistic Missile Defense Capabilities," *Defender* 5(1), 2009, pp. 2–9, available online at www.raytheon.com (accessed 30 August 2009); and Raytheon, "Raytheon Standard Missile-3 Block IIA Program Achieves Key Milestone," news release, 17 June 2009, available online at investor.raytheon.com (accessed 30 August 2009).

74. The general introduction here draws on Theresa Hitchens, "Space Wars—Coming to a Sky Near You?" *Scientific American*, 18 February 2008, available online at www.sciam.com (accessed 1 July 2008); Theresa Hitchens, Victoria Samson, and Sam Black, "Space Weapons Spending in the FY 2008 Defense Budget," 21 February 2007,

and also "Space Weapons Spending in the FY 2009 Defense Budget," March 2008, both available online from the Center for Defense Information at www.cdi.org (accessed 1 July 2008); Tellis, "China's Military Space Strategy," pp. 53–59; and Wilson, "Threats to United States Space Capabilities," esp. ¶4.

75. See Elaine M. Grossman and Keith J. Costa, "Small, Experimental Satellite May Offer More Than Meets the Eye," *Inside the Pentagon*, 4 December 2003, available online at www.globalsecurity.org (accessed 1 July 2008); Leonard David, "Military Micro-Sat Explores Space Inspection, Servicing Technologies," *Space.com*, 22 July 2005, available online at www.space.com (accessed 1 July 2008); and *Report of the Commission to Assess United States National Security Space Management and Organization*, pp. 20–21.

76. Theresa Hitchens, "Space Wars—Coming to a Sky Near You?"

77. E-mail correspondence, Theresa Hitchens, director, Center for Defense Information, Washington, DC, 8 May 2006.

78. See generally Ben Ianotta, "In-Orbit Satellite Servicing Designers Stick with Human Control," *Space News*, 16 May 2005; Paula Shaki Trimble, "Robotic Spacecraft May Service Future U.S. Spy Satellite," *Space News*, 25 October 1999; both available online at www.space.com (accessed 1 July 2008).

79. On Europe see, European Space Agency (ESA), "ConeXpress-Orbital Life Extension Vehicle (CX-OLEV), 18 June 2008, available online at telecom.esa.int (accessed 1 July 2009). For general information on Orbital Satellite Services see the official Web site at www.orbitalsatelliteservices.com (accessed 1 July 2008); and Orbital Satellite Services, "Orbital Satellite Services Establishes Its Corproate Headquarters in Sweden," News, 10 September 2008, available online at www.orbitalsatelliteservices.com (accessed 1 July 2009).

80. For the U.S., see Michael A. Dornheim, "Orbital Express to Test Full Autonomy for On-Orbit Service," *AWST*, 5 June 2006; "Brian Berger, "Fender Bender: NASA's DART Spacecraft Bumped into Target Satellite," *Space.com*, 22 April 2005; Brian Berger, "Pentagon Pulls Plug on Orbital Express," *Space News*, 23 July 2007; Jeremy Singer, "Air Force ANGELS: Satellite Escorts to Take Flight," *Tech Wednesday*, 30 November 2005; Leonard David, "Air Force Satellite Shows Off Rendezvous Skills," *Space.com*, 12 September 2005. All *Space News*, *Space.com*, and *Tech Wednesday* articles are available online at www.space.com (accessed 1 July 2008). See also the information on the "Orbital Express Space Operations Architecture," and on the "Front-End Robotic Enabling Near-Term Demonstrations (FREND)," available online at the official Web site of the Defense Advanced Research Projects Agency (DARPA) at www.darpa.mil (accessed 1 July 2008).

81. On Japan's movements, see the overview by Uchū Tsūshin Nettowaaku Guruupu [Smart Satellite Technology Group (SSTG)], "Kidōjyō Hozen Shisutemu ni Kan suru Kenkyū" [Research Related to OMS], available online at sstg.nict.go.jp (accessed

30 August 2009). For the background, see one of the earliest listed works on point by one of Japan's leading space robot scientists, Shinichi Kimura, who was subsequently responsible for Communications Research Laboratory's (CRL) *SmartSat* program. See, for example, Shinichi Kimura et al., "A Control Algorithm of Redundant Space Manipulator for Remote Inspection," 96-C-49 (paper presented at the 11th International Astrodynamics Symposium, Gifu, Japan, 1996), available online at www.kimura-lab.net (accessed 19 July 2009) (title and year only); Shinichi Kimura, "Orbital Maintenance System (OMS)," *Journal of the Institute of Electronics, Information and Communication Engineers* 82(8), 1999, pp. 820–823 (abstract only); and Shinichi Kimura et al., "Preliminary Experiments on Orbital Maintenance System Using Micro-LabSat," *Uchū Kagaku Gijutsu Rengō Kōenkai Kōenshū* 47, 2003, pp. 92–97 (abstract only); both available online at sciencelinks.jp (accessed 19 July 2009). See also Shinichi Kimura et al., "A Concept for a Robotic Servicing for the Next-Generation LEO System," *Proceedings of the 6th International Symposium on Artificial Intelligence and Robotics & Automation in Space: i-SAIRAS 2001*, Canadian Space Agency, St. Huber, Quebec, Canada, 18–22 June 2001, available online at robotics.estec.esa.int (accessed 19 July 2009).

82. See Paul Kallender, "ADEOS Loss May Force Redesign," *Space News*, 14 July 1997; Paul Kallender, "Failure of Adeos Fuels Debate on Size of Satellites," *Space News*, 7 July 1997; and Paul Kallender, "Japan's Adeos-2 Suffers Catastrophic Loss," *Space News*, 11 November 2003. All *Space News* articles are available online at www.space.com (accessed 1 July 2008).

83. Many of the same themes continued after CRL and the Telecommunications Advancement Organization (TAO) were merged as NICT in 2004, out of which the Smart Satellite Technology Group (SSTG) operates today.

84. For a general background on *ETS-VII/Kiku-7*, see "Japanese List Space Technology Goals," *AWST*, 22 August 1983; "Japanese Propose $2.3 Billion for Space Programs," *AWST*, 16 September 1991; Paul Kallender, "*ETS-VII/KIKU-7* Raises Japan's Hopes for Additional Funding," *Space News*, 31 March 1997; Paul Kallender, "Japan's Satellite Success Rests on Launch of *ETS-VII/KIKU-7*," *Space News*, 17 November 1997; Eiichiro Sekigawa and Michael Mecham, "NASDA Plans November Launch for *ETS-VII/KIKU-7*, TRMM," *AWST*, 14 April 1997; "Japan's National Space," *AWST*, 15 December 1997, p. 19; Craig Covault and Eiichiro Sekigawa, "Japanese H-2 Failure Ruins Satcom Research Mission," *AWST*, 2 March 1998; Eiichiro Sekigawa, "Budget Pressure Takes Toll on Hope Program," *AWST*, 13 July 1998, p. 55; "Japan's National Space Development Agency," *AWST*, 7 September 1998, p. 52; "Tokyo First to Achieve Unmanned Docking in Space," *BBC Monitoring Asia Pacific*, 7 July 1998; and "Japan Calls off New Attempt to Dock Orbiting Spacecraft," *Flight International*, 26 August 1998.

85. The basic description of the satellite is from JAXA, available online through www.jaxa.jp (accessed 1 July 2008). For the latest evaluations, see Kazuya Yoshida, "Engineering Test Satellite VII Flight Experiments for Space Robot Dynamics and

Control: Theories on Laboratory Tests Beds Ten Years Ago, Now in Orbit," *International Journal of Robotics Research* 22(5), 2003, esp. pp. 321–323. See also "Japan's *ETS-VII/KIKU-7* Spacecraft," *AWST*, 17 August 1998, p. 25.

86. See, for example, Sumiko Oshima, "Satellites to Perform World's First 'Cosmic Dance' Today," *Japan Times*, 7 July 1998.

87. For assessments of performance, see Paul Kallender, "Japan Turns to U.S. Satellite for Data Link to *ETS-VII/KIKU-7*," *Space News*, 30 March 1998; Paul Kallender, "*ETS-VII/KIKU-7* Problems Unresolved—Satellite Mission Could be Hindered by Computer Snafu," *Space News*, 9 February 1998; Paul Kallender, "*ETS-VII/KIKU-7* Achieved Goals Despite Glitches," *Space News*, 17 August 1998; Paul Kallender, "*ETS-VII/KIKU-7*—Orbital Rendezvous and Robotic Mission," *Japan Space Net*, 2 April 1997; and Paul Kallender, "H2 Launch Delay Imperils Japan's Robotic Mission," *Japan Space Net*, 19 November 1997. All *Space News* articles available online at www.space.com, with *Japan Space Net* articles at www.spacedaily.com, both accessed (1 July 2008). See also Paul Proctor, "Arm Wrestling," *AWST*, 30 March 1998; and Bruce A. Smith, "Experiments Completed," *AWST*, 8 November 1999, p. 21.

88. Some of the problems were as follows: Toshiba had installed the arm 30 degrees off-center from its correct position; after launch, the automatic sun-tracking function on the solar panel failed; right after launch it turned out that the satellite's attitude control was being maintained by thrusters rather than by the reaction wheel; the chaser satellites suffered a series of shutdowns; and there were serious intra-system noise problems that not only raised concerns about the satellite's integration but also its very ability to perform the docking and rendezvous missions.

89. The chaser satellite separated from the passive and stable target satellite by about 6 feet, and the chaser's laser attitude control system then also succeeded in uniting the pair. The second automatic docking test failed, partly because it was ambitious in increasing the chaser approach distance from a mere 2 meters the first time around to 520 meters. NASDA persisted and by the end of August 1998 was rewarded with a successful capture, with the chaser satellite 2,000 meters (6,560 feet) from the target.

90. See Japan Aerospace Exploration Agency (JAXA), "Engineering Test Satellite 'Micro-LabSat'—'Maikuro Rabusatto 1 Go ki—μLabsat [μLabsat: Micro Labsat]'"; and Aerospace Research and Development Directorate/JAXA, "Research on Micro Labsat 1," all available through JAXA at www.jaxa.jp (accessed 1 July 2008).

91. See also Paul Kallender, "Japanese Group Sees Important Role for Microsatellites," *Space News*, 29 April 2003, available online at www.space.com (accessed 1 July 2008); Eiichiro Sekigawa, "H-IIA Launch Puts Japan on Earth Watch," *AWST*, 23 December 2002; and "Young Engineers Construct Working Satellite," *Daily Yomiuri*, 6 June 2004.

92. See especially S. Kimura et al., "Preliminary Experiments on Technologies for Satellite Orbital Maintenance Using Micro-Labsat-I," *Advanced Robotics* 18(2), 2004,

pp. 117–138; and also Hisaya Watanabe, "Space-Environment-Conscious Satellite System" (presentation, MPT's Space Communications Policy Division, Tokyo, 23 April 1997).

93. This section relies primarily on an interview with Shinichi Kimura, NICT, Tokyo, 8 August 2005; and e-mail correspondences with Shinichi Kimura, SmartSat project leader, 2 April 2008 and 7 April 2008.

94. Additional background information is from SmartSat Technology Group, "The SmartSat-1 Experiment Concept," available online at sstg.nict.go.jp (accessed 30 August 2009); Shinichi Kimura et al., "SmartSat-1: On Orbit Experiment Plan Using Mini-Satellite," *Space Japan Review*, no. 48, 2006, pp. 1–4; and Shinichi Kimura et al., "Rendezvous Experiments on Smartsat-1," Space Mission Challenges for Information Technology (SMC-IT 2006), Pasadena, CA, 2006, pp. 374–379, available online at smc-it.jpl.nasa.gov (accessed 13 August 2008).

95. See generally "Mitsubishi Plans Bigger Small Experimental Satellite," *AWST*, 27 January 2003; and Paul Kallender, "Japanese Lab Forms Unit to Develop Microsatellites," *Space News*, 14 October 2002, available online at www.space.com (accessed 1 July 2008).

96. Strategic Headquarters for Space Policy (SHSP), *Uchū Kihon Keikaku*, esp. pp. 24–25.

97. Committee on Promotion of Outer Space Development and Use (CPSDU), "Uchū Kaihatsu Riyō ni kan suru Kihon Hōshin ni Tsuite [Concerning the Basic Policy on Space Development and Utilization], 15 January 2009, esp. pp. 6–7, available online at www.mod.go.jp (accessed 27 July 2009), pp. 6–7.

98. For *MUSES-C/Hayabusa*, see Japan Aerospace Exploration Agency (JAXA), "Asteroid Explorer 'Hayabusa' (MUSES-C)"; Institute of Space and Astronautical Science (ISAS), "Asteroid Explorer Hayabusa Information"; and JAXA, "Shōwakusei Tansaki 'Hayabusa' no Genzai no Jyōkyō ni Tsuite" [Concerning the Present Condition of Asteroid Probe 'Hayabusa'], press release, 4 February 2009; all available online through JAXA at www.jaxa.jp (accessed 1 July 2009). See also Jun'Ichiro Kawaguchi et al., "Hayabusa—Its Technology and Science Accomplishment Summary and Hayabusa-2," *Acta Astronautica* 62(10–11), 2008, pp. 649–647; Kazuko Shiibashi, "Nervous Descent," *AWST*, 28 November 2005, p. 36; and "New Hope for Return of Hayabusa Space Probe," *Daily Yomiuri*, 12 February 2009.

99. Tellis, "China's Military Space Strategy," p. 45.

100. See Christopher W. Hughes, "'Super-sizing' the DPRK Threat: Japan's Evolving Military Posture and North Korea," *Asian Survey* 49(2), 2009, esp. p. 293.

Chapter 6

1. For the background on the *H-IIB*, see Japan Aerospace Exploration Agency (JAXA), "H-IIB Roketto" [H-IIB Launch Vehicle], and the latest posted developments

at www.jaxa.jp (accessed 29 August 2009); "Kokusan Roketto Tooi Kido" [Remote Orbit for Domestic Rockets], *Asahi Shinbun*, 28 February 2008; Michael Mecham and Frank Morring, Jr., "Simplified Lift: JAXA Looks to H-IIB Design as Backup for HTV Launches," *Aviation Week & Space Technology* (hereafter, *AWST*), 28 November 2005, p. 66; and "Japan Unveils First HTV," *AWST*, 28 April 2008, p. 18.

2. Japan Aerospace Exploration Agency (JAXA)/Mitsubishi Heavy Industries (MHI), "H-IIB Roketto Shikenki no Uchiage ni Tsuite" [Concerning the Launch of the H-IIB Rocket Test Vehicle], press release, 8 July 2009.

3. See Japan Aerospace Exploration Agency (JAXA), "H-IIB Roketto Enjin Nenshō Shiken" [H-IIB Rocket Engine Combustion Tests], for the series of tests performed on the LE-7A and the LE-5B engines between 2007 and 2009.

4. As indicated, information in the remainder of this section draws on interviews with MHI officials, Tokyo, 5 August 2005.

5. See generally Michael Mecham, "Robotic Roll Call," *AWST*, 4 June 2007, p. 17; and Frank Morring, Jr., "First Principles: Spacefaring Nations, Scientists Set Guidelines for Exploration," *AWST*, 11 June 2007, p. 32.

6. Bradley Perrett and Kazuki Shiibashi, "Japanese Hope: More Than 20 Years in Development, Kibo Is Approaching Launches to Send It to ISS," *AWST*, 22 October 2007, p. 51; Kazuki Shiibashi, "Overtaken by Events: Part of Japan's ISS Effort Probably Won't Be Fully Used," *AWST*, 14 January 2008, p. 50; JAXA, "Launch Result of the Kaguya (SELENE) by the H-IIA Launch Vehicle No. 13 (H-IIA F13)," available online at www.jaxa.jp (accessed 24 June 2008); "Space Laureate," *AWST*, 17 March 2008, p. 45; and Kazuki Shiibashi, "Moonward Ho: Japan's Selene Spacecraft Marks the Start of an Intense Era of International Lunar Study," *AWST*, 24 September 2007, p. 83.

7. David Bond, "China First," *AWST*, 24 September 2007, p. 31; "Yang Liwei, China's First Astronaut in Space," and "President Hu Hails Successful Launch of Shenzhou V," both *Xinhua News Agency*, 15 October 2003, and both available online at www.china.org.cn (accessed 24 June 2008); and also Paul Kallender, "Official Calls for Japan to Establish Manned Space Program," *Space News*, 26 February 2003, available online at www.space.com (accessed 24 June 2008).

8. Following JAXA and other sources we too use *J-1* and *J-I* (or *J-2* and *J-II* etc.) interchangeably. For earlier criticisms of the *J-I*, see Paul Kallender, "Japan's NASDA Shoots for Advanced J-1 Launcher," *Space News*, 2 June 1997; Paul Kallender, "J-1 Updates May Involve Non-Japanese Suppliers," *Space News*, 26 January 1998; Paul Kallender, "Critical Report Hammers J-1 Program Costs," *Space News*, 18 May 1998; "Study Blasts Rocket Plans," *Daily Yomiuri*, 27 April 1998; Michael Meecham, "Japan Eyes Low-Cost J-1," *AWST*, 2 June 1997, p. 67; and Eiichiro Sekigawa, "Japanese Watchdog Agency Questions Need for J-1," *AWST*, 11 May 1998, p. 41.

9. Interview, Akihiro Fujita, director of STA's Aeronautics and Space Development Division, Tokyo, 6 May 1998; and Science and Technology Agency, "J-1 Roketto

Kairyōgata ni Tsuite" [Concerning an Improved Version of the J-1 Rocket] (Tokyo: STA, 16 May 1997), esp. table 1.

10. Joseph C. Anselmo, "Two Launches and Out?" *AWST*, 1 June 1998, p. 17; Joseph C. Anselmo, "Space Alliance?" *AWST*, 15 June 1998, p. 21; "State, Consortium to Link on Rocket Development," *Japan Times*, 4 December 2001; and Paul Kallender and Warren Ferster, "Japan Plans Russian, U.S. Parts for J1 Upgrade," *Space News*, 3 August 1998.

11. In combination with Chapter 3, additional updated information here on GALEX and the *GX* rocket is from Koji Sato and Yoshirou Kondou, "Overview of GX Launch Services by GALEX," *Acta Astronautica* 59(1–5), 2006, pp. 381–391; Galaxy Express Corporation, "A Case Study of International Cooperation—GX Rocket," prepared for AAS Seminar, 6 June 2005, available online at csis.org (accessed 27 July 2009); and JAXA, "LNG Propulsion System Project," available online at www.jaxa.jp (accessed 23 July 2008). See also Ed Kyle, "New Launchers—Galaxy Express GX," *Space Launch Report*, 8 May 2008, available online at www.spacelaunchreport.com (accessed 27 July 2009).

12. Interview, GALEX official, November 2002.

13. In an important move formalized in December 2006, Lockheed Martin and Boeing combined their expendable launch vehicle business by forming a joint venture called United Launch Alliance (ULA). According to the official press release, ULA combines the production, engineering, test, and launch operations associated with U.S. government launches of Boeing's Delta and Lockheed's Atlas rockets. ULA, not Lockheed Martin, was listed formally as part of the GALEX consortium. See Boeing, "Boeing and Lockheed Martin Complete United Launch Alliance Transaction," press release, 1 December 2006, available online at www.boeing.com (accessed 22 July 2008).

14. Sato and Kondou, "Overview of GX Launch Services," pp. 386–388; GALEX, "Case Study," slide 7.

15. Interviews, IHI officials, Tokyo, 2 September 2002 and 28 October 2002.

16. Unless otherwise indicated, the remainder of this section relies on Eiichiro Sekigawa and Michael Mecham, "Japan Shelves Support of Private GX Booster," *AWST*, 15 July 2002, p. 31; Eiichiro Sekigawa and Michael Mecham, "Mitsubishi Assumes Control of Japan's H-IIA Launcher," *AWST*, 2 December 2002, p. 51; "Lockheed Martin," *AWST*, 6 January 2003, p. 16; "Japan Will Develop Liquid Natural Gas-Fueled Second Stage Engine," *AWST*, 24 March 2003, p. 21; Keiko Chino, "GX Rocket Plans Up in the Air," *Daily Yomiuri*, 16 November 2006; Amy Svitak, "Proliferation Concerns Prompt U.S. Scrutiny of Launch Ventures," *Space News*, 23 September 2002; Paul Kallender, "Japan Revives Work on Galaxy Express Rocket Program," *Space News*, 24 March 2003; Paul Kallender, "Japan's Proposed Space Budget Would Reverse Years of Decline," *Space News*, 13 September 2005; "Editorial: Japanese Space Program

Needs Focus," *Space News*, 19 September 2005; "Editorial: Cancel the Galaxy Express Project," *Space News*, 30 October 2006; and Paul Kallender, "Japan's GX Rocket Debut Delayed Another Five Years," *Space News*, 30 October 2006. All *Space News* articles are available online at www.space.com (accessed 24 June 2008).

17. Michael P. Kleiman, "Looking for Responsive, Low-Cost Space Lift," *Air Force Print News Today*, 16 February 2006, available online at www.afmc.af.mil (accessed 23 July 2008).

18. Information on the Atlas launch vehicles is from the official Web site of Lockheed Martin, available online at www.lockheedmartin.com (accessed 23 July 2008). Additional information is from Craig Covault, "Atlas III Flight Tightens LockMart, Boeing Faceoff," *AWST*, 29 May 2000, p. 29; and Todd Halvorson, "First-Ever Atlas 3 Lifts Off," *Space.com*, 24 May 2000, available online at www.space.com (accessed 24 July 2008).

19. Interview, space analyst, Tokyo, 9 October 2002.

20. See, specifically, Jiyū Minshūto Seimu Chōsakai, Uchū Kaihatsu Tokubetsu Iinkai, *Aratana Uchū Kaihatsu Riyō Seido no Kōchiku ni Mukete: Heiwa Kokka Nihon Toshete no Uchū Seisaku (An)* [Policy Affairs Research Council of the Liberal Democratic Party, Special Committee on Space Development, Toward the Establishment of a New Space Development and Utilization System: Japanese Space Policy as a Peaceful Nation (Proposal)], September 2006, pp. 21–22; and also Sato and Kondou, "Overview of GX Launch Services," pp. 388, 391.

21. JAXA, *Jinkō Eisei, Roketto no Genjyō, Kadai, Tenbō* [The Present Condition, Challenges, and Prospects of Satellites and Rockets] (Tokyo: JAXA, 4 November 2008), p. 13, reference attachments pp. 8, 9; and SHSP Secretariat, *Wagakuni ni Okeru Jinkō Eisei-Roketto no Kaihatusu-Riyō Genjyō ni Tsuite* [The State of Development and Utilization of Japan's Satellites and Rockets] (Tokyo: SHSP Secretariat, 4 November 2008), p. 9.

22. Unless otherwise indicated, the information on the *ASR/Epsilon* draws heavily on Yasuhiro Morita et al., "Research on an Advanced Solid Rocket Launcher in Japan," pp. 1–6, and Atsushi Mayumi et al., "The Advanced Solid Rocket for Various Small-sat Missions," pp. 1–6, both papers presented at the 26th International Symposium on Space Technology and Science, 2–8 June 2008, Hamamatsu City, Japan, available online at www.senkyo.co.jp (accessed 27 July 2009). See also Morita Yasuhiro, "Saisentan no Gijutsū to Saiko no Chimu o Kessoku Shite" [Cutting-Edge Technology Unites With the Best Team], Interview No. 16, 17 February 2006; and Morita Yasuhiro, "Kotai Roketto no Kenkyū—Sekai Ichi kara Sekai Ichi e no Chōsen" [Solid Rocket Research: From the World's Number 1 to the Challenge of Staying the World's Number 1], October 2007, available online via www.jaxa.jp (accessed 7 June 2008).

23. Some sources suggest that the initial development costs were set at around $100 to $120 million. See Keiko Chino, "Rocket Plans Lack Direction," *Daily Yomiuri*,

5 August 2006; and Keiko Chino, "Lowering Cost Key to Post-M5 Rocket Success," *Daily Yomiuri*, 6 September 2007.

24. See Office of Science and Technology Policy (OSTP), Executive Office of the President, "U.S Space Transportation Policy: Fact Sheet, 6 January 2005," available online at www.ostp.gov (accessed 24 July 2008); Richard Dalbello, "Op ed: Putting the 'Operational' in Operationally Responsive Space," *Space News*, 10 April 2006, available online at www. space.com (accessed 24 July 2008); Kendall K. Brown, "Is Operationally Responsive Space the Future of Access to Space for the U.S. Air Force?" *Air & Space Power Journal* 20(2), 2006, pp. 11–18; and Les Doggrell, "Operationally Responsive Space: A Vision for the Future of Military Space," *Air & Space Power Journal* 20(2), 2006, pp. 42–49. On Japan's SOD initiative, see Norihiko Saeki, "COTS Policy & 'Space on Demand' in Japan" (Tokyo: METI, 29 October 2007).

25. In keeping with the SOD initiative, there may be moves toward developing both submarine- and air-launch capabilities. We know of no public submarine launch studies. However, the Institute for Unmanned Space Experiment Free Flyer (USEF), IHI, and ISAS have produced a preliminary study for a Japanese equivalent of the Minotaur LV. The research concept is for a family of vehicles 50 to 300 kilograms to a 500-kilometer LEO that can be launched from a Boeing 747 based on ISAS's *M-3S*, but possibly updating it with *M-V* and later technologies for a low-cost, fast-launch system. See, for example, Seiji Matsuda et al., "An Affordable Micro Satellite Launch Concept in Japan" (paper presented at the 6th Responsive Space Conference, Los Angeles, CA, 28 April–1 May 2008), pp. 1–6, available online at www.responsivespace.com (accessed 3 August 2009).

26. "Japan Mulls New Missile Defense System: Report," Reuters, 4 July 2009, available online at www.reuters.com; Bradley Perett, "Japan Considering THAAD Missile Defense," Aviationweek.com, 6 July 2009, available online at www.aviationweek.com; and "Japan Hints at Buying THAAD Missiles," UPI.com, 7 July 2009, available online at www.upi.com (all accessed 25 July 2009). The reports cited a Japanese newspaper, *Mainichi*, which had not specified its sources.

27. Michael O. Lavitt, "Cooperative TMD Proposal," *AWST*, 4 October 1993, p. 15; Michael A. Dornheim, "'Theater-Wide' Missile Defense Appealing, Controversial, Difficult," *AWST*, 3 March 1997, p. 62; Joseph C. Anselmo, Robert Wall, and Eiichiro Sekigawa, "Missile Test Extends North Korea's Reach," *AWST*, 7 September 1998, p. 56; Robert Wall, "U.S., Japan Agree on Cooperative Missile Defense," *AWST*, 23 August 1999, p. 46; and Eiichiro Sekigawa, "Strategy Confirmed," *AWST*, 29 September 2003, p. 60.

28. "Announcement of Strategic Defense Initiative—President Reagan," 23 March 1983, available online at www.mda.mil (accessed 28 July 2008).

29. Paul Mann, "Reagan Rules Out Ending SDI Efforts in Exchange for Soviet Missile Cuts," *AWST*, 30 September 1985, p. 93.

30. "President Announces Progress in Missile Defense Capabilities, Statement by the President," 17 December 2002, available online at www.whitehouse.gov (accessed 28 July 2008).

31. The basic data and information on BMD systems and related innovations discussed here is from the official site of the U.S. Department of Defense (USDOD) and Missile Defense Agency (MDA), *Missile Defense Worldwide—BMDS Booklet*, 5th ed. (Washington, DC: MDA, n.d.), available online at www.mda.mil (accessed 28 July 2008). The Strategic Defense Initiative Organization (SDIO) started in 1984, was redesignated the Ballistic Missile Defense Organization (BMDO) in 1993, and became the MDA in 2002. For consistency's sake, reference is made to only MDA throughout. See also general overviews and controversies about MD technologies in Joan Johnson-Freese, *Space as a Strategic Asset* (New York: Columbia University Press, 2007), pp. 114–131; Robert Wall, "Collision Course; Established Programs Seen Threatened by Kinetic-Energy Interceptor," *AWST*, 19 April 2004, p. 36; Robert Wall, "Shipping Out; Continued Upgrades Are Planned for Aegis, Standard Missile SM-3 Components," *AWST*, 28 June 2004, p. 51; and Amy Butler, "The Speed of Light; Directed Energy Is Just One of Many Technologies Being Developed for Missile Defense," *AWST*, 2 May 2005, p. 48.

32. Basic data and information on Japan and BMD discussed here is drawn from: Hideaki Kaneda, Kazumasa Kobayashi, Hiroshi Tajima, and Hirofumi Tosaki, *Japan's Missile Defense: Diplomatic and Security Policies in a Changing Strategic Environment* (Tokyo: The Japan Institute of International Affairs, March 2007), esp. pp. 53–82; Takahashi Sugio, "Briefing Memo: Emerging Missile Defense Issues," *The National Institute for Defense Studies News* 114, August and September 2007, pp. 1–7; Ken Jimbo, "A Japanese Perspective on Missile Defense and Strategic Coordination," *Nonproliferation Review*, Summer 2002, pp. 56–62; Andrew L. Oros, *Normalizing Japan: Politics, Identity and the Evolution of Security Practice* (Stanford, CA: Stanford University Press, 2008), pp. 149–169; and Michael D. Swaine, Rachel M. Swanger, and Takashi Kawakami, *Japan and Ballistic Missile Defense* (Santa Monica, CA: RAND Corporation, 2001).

33. Ministry of Defense (MOD), "Space-Related Defense Policies and Future Topics for Consideration" (Tokyo: MOD, November 2003), slide 2.

34. See Missile Defense Agency, *Missile Defense Worldwide*, esp. pp. 7–10.

35. Eiichiro Sekigawa, "Panel Urges Overhaul of Japan's Military," *AWST*, 22 August 1994, p. 59; Paul Proctor, "Missile Defense Quandary," *AWST*, 21 August 1995, p. 11; Paul Mann, "Economic Woes Shadow Japan's Missile Defense," *AWST*, 11 March 2002, p. 55; and "Mitsubishi to Sell Lockheed Missile," *Nikkei Weekly*, 28 November 1994, p. 9.

36. Nao Shimoyachi, "Japan Approves Plan for Missile Defense," *Japan Times*, 20 December 2003.

37. Brendan M. Greeley, Jr., "Army Missile Intercept Success Spurs SDI Theater Defense Study," *AWST*, 29 September 1986, p. 22.

38. "U.S. Suggests Japan Develop Native SDI," *Japan Economic Journal*, 2–9 January 1988.

39. "Mitsubishi Heavy Submits SDI-Related Research Bid," *Japan Economic Journal*, 13 August 1988; and "Multinational Teams," *AWST*, 21 November 1988, p. 13.

40. Michael O. Lavitt, "Cooperative TMD Proposal," *AWST*, 4 October 1993; and Takao Hishinuma, "U.S. to Press Japan to Build Missile Shield," *AWST*, 8 November 2002, p. 1.

41. See generally, David Hughes, "U.S., Japanese Team on Missile Defense," *AWST*, 28 November 1994, p. 22; "Mitsubishi to Sell Lockheed Missile," *Nikkei Weekly*, 28 November 1994, p. 9; and "Key Contract Won for Missile Defense," *Nikkei Weekly*, 13 September 1999, p. 6.

42. "New Defense Alliance Goes on Offensive—Mitsubishi Electric, Lockheed Martin Said to Be Seeking Contract on Missile-Interception System Pushed by U.S.," *Nikkei Weekly*, 7 September 1998, p. 7.

43. Paul Mann, "Missile Defense Boosted, Despite Weak Management," *AWST*, 26 October 1998, p. 34.

44. See U.S. Department of Defense, Memorandum for Correspondents, No. 134-M, 16 August 1999, available online at www.defenselink.mil (accessed 15 January 2008); and also James R. Asker, "Wedding Plans," *AWST*, 11 January 1999, p. 425.

45. "Japan Firms Picked for Missile-Defense Project," *Nikkei Weekly*, 10 January 2000, p. 9.

46. Sotaro Suzuki, "Industry Targets U.S. Missile Defense," *Nikkei Weekly*, 25 June 2001.

47. Robert Wall, "Japan Commits Funds for Missile Defense," *AWST*, 21 July 2003, p. 43.

48. "U.S. Calls on Japan to Shield It from Missiles," *Japan Times*, 17 May 2007; and "Change in Interpretation Urged to Allow Collective Self-Defense," *Japan Times*, 26 June 2008.

49. Michael Mecham, Frank Morring, Jr., and Robert Wall, "All Together: Japanese Forces Integrate Command Authority to Meet Regional Threats," *AWST*, 31 October 2005, p. 37.

50. "Japan Is Examining Ways to Ease," *AWST*, 24 November 2003, p. 17; and Eiichiro Sekigawa, "New Realities: In a Major Realignment, Japan Completes Its Defense Policy Review," *AWST*, 20 December 2004, p. 38; and "Japan, U.S. Agree on Joint Development of Ballistic Missile Defense System," *BBC Monitoring Asia Pacific*, 23 June 2006.

51. See "Japan's Dilemmas," *AWST*, 16 June 2008, p. 53; "The U.S. and Japan Have Completed," *AWST*, 13 March 2006, p. 18; "Japan Tests Aegis ABM System," *AWST*, 24 December 2007, p. 12; Henry S. Kenyon, "Japan Acquires Missile Defense Shield," *SIGNAL*, March 2008, available online at www.afcea.org (accessed 29 July 2008); and "Tokyo Defended," *AWST*, 7 April 2008, p. 18.

52. "Japan Tests Aegis ABM System," *AWST*, 24 December 2007, p. 12.

53. See, for example the remarks attributed to then JDA Director-General Shigeru Ishiba, who pronounced that it was worth considering whether Japan should have the capability for attacking even a missile base in a hostile country. "Spy Satellites for Peaceful Use," *Japan Times*, 3 April 2003; and David McNeill, "Japan Warns That It Will Attack If North Korea Aims Missile," *Independent* (London), 15 September 2003.

54. Craig Covault, "Space Control: Chinese Anti-Satellite Weapon Test Will Intensify Funding and Global Policy Debate on the Military Uses of Space," *AWST*, 22 January 2007, p. 24; and Amy Butler, David A Fulghum, and Craig Covault, "Missile Fallout: U.S. Insists SM-3 Is Geared for Missile Defense, Not Asat," *AWST*, 3 March 2008, p. 28.

55. Dave Ahearn, "Japanese May Contribute Technology to ABL; Purchase Might Be Considered," *Defense Daily*, 26 September 2007; David A. Fulghum and Robert Wall, "Assumption Angst," *AWST*, 13 February 2006, p. 32; "Japan Advised to Preempt North Korean Missile Attack," *Spacewar*, 25 March 2004, available online at www.spacewar.com (accessed 13 July 2008); "Japan to Consider Joint Study on Airborne Anti-Missile Laser System," *SpaceDaily*, 10 January 2005; and "Japan Develop Laser Weapons Amid North Korean Threat," *SpaceDaily*, 13 May 2007, both available online at www.spacedaily.com (accessed 13 July 2008).

56. Johnson-Freese, *Space as a Strategic Asset*, pp. 7, 101, 107, 125.

57. See, for example, Nao Shimoyachi, "Missile Shield Project Ignites Bidding War—Japan Defense Firms See 1 Trillion Yen Project as Chance to Build Industry," *Japan Times*, 18 November 2004; "Japan Defense Industry Buoyed by Plan to Build Advanced Patriot Missiles," *Nikkei Weekly*, 29 November 2004; and "Japan, U.S. Agree on Joint Radar Research," *BBC Monitoring Asia Pacific*, 5 April 2006.

58. The basic information on the key new space-related technologies discussed here relies on Yoshikazu Miyazawa, "Current Status of Japanese Aerospace Programs—Focusing on the High Speed Flight Demonstration" (Tokyo: JAXA, The Institute of Space Technology and Aeronautics, 2004), pp. 1–12; and Shinji Ishimoto et al., "Flight Demonstrator Concept for Key Technologies Enabling Future Reusable Launch Vehicles," *Acta Astronautica* 57(2–8) 2005, pp. 438–443.

59. In addition to the general information available through the Space Transportation System Research and Development Center (STSRDC) at JAXA that is available online at www.jaxa.jp (accessed 30 August 2009), see the general and Japan-specific information in Marco Antonio Caceres, "Many RLVs Still a Distant Dream," *AWST*, 15 January 2001, p. 148; Eiichiro Sekigawa, "Japan to Stress Reusable Vehicles," *AWST*, 8 August 1994, p. 64; Bruce A. Smith, "Space Unit Grounded," *AWST*, 18 January 1999, p. 17; Stanley W. Kandebo, "Japanese Making Rapid Strides in Hypersonic Technologies," *AWST*, 16 December 1991, p. 60; and "NASP Should Become Civilian Global Program to Survive, Japanese Specialist Suggests," *AWST*, 20 April 1992.

60. "First H-2 Payload Aids HOPE Shuttle," *AWST,* 31 January 1994, p. 53; Paul Kallender, "Japan Cuts Spending; HOPE Killed," *Space News,* 21 July 1997; Paul Kallender, "NASDA Withdraws HOPE-X Funding—Shelves Program for a Year to Give Money to More Pressing Project," *Space News,* 29 June 1998; Paul Kallender, "Launch Delay Opens Door for HOPE-X Tests," *Space News,* 17 August 1998; Eiichiro Sekigawa, "Japan Shuttles Forth, Even Without Hope," *AWST,* 16 September 2002, p. 63; and "Japan's Hope-X Mini-Shuttle Model Crashes During Test," *AWST,* 7 July 2003, p. 16. All *Space News* articles are available online at www.space.com (accessed 24 June 2008).

61. See Japan Aerospace Exploration Agency (JAXA), "NASDA History," and STSRDC's "Activities in the Past" for brief overviews of the main reentry projects, available online at www.jaxa.jp (accessed 30 August 2009).

62. See, for example, Koji Shimura et al., "Current Activities for Future Reusable Space Transportation Systems at MHI" (paper presented at the 56th International Astronautical Conference [IAC], Fukuoka, Japan, 17–21 October 2005), pp. 1–6.

63. Michael A. Taverna, "Europe Begins Plotting Strategy for RLVs," *AWST,* 12 April 1999, p. 73; Michael A. Taverna, "France May Join Japan's Hope-X," *AWST,* 7 February 2000, p. 40; Bruce A. Smith, "Added Cooperation," *AWST,* 6 November 2000, p. 23; Frank Morring, Jr., "Europe and Japan Have RLV Research Plans," *AWST,* 15 October 2001, p. 46; and Frank Morring, Jr., "Japan, France Start Spaceplane Tests," *AWST,* 16 June 2003, p. 49.

64. Japan Aerospace Exploration Agency (JAXA), "Flight Trial Result of Scaled Experimental Supersonic Transport (SST)," press release, 10 October 2005.

65. The brief survey below draws on the following: "Hypersonic Technology," *AWST,* 10 October 1988, p. 38; "Editorials: Restore National Aero-Space Plane Funds," *AWST,* 1 May 1989, p. 15; Stanley W. Kandebo, "Japanese Making Rapid Strides in Hypersonic Technologies," *AWST,* 16 December 1991, p. 60; Michael O. Lavitt, "Industry Outlook: Japanese Scramjet," *AWST,* 23 May 1994, p. 11; Michael A. Dornheim, "Australian Scramjet Flight Should Bolster Research Database," *AWST,* 5 August 2002; "'National Aerospace Initiative' Pushes Dual-Use Technology," *AWST,* 20 May 2002, p. 32; Ann Finkbeiner, "Hypersonics Redux," *AWST,* 30 January 2006, p. 510; Guy Norris, "Hyper Test," *AWST,* 23 July 2007, p. 23; Guy Norris, "Going Global," *AWST,* 14 July 2008, p. 128; and Paul Kallender, "Restoring Faith in Space—Profile: Keiji Tachikawa, President of Japan Aerospace Exploration Agency," *Space News,* 27 June 2005, available online at www.space.com (accessed 1 February 2008).

66. Interview, Takashi Ishikawa, director, Aviation Program Group, JAXA, "The Future of Environmentally Friendly Aviation Technology: Clean Engines and Supersonic Transport—Leading Japan's Aviation Industry," 22 February 2008, available online at www.jaxa.jp (accessed 30 July 2009).

67. Japan Aerospace Exploration Agency (JAXA), "Flight Test Results for Scramjet Engine (Follow-up Report #3)," press release, 19 April 2006.

68. See, generally, Michael Dornheim, "Missiles Lead Hypersonics Revival," *AWST*, 13 October 1997, p. 62; "Hypersonic Missiles Aimed at Deep Targets," *AWST*, 6 January 1997, p. 44; "'National Aerospace Initiative' Pushes Dual-Use Technology," *AWST*, 20 May 2002, p. 32; Douglas Barrie, "Speed Merchants: Latest British Scramjet Test Paves Way for Advanced Weapons Propulsion R&D," *AWST*, 3 April 2006, p. 32.

69. See Strategic Headquarters for Space Policy (SHSP), *Uchū Kihon Keikaku; Nihon No Eichi ga Uchū o Ugokasu* [Basic Space Plan: Wisdom of Japan Moves Space] (Tokyo: SHSP, 2 June 2009), pp. 19–20, 28–29. For a description and details of the *QZSS* see Japan Aerospace Exploration Agency, "Jyun Tenchō Eisei Shisutemu" [Quasi-Zenith Satellite System]; JAXA project manager Koji Terada, "Itsumo Nihon no Maue [Tenchō] ni aru Eisei ni Akusesu Kanō" [The Possibility of Constantly Accessing Satellites Directly Overhead (Zenith) Japan]; and also Interview, Koji Terada, "Establishment of a Seamless Positioning System," 20 October 2008. All JAXA items are available online at www.jaxa.jp (accessed 29 July 2009).

70. Bruce Smith, "Space-Based Positioning, Navigation & Timing Policy (The Tension Between Military and Civil Requirements)," *Strategy Research Report* (Carlisle, PA: United States Army War College, 15 March 2006), pp. 1–17; and also similar dual-use themes in "Japanese GPS Project Eyes Launch in FY08," *Nikkei Weekly*, 14 March 2005. See also some Japanese concerns relative to other Asian players such as China in Michael A. Taverna, "New Compass Heading," *AWST*, 28 April 2008, p. 34.

71. The White House, Office of the Press Secretary, "Joint Statement by the Government of the United States of America and the Government of Japan on Cooperation in the Use of the Global Positioning System," press release, 22 September 1998. For continuing cooperation, see Ministry of Foreign Affairs (MOFA), "United States–Japan Consultations on the Use of the Global Positioning System," joint announcement, 18 November 2004; and MOFA, "United States–Japan GPS Cooperation," joint announcement, 24 May 2007, both available online at www.mofa.go.jp (accessed 26 July 2008).

72. Interview, Hiroshi Kimura, senior executive director, Satellite Positioning Research Center (SPAC), 21 May 2007.

73. Interview, Teruhisa Tsujino, special researcher, National Institute of Science and Technology Policy (NISTEP), Ministry of Education, Culture, Sports, Science and Technology (MEXT), Tokyo, 21 September 2005.

74. Some of the historical trajectory and facts below rely on Paul Kallender, "Japanese Agency Thwarts GPS Plan," *Space News*, 9 September 1996; Paul Kallender, "Japan Resumes Research into Satellite Navigation," *Space News*, 14 April 1997; Paul Kallender, "Japan, Korea, Australia Mull Joint Satellite Plan," *Space News*, 7 December 1998; Paul Kallender, "Japanese Industry Pushes GPS Enhancement System," *Space News*, 30 September 2002; Paul Kallender, "Gearing Up for a Challenge," *Space News*, 3 March 2003; Paul Kallender, "Japan Aims for Operational Military Space

Systems by 2006," *Space News*, 25 August 2003; Paul Kallender, "Impasse Over Japan's QZSS System Persists," *Space News*, 1 November 2004; as well as documentation provided by Toshihiro Matsui, director of MEXT's Space Utilization division covered in Paul Kallender, "Japanese Quasi-Zenith Satellite System May Face Delays," *Space News*, 24 August 2004; Paul Kallender-Umezu, "Japanese Government Commits to Funding 1st of Three QZSS Satellites," *Space News*, 28 May 2007. All *Space News* articles are available online at www.space.com (accessed 1 August 2008). See also Frank Morring, Jr., "Bureaucratic Blockage," *AWST*, 4 July 2005, p. 17; James R. Asker, "Think Again," *AWST*, 18 October 2004, p. 25; Frank Morring, Jr., "Japanese GPS," *AWST*, 21 March 2005, p. 17; and Ivan G. Petrovski et al., "QZSS—Japan's New Integrated Communication and Positioning Service for Mobile Users," *GPS World*, June 2003, pp. 1–6 (online version), available online at www.gnss.co.jp (accessed 4 August 2009).

75. See, for example, interview with Dr. Hiroshi Kimura, president and CEO, Advanced Space Business Corporation (ASBC), *Space Japan Review*, No. 43, October/November 2005, pp. 6–7; and "Japan Looks to the Sky to Find Way Around the Ground," *Nikkei Weekly*, 18 March 1996.

76. Interview, Mitsugi Chiba, director, Space Policy Division, STA, Tokyo, April 1997.

77. On corporate interests, see also "Kajima to Develop Precision GPS System," *Japan Economic Journal*, 12 November 1988; Yasushi Nakata, "Mazda Unveils Navigation System," *Japan Economic Journal*, 23 June 1990; "Satellite-Based Navigation Booming," *Daily Yomiuri*, 24 May 1994; Sam Silverstein, "Melco Plans Telecommunications Satellite Fleet," *Space News*, 28 January 2002; "Private Sector Heightens Role in Space," *Nikkei Weekly*, 5 August 2002; Keiko Chino, "Govt Loses Way on GPS Bill," *Daily Yomiuri*, 15 July 2006; Keiko Chino, "Space Review Must Be Clear-Cut," *Daily Yomiuri*, 28 February 2006; and Keiko Chino, "Interest Grows in Space Program," *Daily Yomiuri*, 3 October 2006.

78. Keidanren, "Sōgōteki na Uchyū Kaihatsu-Riyō Sesisaku no Kakuritsu to Uchyū Sangyō no Kiban Kyōka-Sangyōka no Suishin" [Toward the Establishment of Comprehensive Space Development-Utilization Policies, the Strengthening of the Space Industry Base, and the Promotion of Industrialization], policy proposal, 6 July 1999, ¶3.3.2; Keidanren, "Jidai no Sangyō no Kiban Zukuri ni Muketa Kenkyū Kaihatsu no Suishin ni Tsuite—Jyūten Kenkyū Kaihatsu Kadai ni Tsuite" [About the Promotion of Research and Development Oriented Toward the Construction of the Frontier Industry Base: Concerning the Challenges in the Main Research and Development], policy proposal, 21 May 2002, ¶4; Nippon Keidanren, "Oshirase: 'Jyun Tenchō Shisutemu Suishin Kentōkai' no Shinsetsu ni Tsuite" [Concerning the Newly Established "Promotion and Investigative Committee for the QZSS"], Announcement, July 2002; and Nippon Keidanren, "'Jyun Tenchō Eisei Shisutemu Suishin

Kentōkai' no Katsudō ni Tsuite—Jyun Tenchō Eisei Shisutemu to wa" [Concerning the Activities of the Promotion and Investigative Committee for the QZSS—The QZSS System], policy proposal, August 2002.

79. Hideto (Duke) Takahashi, "Japanese Regional Navigation Satellite System 'The JRANS Concept,'" *Journal of Global Positioning Systems* 3(1), 2004, pp. 259–264.

80. Satellite Positioning Research and Application Center (SPAC), "SPAC ni Tsuite" [About SPAC], available online at www.eiseisokui.or.jp (accessed 27 July 2009); and Shigeru Matsuoka, "Service Status of QZSS" (paper presented to the 15th session of the Asia-Pacific Regional Space Agency Forum [APRSAF-15], Hanoi & Ha Long Bay, Vietnam, 10 December 2008), slides 1–24, available online at www.aprsaf.org (accessed 29 July 2009).

81. Melco, "Mitsubishi Electric to Exhibit at Telecom 2003," press release, 7 October 2003, available online at global.mitsubishielectric.com (accessed 29 July 2009); and Makoto Asaba (Melco), "How Can SatCom Reduce the Broadband Access Cost?" Space Systems Department Kamakura Works No. MEL-SE-05-027, paper presented [APRSAF-12], Kitakyushu, Japan, 11–13 October 2005, esp. slides 21–24.

82. Interview, Hiroshi Kimura, senior executive director, Satellite Positioning Research Center (SPAC), 21 May 2007.

83. On the government roles, see Tsutomu Shigeta and Takashi Miyoshi, "NASDA's Activities and Roles in Promoting Satellite Utilization Experiments," *Acta Astronautica* 54(3) 2004, pp. 161–162; Hideto (Duke) Takahashi, "Japanese Regional Navigation Satellite System 'The JRANS Concept,'" p. 259; Institute for Unmanned Space Experiment Free Flyer (USEF), "ASER," available online at www.usef.or.jp (accessed 29 July 2009); and Japan Aerospace Exploration Agency (JAXA), "Jyun Tenchō Eisei Shisutemu" [Quasi-Zenith Satellite System], available online at www.jaxa.jp (accessed 29 July 2009).

84. See Geographical Survey Institute (GSI), "Policy Planning" for the text of the law, available online at www.gsi.go.jp (accessed 28 July 2009); Hiroshi Murakami, "New Legislation on NSDI in Japan: 'Basic Act on the Advancement of Utilizing Information,'" *Bulletin of the Geographical Survey Institute* 55, March 2008, pp. 1–2; and Satellite Positioning Research and Application Center (SPAC), "SPAC ni Tsuite" [About SPAC], available online at www.eiseisokui.or.jp (accessed 27 July 2009).

85. Interview, Satoshi Tsuzukibashi, deputy director of Space and Technology Policy and Office of Defense Production Committee, Nippon Keidanren, Tokyo, 23 May 2007.

86. Matsuoka, "Service Status of QZSS," slide 1.

87. Interview, Kanae Kurata, unit chief, Space Development and Utilization Division, Research and Development Bureau, MEXT, Tokyo, 28 April 2007.

88. Petrovski et al., "QZSS—Japan's New Integrated Communication and Positioning Service," p. 1.

89. Unless otherwise indicated, the following draws on Frank Morring, Jr., "Japanese Recce," *AWST*, 24 January 2005, p. 15; Paul Kallender-Umezu, "Japan Plans to Launch 3 Satellites," *Defense News*, 13 September 2004, p. 28; "Japan Eyes Production of Smaller 4th-Generation Spy Satellite," *Xinhua General News Service*, 10 January 2005; Paul Kallender, "Japan Contemplates Second-Generation Surveillance Craft," *Space News*, 16 September 2002; and Paul Kallender-Umezu, "Japan Mulls Options Following Radar Satellite Failure," *Space News*, 9 April 2007. All *Space News* articles are available online at www.space.com (accessed 1 August 2008).

90. Interview, Yasuhiro Itakura, research officer, CSIC, Tokyo, 30 August 2005; and Paul Kallender, "Japan's Proposed Space Budget Would Reverse Years of Decline," *Space News*, 12 September 2005.

91. Interview (follow-up), Yashuhiro Itakura, research officer, CSIC, Tokyo, 19 April 2006.

92. Unless otherwise indicated, all background information and details on the various technology experiments in this section are from JAXA, available online at www.jaxa.jp, and checked where possible against those in *Encylopedia Astronautica*, available online at www.astronautix.com; and NASA's National Space Science Data Center (NSSDC), available online at nssdc.gsfc.nasa.gov (all accessed 27 August 2009).

93. Yasuhiro Morita et al., "Demonstrator of Atmospheric Reentry System with Hyperbolic Velocity—DASH," *Acta Astronautica* 52(1), 2003, pp. 29–39; Yasuhiro Morita et al., "Development Status of the High-Speed Reentry System—DASH," *Acta Astronautica* 53(12), 2003, pp. 971–981; and Japan Aerospace Exploration Agency (JAXA), "Kōsoku Saitotsunyū Jikkenki (DASH) no Kekka ni Tsuite" [Concerning the DASH Results], press release, 4 February 2002.

94. JAXA, "Catalogue of ISAS Missions: EXPRESS—Mission Profile," available online through ISAS at www.jaxa.jp (accessed 27 August 2009); and Takashi Abe et al., "Reentry Technology Experiment on the First Mission of Reentry Capsule 'Express,'" *International Symposium on Space Technology and Science* (18th), vols. 1 and 2, Kagoshima, Japan, 17–22 May 1992, pp. 1383–1388 (abstract only).

95. JAXA, "Past Projects: Orbital Re-entry Experiment 'OREX,'" available online at www.jaxa.jp (accessed 27 August 2009); JAXA, "Activities in the Past: OREX," available online through STSRDC at www.jaxa.jp (accessed 27 August 2009); and "First H-2 Payload Aids Hope Shuttle," *AWST*, 31 January 1994, p. 53.

96. See, for example, Selig S. Harrison, "Missile Capabilities in Northeast Asia: Japan, South Korea and North Korea," in appendix III, Unclassified Working Papers, *Rumsfeld Commission Report*; and Shaun Burnie, research director, speech (Greenpeace International Nuclear Program, Foreign Correspondents' Club of Japan, 2 August 2005). Based on JAXA information, we would add that OREX in fact demonstrated a wide variety of technologies under the reentry technology rubric. These included

accurate reentry from orbit using GPS and an onboard computer to control the vehicle and the firing of retro-rockets; aerodynamic heating and testing of heat-resistant materials, including ceramic fiber-hardened heat-resistant tiles and a very thin (4-millimeter-thick) carbon-carbon nose cone; and measurements of drag and aerodynamic characteristics and altitude determination. As discussed, OREX is in fact only one of a series of reentry experiments and technology demonstrators for warhead reentry systems conducted under other technology programs (such as RLVs) with a variety of stated goals over the last fifteen years.

97. JAXA, "Past Projects: Hypersonic Flight Experiment 'HYFLEX,'" available online at www.jaxa.jp (accessed 27 August 2009); JAXA, "Activities in the Past: HYFLEX" available online through STSRDC at www.jaxa.jp (accessed 27 August 2009); and Michael Mecham, "Last-minute Sinking of HYFLEX Mars J-1 Debut," *AWST*, 19 February 1996, p. 27.

98. Institute for Unmanned Space Experiment Free Flyer (USEF), "USERS," available www.usef.or.jp (accessed 1 August 2008); Koichi Ichiji et al., "Unmanned Space Experiment Recovery System (USERS)—Establishment of Standardized Bus System and the Self-Return from the On-Orbit Operation," IEICE Transactions on Communications, J88-B(1), 2005, pp. 117–130 (abstract only); Koichi Ijichi et al., "The USERS System Design & On-Orbit Requirements," 1998, available online at track.sfo.jaxa.jp (accessed 27 July 2009); Ulrich M. Schottle, "Overview on Past and Present Projects in Germany to Flight-Test Reentry Technologies" (USERS Workshop, Tokyo, Japan, 8 October 2004), esp. slides 10–12 (on USERS); "First Japanese Reusable Spacecraft to Carry Science, Technology Payload," *AWST*, 20 August 1990, p. 70; Craig Covault, "Japanese Technology Missions Readied," *AWST*, 7 December 1998, p. 55; and "Satellite Returns to Earth Successfully," *Daily Yomiuri*, 31 May 2003.

99. Jun'ichiro Kawaguchi et al., "MUSES-C, Its Launch and Early Orbit Operations," *Acta Astronautica* 59(8–11) 2006, pp. 669–678; Institute of Space and Astronautical Science/Japan Aerospace Exploration Agency, "Hayabusa Arrived at Itokawa," 14 September 2005, available online at www.jaxa.jp (accessed 29 July 2009); and Kazuki Shiibashi, "Journey Begins; Crippled Japanese Asteroid Hunter on Its Way Back to Earth in 2010," *AWST*, 30 April 2007. For information on the capsule, see Jun'ichiro Kawaguchi, "Hayabusa's Return Journey to Earth—The Final Stage: The Age of Solar Exploration," 24 May 2007, available online at www.jaxa.jp (accessed 29 July 2008); and National Space Science Data Center, "Hayabusa: NSSDC ID 2003-019A," available online at nssds.gsfc.nasa.gov (accessed 27 August 2009).

100. Unless otherwise indicated, the background and information draws on N. Usui, "U.S., Japan Discuss Sharing Missile Warnings," *Space News*, 23–29 January 1995; "Collective-Defense Ban Seen Keeping Japan out of Missile First-Alert Loop," *Japan Times*, 10 June 2005; Martin Sieff, "U.S. Japan to Integrate BMD IT Networks," *Washington (UPI)* 18 January 2006; Tetsuo Hidaka and Koichi Yasuda, "Better Spy

Satellite System Needed; Reliance on U.S. Intelligence on Missile Launch Shows Need for Improvement," *Daily Yomiuri*, 31 July 2006; and Taylor Dinerman, "Is the Japanese-U.S. Missile Defense Program Changing the Asian Military Balance?" *Space Review*, 10 April 2006, available online at www.thespacereview.com (accessed 31 January 2008).

101. See Strategic Headquarters for Space Policy (SHSP), *Uchū Kihon Keikaku* [Basic Space Plan] (Tokyo: SHSP), pp. 5–6, 20–21, and 27. See also Paul Kallender-Umezu, "North Korean Missile Test Puts Focus on Japanese Space Policy," *Space News*, 13 April 2009; and Paul Kallender-Umezu, "Missile Warning System at Forefront of Japan's New Space Policy," *Space News*, 8 June 2009.

102. The following draws on an interview, Masanori Homma, director, Program Management and Integration Department, Office of Space Applications, JAXA, Tokyo, April 2006.

103. Interview, Japanese government official (anonymity requested), Tokyo, 25 April 2006.

104. Richelson, *America's Space Sentinels*, pp. 7–8, 65–66, 237 (appendix A); and Johnson-Freese and Gatling, "Security Implications of Japan's Information Gathering Satellites (IGS) System," pp. 544–545.

105. E-mail correspondence, Norihiko Saeki, deputy director, Space Industry Office, METI, Tokyo, 17 July 2009.

106. For *ASNARO*, see Institute for Unmanned Space Experiment Free Flyer (USEF), "ASNARO," available online at www.usef.or.jp (accessed 27 July 2009).

107. Additional information here is from Paul Kallender-Umezu, "Japan Moving Ahead with Smaller Earth Imaging Satellites," *Space News*, 10 August 2009; Paul Kallender-Umezu, "Missile Warning System at Forefront of Japan's New Space Policy," *Space News*, 8 June 2009; and Goodrich Corp., "Goodrich to Support Japan's Next Generation Advanced Observation Satellite," 11 February 2009, available online at phx.corporate-ir.net (accessed 27 July 2009).

108. Saeki, "COTS Policy & 'Space on Demand' in Japan," esp. slides 10, 14–18.

109. Ministry of Defense (MOD), *Nihon no Bōei* (Tokyo: MOD, 2009), p. 105; CPSDU, "Uchū Kaihatsu Riyō ni kan suru Kihon Hōshin ni Tsuite [Concerning the Basic Policy on Space Development and Utilization], 15 January 2009, pp. 1–13, available online at www.mod.go.jp (accessed 27 July 2009); Committee on Promotion of Outer Space Development and Use (CPSDU), "Uchū Kaihatsu Riyō ni kan suru Kihon Hōshin ni Tsuite (Gaiyō)" [Concerning the Basic Policy on Space Development and Utilization (Outline)], 15 January 2009, available online at www.mod.go.jp (accessed 27 July 2009); and MOD, "Space-Related Defense Policies and Future Topics for Consideration" (Tokyo: MOD, November 2008), slides 2–7. See also Paul Kallender-Umezu, "Japan Military Space Guidelines Identify Capabilities but Lack Planning Specifics," *Space News*, 16 February 2009.

110. For a general overview, see "The Space Race Now Includes Engineering Students from Six Universities in Japan," *International Herald Tribune (Herald Asahi)*, 5 December 2007; N. Gopal Raj, "Satellites Built by Universities," *Hindu*, 6 May 2008; Leonard David, "Cubesat: Tiny Spacecraft, Huge Payoffs," *Tech Wednesday*, 8 September 2004, available online at www.space.com (accessed 25 March 2006); Paul Kallender, "Failure of Adeos Fuels Debate on Size of Satellites," *Space News*, 7 July 1997; Paul Kallender, "Japanese Lab Forms Unit to Developing Microsatellites," Space News, 14 October 2002; and Paul Kallender, "Japanese Group Sees Important Role for Microsatellites," *Space News*, 29 April 2003. All *Space News* articles are available online at www.space.com (accessed 25 March 2006).

111. The work of UNISEC members throughout Japan can be conveniently tracked at the NPO's Web site at www.unisec.jp (accessed 28 July 2009), which is used for the brief description below.

112. Interview, Associate Professor Shinichi Nakasuka, Department of Aeronautics and Astronautics, University of Tokyo, 25 March 2003. Nakasuka is a founding member of the Space Open Laboratory (SOL) and a prime mover behind UNISEC and especially the miniaturization of satellites.

113. See the description of the lab's work at lss.mes.titech.ac.jp (28 July 2009).

114. General background is from Keiko Chino, "Minisatellite Project Blasts Off—METI Hopes Small, Cheap Satellites Can Boost Domestic Space Industry," *Daily Yomiuri*, 15 December 2007; and Michael Mecham, "Smallsats Open Regional Doors," *AWST*, 7 December 1998, p. 70.

115. On JAXA and STDRC, see Hidekazu Hashimoto (director, STDRC), "Our Missions: STDRC—'Micro-LabSat,'" available online at www.jaxa.jp (accessed 27 July 2009); and Matsuaki Kato et al., "Road Map of Small Satellite in JAXA" (paper presented at the 56th International Astronautical Conference [IAC], Fukuoka, Japan, 17–21 October 2005), pp. 1–7.

116. Tomoaki Toda et al., "Design of Onboard Communication Systems for Formation Flight SCOPE Mission" (paper presented at the 56th IAC, Fukuoka, Japan, 17–21 October 2005), pp. 1–2 (abstract only).

117. Shin-ichiro Nishida et al., "Development Status of an Active Space Debris Removal System" (paper presented at the 56th IAC, Fukuoka, Japan, 17–21 October 2005), pp. 1–7; and United Nations Office for Outer Space Affairs, "Report on Space Debris Related Activities in Japan (for UNCOPUOS/STSC)," February 2009, available online at www.oosa.unvienna.org (accessed 27 July 2009).

118. See Kazuki Hayashi (MHI), "A New Approach for Spacecraft Rendezvous and Fly-around Control on Elliptic Orbits" (paper presented at the 56th IAC, Fukuoka, Japan, 17–21 October 2005), pp. 1–2, 11; and Toshiyuki Nakamura et. al., "The *μ-LabSat* Design and Development Status in Japan," *Proceedings of the Euro-Asia Space Week on Cooperation in Space—"Where East & West Finally Meet,"* 23–27 November 1998, Sin-

gapore (ESA SP-430, February 1999), available online via the Smithsonian Astrophysical Observatory (SAO)/NASA Astrophysics Data System (ADS) at adsabs.harvard.edu (accessed 29 August 2009).

119. Keiichi Okuyama, faxed correspondence No. 4443, KHI Gifu Works, Defense Systems Design Department, 20 March 2003.

120. See generally Larry Dickerson, "UAVs on the Rise," *AWST*, 15 January 2007; David Bond, "Hawk Excursion," *AWST*, 13 March 2006, p. 23; David A. Fulghum, "Space-RAAM," *AWST*, 21 May 2007, p. 31; and Bill Sweetman, "Rising Ambitions," *AWST*, 18 February 2008, p. 70.

121. David A. Fulghum, "Japan's Dilemmas," *AWST*, 16 June 2008, p. 53.

122. "Yamaha Tied to Sensitive China Exports," *Japan Times*, 24 January 2006; "Yamaha Linked to Firms with Ties to China Army," *Japan Times*, 29 January 2006; "Yamaha Helped METI Craft Export Rules," *Japan Times*, 25 February 2007; and "Yamaha Motor Gets Chopper Sale Ban," *Japan Times*, 12 May 2007.

123. Edward H. Phillips, "UAV Decision," *AWST*, 16 May 2005, p. 17; and Sunho Beck, "Missile Watch—Japan Is Working on Its Own Solution to Long-Endurance Surveillance of North Korea," *AWST*, 18 February 2008, p. 42.

124. Although the aircraft has not been named, it is reportedly part of the Future Unmanned Aircraft Systems Study being conducted by MOD's Technology Research and Development Institute, and possibly has links to another research effort entitled the Endurance Unmanned Aircraft Systems Technology Project.

125. Keiko Chino, "Minisatellite Project Blasts Off—METI Hopes Small, Cheap Satellites Can Boost Domestic Space Industry," *Daily Yomiuri*, 15 December 2007.

Chapter 7

1. See, for example, Johnson-Freese, *Space as a Strategic Asset*, p. 30; also Joan Johnson-Freese, Naval War College, Testimony Before the U.S.-China Economic and Security Review Commission: China's Military Modernization and Cross-Strait Balance," 15 September 2005.

2. Johnson-Freese, *Space as a Strategic Asset*, pp. 33–34 and esp. table 2.1.

3. See Pyle, *Japan Rising*, pp. 24, 41–65, for a historical overview; but also see the richness and diversity in contemporary domestic security policy preferences across neoautonomists, normal nationalists, pacifists, and middle-power internationalists as depicted by Samuels, *Securing Japan*, pp. 109–132.

4. For a brief summary of the hitherto self-binding restrictions on Japan's security policymaking, see Christopher Hughes, *Japan's Re-Emergence*, pp. 31–40, esp. table 1; and Kenneth B. Pyle, "Author's Response: The Primacy of Foreign Policy in Modern Japan," *Asia Policy*, no. 4, July 2007, p. 209. The strictures include the following: not to acquire power projection capabilities, not to dispatch SDF forces abroad, not to engage in collective defense, not to exceed one percent of GNP on defense expenditures,

not to export arms, not to share military technology, not to breach the non-nuclear principles, and (of specific interest to us) not to violate the peaceful uses of space.

5. See, most recently, Andrew L. Oros, *Normalizing Japan*, pp. 1–21. The central tenets generally include no traditional armed forces, no use of force except in self-defense, and no participation in foreign wars.

6. 1947 Japanese Constitution, Chapter II (Renunciation of War), Article 9.

7. In early 1946 during the extraordinarily swift and entirely American "constitutional convention" responsible for the text of the new constitution, the drafters of Article 9, principally Colonel Charles Kades, reasoned that every nation had to have the right to preserve its own security against internal and external threats—a right which left the possibility of rearmament in the service of that goal a distinct and from then on a controversial possibility. See John W. Dower, *Embracing Defeat: Japan in the Wake of World War II* (New York: W.W. Norton & Company/The New Press, 1999), esp. p. 369.

8. Unless otherwise indicated, the constitutional discussion draws on a combination of the following: Richard J. Samuels, "Politics, Security Policy, and Japan's Cabinet Legislation Bureau: Who Elected These Guys, Anyway?" JPRI Working Paper No. 99, March 2004, pp. 1–19; Christopher W. Hughes, "Why Japan Could Revise Its Constitution and What It Would Mean for Japanese Security Policy," *Orbis* 50(4), 2006, pp. 725–744; J. Patrick Boyd and Richard J. Samuels, *Nine Lives? The Politics of Constitutional Reform in Japan*, Policy Studies 19 (Washington, DC: East-West Center, 2005), pp. 1–77; Richard J. Samuels, "Japan's Goldilocks Strategy," *Washington Quarterly* 29(4), Autumn 2006, pp. 111–127; and Asahi.com, "Self-Defense Forces: New 'Basic Law on Peace and Security' Should Define the Role of the SDF," 23 May 2007, available online at www.ashai.com (accessed 20 June 2007).

9. The first formal interpretation paved the way for subsequent confusion on pinning down the quantitative measure of "war potential," because Article 9 was construed as not banning military capabilities that fell short of the ability to "conduct modern warfare" and that could also be used to defend Japan from direct attack. The second formal interpretation reconstrued Article 9's ban on "war potential" to any military capability that was in excess of the "minimum necessary level" to defend Japan from direct attack—thereby limiting force levels to those necessary to defend the homeland and also limiting the use of force to the defense of the homeland. Textual controversy persists on the "minimum necessary level" standard to this day. Meanwhile, real-world applications of this limiting standard are becoming even more confusing, such as the JSDF transporting armed American troops (and possibly additional weapons) in the Iraq war or flying jointly produced, U.S.–Japan F-2s directly from Northern Japan to Guam for 1,700 miles without refueling to carry out live-bombing runs as part of its annual exercises with the U.S. Air Force. See, for example, Norimitsu Onishi, "Bomb by Bomb, Japan Sheds Military Restraints," *New York Times*, 23 July 2007. Overall, as put succinctly by former Prime Minister Koizumi,

who called for squarely confronting the reality of the JSDF with the text of Article 9 in constitutional revision, it is meaningless to continue to think of a JSDF without war potential. See "Jieitai, Senryōku De Nai to iu Kangae Wa 'Mekura'" [It Would Be 'Blind' Not to Think of SDF as War Potential], *Asahi Shinbun*, 22 July 2007.

10. "Towareru Kenpō to Jietai" [Questioning the Constitution and the SDF], *Asahi Shinbun*, 10 July 2007. Under ad hoc laws, Japanese troops have been dispatched abroad to full-blown combat areas, and the dawning necessities of having to defend U.S. personnel and assets in military operations overseas that also affect Japanese national security are not lost on anyone. The strict self-defense posture of the JSDF has, in incremental steps, moved from protection of the homeland to engagement at the regional—and from there, exercise at the global, stage. At the regional level, a 1996 regional declaration, affirming the U.S.-Japan security alliance transformed into law through the 1999 U.S.-Japan Defense Guidelines, allows Tokyo to expand its security role from the homeland to "areas surrounding Japan." Similarly, a series of legislative measures have been behind the out-of-area deployment of JSDF since the early 1990s, which have brought Japanese forces closer to the heart of military conflict around the world: The International Peacekeeping Peace Cooperation Law of June 1992, enabled the dispatch of JSDF on limited noncombat peacekeeping missions of the United Nations; the Antiterrorism Special Measures Law of October 2001 enabled the dispatch of JSDF to the Indian Ocean for noncombat support of U.S.-led operations in Afghanistan; and the Iraqi Reconstruction Law in July 2003 enabled the dispatch of JSDF on the ground in Iraq on noncombat reconstruction missions in support not just of the United States but also the international community's war on terrorism.

11. Given the uncertainties regarding the extent to which JSDF can engage abroad under existing interpretations (whether collective self-defense, in which Japan uses force to aid an ally like the United States on a case-by-case basis in its own self-defense, or collective security, in which Japan aids other countries under the mandate of the United Nations), the government has set up a panel to clarify the scenarios under which Japan can exercise force under its right of collective self-defense. At least one government panel has already suggested that Japan needs to change its exclusively defense-only posture and to drop the collective self-defense posture altogether. See on these debates the following sources: "Shūdan Anzen Hoshō—Nihon wa" [Collective Security and Japan], *Asahi Shinbun*, 20 March 2007; "Shūdanteki Jieiken Kenkyū—'Nichū Kankei ni Eikyō'" [Research on Collective Self-Defense—'Impact on Sino-Japanese Relations'], *Asahi Shinbun*, 18 May 2007; and Jun Hongo, "Defense-Only Posture Needs Reviewing: Panel," *Japan Times*, 5 August 2009; and Masami Ito, "New Panel Mulls New Security Parameters," *Japan Times*, 19 February 2010. Additionally, for a summary of the different positions on offensive and defensive realists in the context of overseas JSDF deployment, see Paul Midford, *Japanese Public Opinion and the*

War on Terrorism: Implications for Japan's Security Strategy, Policy Studies 27 (Washington, DC: East-West Center Washington, 2006)," pp. 49–51.

12. For overviews, see Robert Pekkanen and Ellis S. Krauss, "Japan's 'Coalition of the Willing' On Security Policies," esp. pp. 429–431; Richard J. Samuels, "Securing Japan: The Current Discourse," pp. 135–137; and Hironori Sasada, "Youth and Nationalism in Japan," *SAIS Review* 26(2), 2006, pp. 109–122.

13. James Brooke, "After Failures, Space Efforts in Japan Gets a Lift," *New York Times,* 27 February 2005.

14. For information on the United States, the following draws on Department of Defense (DOD), Office of the Undersecretary of Defense for Acquisition, Technology, and Logistics, *Report of the Defense Science Board/Air Force Scientific Advisory Board Joint Task Force on Acquisition of National Security Space Programs* (Washington, DC: U.S. Department of Defense, May 2003), pp. 1–2, 12; Office of Science and Technology Policy (OSTP), Executive Office of the President, "U.S. Space Transportation Policy: Fact Sheet, 6 January 2005," available from OSTP at www.ostp.gov (accessed 8 June 2007); Leonard David, "U.S. Security Depends on Space Assets, Says Air Force Undersecretary," *Space.com,* available online at www.space.com, 20 February 2002 (accessed 7 June 2007); and Johnson-Freese, *Space as a Strategic Asset,* esp. pp. 1–26.

15. Les Doggrell, "Operationally Responsive Space: A Vision for the Future of Military Space," *Air & Space Power Journal* 20(2), 2006, pp. 42–49.

16. See Norihiko Saeki, "COTS Policy & 'Space on Demand' in Japan" (Tokyo: METI, 29 October 2007).

17. See, in general, Steven Berner, "Japan's Space Program: A Fork in the Road?" RAND National Security Research Division, Santa Monica, CA, 2005, available online at www.rand.org (accessed 22 December 2006), p. 11.

18. As discussed in Chapter 4, see Central Intelligence Agency (CIA), "Prospects for the Worldwide Development of Ballistic Missile Threats to the Continental United States" (Washington, DC: CIA, 17 November 1993), available online at www.fas.org (accessed 30 April 2008).

19. Unless otherwise indicated, the following discussion on Japan and nuclear weapons draws on Yuri Kase, "The Costs and Benefits of Japan's Nuclearization: An Insight into the *1968/70 Internal Report,*" *Nonproliferation Review,* Summer 2001, pp. 55–68; Michael J. Green and Katsuhisa Furukawa, "New Ambitions, Old Obstacles: Japan and Its Search for an Arms Control Strategy," *Arms Control Today,* July/August 2000, available online at www.armscontrol.org (accessed 21 July 2007), pp. 1–12 (online version); Matake Kamiya, "Nuclear Japan: Oxymoron or Coming Soon?" *Washington Quarterly* 26(1), Winter 2002–2002, pp. 63–75; Llewelyn Hughes, "Why Japan Will Not Go Nuclear (Yet)," *International Security* 31(4), 2007, pp. 67–96; Frank Barnabie and Shaun Burnie, "Thinking the Unthinkable: Japanese Nuclear Power and Proliferation in East Asia," *Japan Focus,* 8 September 2005, available online at www

.japanfocus.org (accessed 2 August 2007), pp. 1–14 (online version); Dan Plesch, "Without the UN Safety Net, Even Japan May Go Nuclear," *Guardian*, 28 April 2003, available online at www.guardian.co.uk (accessed 24 October 2006); and Gavan McCormack, "Japan as a Plutonium Superpower," *Japan Focus* 50, 18 December 2007, available online at japanfocus.org (accessed 21 June 2008).

20. For the following, in addition to above sources, see "The Missile Technology Control Regime," available online at www.mtcr.info (accessed 1 August 2007); "Japan's Position on the Millennium Summit and Millennium Assembly" (55th Session of the General Assembly; expressing Japan's commitment to the MTCR), August 2000, ¶2.1.d, available online at www.mofa.go.jp (accessed 1 August 2007); "Statement by the Press Secretary/Director-General for Press and Public Relations, Ministry of Foreign Affairs, on the International Code of Conduct Against Ballistic Missile Proliferation," 26 November 2002, available online at www.mofa.go.jp (accessed 1 August 2007); Japan Atomic Energy Commission, *White Paper on Nuclear Energy 2006* (Tokyo: JAEC, March 2007), pp. 12, 14–18; succession of nuclear disarmament and non-proliferation resolutions submitted by Japan to the United Nations, available at www.mofa.go.jp (accessed 1 August 2007); and Ministry of Foreign Affairs (MOFA), "Japan to Host the Proliferation Security Initiative Maritime Interdiction Exercise," August 2004, available online at www.mofa.go.jp (accessed 1 August 2007).

21. These include the 1955 Atomic Energy Law, which limits nuclear energy use to peaceful purposes; the Three Non-Nuclear Principles from December 1967, which were formalized as a resolution in the Diet in November 1971 (not to possess, produce, or introduce nuclear weapons); the Non-Proliferation Treaty (NPT) signed in 1970, ratified much later in 1976, and extended indefinitely in 1995; the Missile Technology Control Regime (MTCR) in 1987, which Japan helped found as an informal agreement that focuses on controlling the proliferation of delivery systems for weapons of mass destruction (WMD) primarily through export controls; the International Code of Conduct against Ballistic Missile Proliferation (ICOC), which sets up politically (but not legally) binding principles on point; the IAEA comprehensive safeguards agreements in December 1977 (Obligation under Article 3 of the NPT); the IAEA Additional Protocol in December 1999; the ratification of the Comprehensive Test Ban Treaty (CTBT) in July 1997, which seeks to prohibit all nuclear test explosions in any environment, but which has not yet come into force (and was rejected by the U.S. Senate in 1999); the Fissile Material Cut-Off Treaty (FMCT), currently in negotiation, which seeks to ban the production of weapons material and may also require the destruction of existing stocks; peaceful nuclear energy cooperation agreements with a set of countries; as well as a succession of resolutions submitted to the United Nations. Japan is also actively involved in the Proliferation Security Initiative (PSI) for the purpose of practically stopping or impeding the flow of WMD and hosted the first Asian maritime interdiction exercise under it in October 2004.

22. Eric Heginbotham and Richard J. Samuels, "Japan's Dual Hedge," *Foreign Affairs* 81(5), 2002; Eiichi Katahara, "Japan's Plutonium Policy: Consequences for Nonproliferation," *Nonproliferation Review*, Fall 1997, pp. 53–61; and Ariel E. Levite, "Never Say Never Again: Nuclear Reversal Revisited," *International Security* 27(3), 2002–2003, esp. pp. 69–73.

23. See, for example, "Nuclear Arsenal Deemed Infeasible in '81," *Japan Times*, 3 October 2004; and Tamura Hideo, "Kakudantō Shisaku ni 3 Nen Ijyō—Hiyō 2000–3000 Oku Yen—Seifu Naibu Bunsho" [Internal Government Document: 3 or More Years to Manufacture Nuclear Warhead, Cost 2000–3000 Oku Yen], *Sankei Shinbun*, 25 December 2006, available online at www.sankei.co.jp (accessed 6 August 2007).

24. Some of the technical information on nuclear weapons technology is from U.S. Department of Defense (USDOD), *The Militarily Critical Technologies List—Part II: Weapons of Mass Destruction Technologies* (Washington, DC: Office of the Under Secretary of Defense for Acquisition and Technology, February 1998), pp. II-5–13.

25. See Japan Atomic Energy Commission, *White Paper on Nuclear Energy 2006* (Tokyo: JAEC, March 2007), pp. 9–10, 13. With fifty-five nuclear power plants (the third highest number of plants operational in the world at the end of 2006), Japan's nuclear energy did in fact account for about a third of the country's total electricity generation in 2005. See also Shinichi Mizumoto, "The Challenges and Directions for Nuclear Energy Policy in Japan—Japan's Nuclear Energy National Plan" (Tokyo: METI, Agency for Natural Resources and Energy, May 2007), esp. slides 3–4, for an emphasis on reducing dependence on oil in the face of high foreign competition. For views opposing the advance of nuclear power technology based on safety and weapons concerns, see efforts and statements by Greenpeace, "Dual Capable Nuclear Technology," available online at archive.greenpeace.org (accessed 1 August 2007). According to Greenpeace, in striking for nuclear civil power, nuclear power plants also include those with uranium conversion and enrichment facilities, research and power reactors, and reprocessing plants, all of which can produce fissile materials necessary for nuclear warheads. See also statements by other organized groups such as Plutonium Action Hiroshima, "Rokkasho Shori Kōjyō wa Sekai e Kyōi" [Rokkasho Reprocessing Plant Is Threat to the World], 10 December 2004, available online at www.nirs.org (accessed 1 August 2007).

26. See the historical overview and technical details in Selig S. Harrison, "Japan and Nuclear Weapons," in *Japan's Nuclear Future*, esp. pp. 3–44. See also Greenpeace, "Greenpeace Calls for Cancellation of 'Peaceful' Nuclear Cooperation with Japan as Special Diet Committee Debates Nuclear Weapons Policy," 10 June 2002, available online at archive.greenpeace.org (accessed 28 July 2007); and Henry Sokolski, "After Iran: Back to the Basics on 'Peaceful' Nuclear Energy," *Arms Control Today*, April 2005, available online at www.armscontrol.org (accessed 2 August 2007).

27. Although the prototype Monju in Tsuruga was shut down and remains idle after an accident in 1995, the government was subsequently given Supreme Court backing for reopening it. Since then its status has been a work in progress. In the meantime, the government continued research on ways to better reprocess spent fuel (such as that generated by Monju) at the Recycle Equipment Test Facility in Tokai, which could also produce a steady supply of supergrade. The point about supergrade plutonium is that very little is required to produce nuclear warheads (possibly 800 to 900 grams); it is thus especially suitable for miniaturized nuclear warheads like MIRV-type ICBMs. See McCormack, "Japan as Plutonium Superpower"; "Just in Case Monju Is Restarted: ¥350 Million Spent on Idle Nuclear Site," *Japan Times*, 15 April 2008; "Monju Fast-Breeder Nuclear Reactor May Reopen," *Japan Times*, 19 November 2000; "Japan's Top Court Gives OK to Reopen Monju Fast Breeder Reactor," *Space Daily*, 30 May 2005, available online at www.SpaceDaily.com (accessed 21 June 2008); and Selig S. Harrison, "Missile Capabilities in Northeast Asia: Japan, South Korea and North Korea," in appendix III, Unclassified Working Papers, *Rumsfeld Commission Report*.

28. Interview with Shaun Burnie, director for Greenpeace International Nuclear Campaigns, Tokyo, 2 August 2005; Greenpeace, "Japan to Begin Operation of World's Most Expensive Nuclear Facility," press release 30 March 2006, available online at www.greenpeace.or.jp (accessed 28 July 2007); Citizens' Nuclear Information Center (CINC), "Rokkasho Produces First MOX Powder," media release, 17 November 2006, available online at cnic.jp (accessed 1 August 2007); CNIC, "Japan's Plutonium Use Plan—2007 Fiscal Year," 23 February 2007, available online at cnic.jp (accessed 1 August 2007); Yuichi Kaido, "Japan's Reprocessing Policy and Nuclear Proliferation in Asia," *Nuke Info Tokyo* 93, January/February 2003, available online at cnic.jp (accessed 1 August 2007); and Hideyuki Ban (co-director, CNIC), Aileen Mioko Smith (director, Green Action), and Atsuko Nogawa (nuclear campaigner, Greenpeace Japan), "Letter Sent to IAEA re. Japan Atomic Energy Commission Approval of Faulty Plutonium Utilization Plan," 3 February 2006, available online at cnic.jp (accessed 1 August 2007).

29. The assessment by one environmental group in Kyoto, Japan—namely, Green Action—which has been monitoring Japan's plutonium program in light of its purported aim of meeting Japan's electric power needs since the early 1990s, is certainly sobering. In 2003, Green Action claimed that virtually none of Japan's reactors were providing electricity. Because, according to Green Action, not a single light bulb in Japan was actually being directly lit by the Japanese government's plutonium program, it used the support of statements by Nippon Keidanren leaders to state that this program was a huge economic drain. See Aileen Mioko Smith, director, "Briefing on Current Status of Japanese Plutonium Program (Letter Sent to Legislative Assembly, Republic of Panama)," 17 March 2003, available online at www.greenaction-japan.org (accessed 8 August 2007).

30. CNIC, Green Action, Greenpeace Japan, Letter to Dr. Mohamed El Baradei and IAEA Board of Directors, "Rokkasho Reprocessing Plant Testing and Operation will Breach Japan's International Commitment Concerning Plutonium—Japan Atomic Energy Commission Accepts Faulty Plutonium Utilization Plan of Japanese Electric Utilities," 3 February 2006, available online at www.greenaction-japan.org (accessed 20 August 2007).

31. Thomas B. Cochran, "The Problem of Nuclear Energy Proliferation," in *Energy and National Security in the 21st Century,* edited by Patrick L. Clawson (Washington, DC: National Defense University Press, 1995), pp. 95–99. By comparison, the Nagasaki bomb (*Fat Man*), which produced a 20-kiloton explosion, was made with 6.1 kilograms of weapons-grade plutonium and was based on a conservative, low-technology design. For further support, see also U.S. Department of Energy, Office of Declassification, *Restricted Data Declassification Decisions 1946 to the Present* (RDD-7), 1 January 2001, Section II.L.33. In 1993, the United States declassified information that about 6 kilograms of plutonium was enough to make one nuclear explosive device. In 1994 further declassification revealed that hypothetically even a mass of 4 kilograms of plutonium or uranium-233 was sufficient for one nuclear explosive device.

32. U.S. Department of Energy, Office of Declassification, *Restricted Data Declassification Decisions 1946 to the Present* (RDD-7), 1 January 2001, Section II.L.25, 26, 29. On the 1976 announcement, see Richard L. Garwin, "Reactor-Grade Plutonium Can Be Used to Make Powerful and Reliable Nuclear Weapons: Separated Plutonium in the Fuel Cycle Must Be Protected as If It Were Nuclear Weapons," 26 August 1998, available online at www.fas.org (accessed 1 August 2007). The IAEA statements are attributed to Hans Blix, then the agency's director-general, at the Nuclear Control Institute, "Japan Can Construct Nuclear Bombs Using Its Power Plant Plutonium," 9 April 2002, available online at www.nci.org (accessed 4 August 2007).

33. U.S. Department of Defense (USDOD), *The Militarily Critical Technologies List—Part II: Weapons of Mass Destruction Technologies* (Washington, DC: Office of the Under Secretary of Defense for Acquisition and Technology, February 1998), pp. II-5–13.

34. See especially Harrison, "Missile Capabilities in Northeast Asia." Harrison identifies at least five ways in which Japan can pursue a nuclear weapons program: producing weapons-grade plutonium in reactors used exclusively for electricity by shutting them down more frequently to reduce the burn-up (irradiation levels) of the fuel to weapons-grade, using experimental new processes (laser-isotope process) to upgrade to weapons-grade plutonium or to produce enriched plutonium, converting uranium-enrichment facilities from the production of low-enriched uranium to enriched uranium, separating supergrade plutonium in the natural uranium blankets of fast-breeder reactors or, most directly, produce weapons-grade plutonium in a reactor designed for that purpose.

35. Kase, "Costs and Benefits of Japan's Nuclearization," p. 55.

36. Thomas B. Cochran, "The Problem of Nuclear Energy Proliferation," in *Energy and National Security in the 21st Century,* edited by Patrick L. Clawson (Washington, DC: National Defense University Press, 1995), p. 96.

37. Matthew L. Wald, "U.S., Criticized for Helping Japan Over Plutonium, Will Stop," *New York Times*, 9 September 1994. In passing, we would like to note that in 1981 the United States acknowledged that the U.S. aircraft carrier *Ticonderoga* did "lose" a hydrogen bomb off the coast of Okinawa (which technically did not return to Japan until 1972), and it is not clear whether it has been "found." See David E. Sanger, "U.S. Confirms It Lost an H-Bomb Off Japan in '65," *New York Times*, 9 May 1989; and Steven R. Weisman, "Tokyo Drops Issue of Lost U.S. Bomb," *New York Times*, 30 December 1989.

38. Marvin Miller, "Japan, Nuclear Weapons, and Reactor-Grade Plutonium" (paper presented to seminar at the Nuclear Control Institute, 27 March 2002), available online at www.nci.org (accessed 21 July 2007), pp. 1–7 (online version).

39. U.S. Department of Defense (USDOD), *The Militarily Critical Technologies List—Part II*, pp. II-5–8.

40. For example, the following individuals and agencies have spoken directly to Japan's nuclear capability: Nobusuke Kishi in 1957, Yasuhiro Nakasone in 1984, Tsutomu Hata in 1994, Ichiro Ozawa in 2002, Yasuo Fukuda in 2002, Shinzo Abe in 2002 and 2006, and so on, and also the CLB in 1957 and the JDA in 1970. For a comprehensive discussion of Japan acquiring or not acquiring an actual nuclear deterrent, as well as specific statements and actions by Japanese actors on nuclear weapons, see L. Hughes, "Why Japan Will Not Go Nuclear (Yet)," pp. 67–96, 83–84; "Japan Can Be a Nuclear Power: Ozawa," *Japan Times*, 7 April 2002; Howard W. French, "Taboo Against Nuclear Arms Is Being Challenged in Japan," *New York Times*, 9 June 2002; and BBC, "'Krasnaya Zvezda' on Japan's 'Wish for Nuclear Weapons of Its Own,'" *BBC Summary of World Broadcasts*, 20 August 1984.

41. This is led by Japan's Council Against Atomic and Hydrogen Bombs (Gensuikyo, established in 1955, and aligned with the Japanese Communist Party) and Japan's Congress Against Atomic and Hydrogen Bombs (Gensuikin, established in 1965, which broke away from Gensuikyo on the principle of tolerating no nuclear testing and aligned with the Democratic Party of Japan, the Social Democratic Party, and the Japanese Trade Union Confederation, Rengo). For the origins, interaction with the central government, and influence of these organized peace movements, see Anthony DiFilippo, "The Politics of Japanese Nuclear Disarmament Initiatives: Where Government Policies and Civil Society Converge and Diverge" (paper presented at the Annual Meeting of the International Studies Association, Portland, Oregon, 27 February 2003). For their continued opposition to nuclear developments worldwide and their international and domestic activities, see Gensuikyo and Gensuikin represented on their respective Web sites at www10.plala.or.jp and www.gensuikin.org, respectively (accessed 27 June 2007). See Tetsushi Kajimoto, "Youth Here Yet to Pick Up the

Peace Torch," *Japan Times*, 3 August 2005; and for a survey assessment, see especially Toshiyuki Kobayashi, "Fading Memories of the Atomic Bomb and Growing Fears of Nuclear War," *NHK Broadcasting Studies*, 2006–2007, available online at www.nhk.or.jp (accessed 20 August 2009), pp. 193–194, 196–198.

42. "Memos Confirm Secret Okinawa Pact," *Japan Times*, 8 October 2007; "Nakasaone Admits U.S. May Have Moved Nuclear Arms Through Japan," *Japan Times*, 17 June 2007. Although this continues to be controversial for the public, there are reports that the Nakasone's Cabinet had formally stated a policy of allowing port calls by nuclear-powered ships in the early 1980s. See "Japan Says It Permits Visits by Nuclear-Powered Carriers," *New York Times*, 9 February 1983.

43. See Sasada, "Youth and Nationalism in Japan" and T. Kobayashi, "Fading Memories."

44. See, for example, Green and Furukawa, "New Ambitions, Old Obstacles," pp. 2–3 (online version); also Harrison, "Japan and Nuclear Weapons," p. 39; and Michiyo Nakamoto and Richard McGregor, "Japan Vows to Stay out of Nuclear Club," *Financial Times*, 21 November 2006.

45. Richard Samuels, *Securing Japan*, esp. p. 47; and L. Hughes, "Why Japan Will Not Go Nuclear (Yet)," p. 84, who points out that generally weapons classes prohibited by constitutional interpretations in Japan did in fact go on to become acceptable due to changed technological conditions. For a brief overview of the trajectory of changes that fly in the face of the 1969 peaceful purposes resolution, see Paul Kallender-Umezu, "Japan Ruling Party to Seek Military Role in Space," *Space News*, 6 April 2006, available online at www.space.com (accessed 21 July 2008).

46. See Green and Furukawa, "New Ambitions, Old Obstacles," p. 4 (online version).

47. Interview, government official, Tokyo, 5 August 2009.

48. For the law itself, see the original bill in the 169th Diet Session, Bill No. 17, "Uchū Kihon Hōan [Basic Space Law]," available online at www.shugiin.go.jp (accessed 21 May 2008), which was approved by the lower and then upper house in May 2008. See also Appendix II of this book for our translation. Unless otherwise indicated, the following draws on "Uchyū Kihon Hō Kyō Seiritsu" [Basic Space Law Passed Today], *Asahi Shinbun*, 21 May 2008; "Shasetsu: Uchū Kihon Hō—Gunji ni wa Meikaku Na Gensoku o" [Editorial: Basic Space Law—Distinct Military Principles], *Asahi Shinbun*, 22 May 2008; "Japan Diet to End Peaceful-Use Space Policy," *Japan Times*, 10 May 2008; Kakumi Kobayashi, "New Space Policy Result of Regional Tensions," *Japan Times*, 22 May 2008; and "Diet Enacts Law on Use of Space for Defense," *Japan Times*, 22 May 2008. In the bill: on the issue of consistency with pacifism, see Article 2; on implications for national security, see especially Articles 3 and 14; on promotion of the industry, see Articles 4, 11, and 16; on the emphasis on autonomy, see Article 15; and on the Space Development Strategy Headquarters, see Article 25 (and Chapter 4 generally).

49. Interview, Satoshi Tsuzukibashi, deputy director, Office of the Defense Production Committee, Nippon Keidanren, Tokyo, 2008.

50. See, specifically, Paul Kallender-Umezu, "Japan Elevates Space Management, Lifts 1969 Ban on Military Satellites," *Space News*, 26 May 2008.

51. Yasunori Matogawa, "On the Basic Space Law," Y. M. Column Archive, 22 May 2008, available online at www.planetary.or.jp (accessed 15 May 2008).

52. See, for example, Akemi Nakamura, "Private Rocketeers Start Small, Think Big," *Japan Times*, 2 November 2004.

53. Interviews with Kazuto Suzuki, 3 September 2009; and see Paul Kallender-Umezu, "Amid Shift in Power, Japan Seeks Space Budget," *Space News*, 7 September 2009.

54. Paul Kallender, "Japan Faces Major Hurdles in Spy Satellite Development," *Defense News*, 12–18 October 1998.

55. See "Japan Appoints First Space Development Minister: Officials," *Space Daily*, 17 June 2008, available online at www.SpaceDaily.com (accessed 19 June 2008); Aiko Hayashi, "Japan's Top Space Advocate Wants Own NASA," Reuters, available online at www.reuteurs.com (accessed 19 June 2008); and Kazuki Shiibashi, "Japan Oks New Space Law," *Aviation Week*, 18 June 2008, available online at www.aviationweek.com (accessed 19 June 2008).

56. With respect to the United States, see Kurt M. Campbell, Christian Beckner, and Yuki Tatsumi, "U.S.–Japan Space Policy: A Framework for 21st Century Cooperation" (Washington, DC: Center for Strategic and International Studies, July 2003), pp. 1–30; and Berner, "Japan's Space Program" esp. pp. 33–37. With respect to China, see Ashley Tellis, "China's Military Space Strategy," *Survival* 49(3), 2007, pp. 41–72; James Brooke, "China in Space: The Rivals; The Japanese Are Impressed, But They Plan to Leap Ahead," *New York Times*, 16 October 2003; Hiroko Tabuchi, "Japan, China in a Race to the Moon with Upcoming Launches," *Japan Times*, 27 August 2007; and "Defense Think Tank Frets China Space Program," *Japan Times*, 27 March 2008.

57. For the discussion here, see Christopher J. Bowie, *The Anti-Access Threat and Theater Air Bases* (Washington, DC: Center for Strategic and Budgetary Assessments, 2002), pp. i–viii, 25–26, 48, 71; Andrew Krepinevich, Barry Watts, and Robert Work, *Meeting the Anti-Access and Area-Denial Challenge* (Washington, DC: Center for Strategic and Budgetary Assessments, 2003), esp. pp. i–iv, 93–95; William S. Murray III and Robert Antonellis, "China's Space Program: The Dragon Eyes the Moon (and Us), *Orbis* 47(4), 2003, esp. pp. 650–651; David Isenberg, "Navies Overseas: China Buys Russian Vessels to Mount Naval Challenge to U.S.," *Navy News Week*, 18 November 2002, available online at www.globalsecurity.org (accessed 27 July 2008); David A. Fulghum and Douglas Barrie, "What's the Threat? U.S. Response to China's Military Growth Is Mixed," *Aviation Week & Space Technology* (hereafter *AWST*), 17 March 2008; and

David A. Fulghum and Douglas Barrie, "Non-War; China Accelerates Focus on Disruption, Asymmetric Tactics," *AWST*, 10 March 2008.

58. See on this point Michael Auslin and Christopher Griffin, *Securing Freedom: The U.S.-Japanese Alliance in a New Era*, A Report of the American Enterprise Institute (Washington, DC: American Enterprise Institute, 2008), p. 35.

59. DPJ, "Minshutō no Seiken Seisaku Manifesto 2009 [The DPJ's Governance and Policy Manifesto 2009], Tokyo, Japan, 27 July 2009, p. 22, available online at www.dpj.or.jp (accessed 28 August 2009); and Mure Dickie and Alec Russell, "Okada Set to Redefine Ties with Washington, *Financial Times*, 18 September 2009.

60. See generally Masahiro Matsumura, "What Does Japan Want from Washington?" *Japan Times*, 16 December 2009; Masami Ito, "New Panel Mulls New Security Parameters," *Japan Times*, 19 February 2010; and "China's New Ambassador Takes Post," *Japan Times*, 2 March 2010.

61. Peter J. Brown, "Japan's Next Chapter in Space Begins," *Asia Times Online*, 9 September 2009, available online at www.atimes.com (accessed 4 February 2010).

62. See Secretariat for Strategic Headquarters for Space Policy (SHSP), "Heisei 22 Nendo Yosan (Seifu Genan) Ni Okeru Uchū Kaihatsu Yosan ni Tsuite (Sokuhōchi) [Space Development Budget in Heisei 22 Budget (Governmet Original Bill) (Preliminary Figures)], Tokyo, 12 January 2010.

63. Kathrin Hille, "China Chief Foresees Military Space Race," *Financial Times*, 4 November 2009.

64. See also Paul Kallender-Umezu, "Japan Elevates Space Management, Lifts 1969 Ban on Military Satellites," *Space News*, 26 May 2008, available online at www.space.com (accessed 29 May 2008).

65. The Council on Security and Defense Capabilities (CSDC), *The Council on Security and Defense Capabilities Report: Japan's Visions for Future Security and Defense Capabilities* (Tokyo: CSDC, October 2004), p. 5.

INDEX

Note: For names corresponding to abbreviations, see list beginning on p. xv. Tables and figures are indicated by *t* or *f*, respectively, following page numbers.

The authorized representative in the EU for product safety and compliance is:
Mare Nostrum Group
B.V Doelen 72
4831 GR Breda
The Netherlands

www.ingramcontent.com/pod-product-compliance
Lightning Source LLC
LaVergne TN
LVHW091515200926
848483LV00005B/422/J
9780804700634